Y0-CAR-225

AVON FREE PUBLIC LIBRARY

Encyclopedia of Human-Animal Relationships

Encyclopedia of Human-Animal Relationships

A Global Exploration of Our Connections with Animals

Volume 1: A–Con

Edited by
Marc Bekoff

GREENWOOD PRESS
Westport, Connecticut • London

Library of Congress Cataloging-in-Publication Data

Encyclopedia of human-animal relationships : a global exploration of our connections with animals / edited by Marc Bekoff.
p. cm.
Includes bibliographical references and index.
ISBN-13: 978-0-313-33487-0 (set : alk. paper)
ISBN-13: 978-0-313-33488-7 (vol 1 : alk. paper)
ISBN-13: 978-0-313-33489-4 (vol 2 : alk. paper)
ISBN-13: 978-0-313-33490-0 (vol 3 : alk. paper)
ISBN-13: 978-0-313-33491-7 (vol 4 : alk. paper)

1. Human-animal relationships—Encyclopedias. I. Bekoff, Marc.
QL85.E53 2007
590—dc22 2007016552

British Library Cataloguing in Publication Data is available.

Library of Congress Catalog Card Number: 2007016552

ISBN–13: 978–0–313–33487–0 (Set)
ISBN–13: 978–0–313–33488–7 (vol. I)
ISBN–13: 978–0–313–33489–4 (vol. II)
ISBN–13: 978–0–313–33490–0 (vol. III)
ISBN–13: 978–0–313–33491–7 (vol. IV)

First published in 2007

Greenwood Press, 88 Post Road West, Westport, CT 06881
An imprint of Greenwood Publishing Group, Inc.
www.greenwood.com

Printed in the United States of America

The paper used in this book complies with the Permanent Paper Standard issued by the National Information Standards Organization (Z39.48-1984).

10 9 8 7 6 5 4 3 2 1

For my parents, who always encouraged the animal in me.

Contents

Alphabetical List of Entries

Culture, Religion, and Belief Systems

List of Entries that Feature Specific Animals

This list includes those entries that focus on particular animals. Other entries also include information about these animals but not to the extent that the following essays do. Please see the index for all mentions of the large number of different animals that are included in these volumes.

Cats

Cats, Large (*See also* Tigers)

Cattle

Sports
Blood Sports and Animals

Chickens

Animals as Food
Veganism
Culture, Religion, and Belief Systems
Chinese Youth Attitudes toward Animals
Judaism and Animals
Ethics and Animal Protection
Compassionate Shopping

Chimpanzees (*See also* Great Apes; Monkeys)

Bonding
Animal Beings I Have Known: A Personal Essay
Chimpanzee and Human Relationships
Communication and Language
Great Apes and Language Research
Nim Chimpsky [sidebar]
Conservation and Environment
The Great Apes of Asian Forests, "The People of the Forest"
Ethics and Animal Protection
Goodall, Jane
The Significance of Jane Goodall's Chimpanzee Studies [sidebar]
The Great Ape Project

Cockroaches

Bonding
Discrimination between Humans by Cockroaches

Corals

Living with Animals
Corals and Humans

Cows. *See* Cattle

Coyotes

Conservation and Environment
Coyotes, Humans, and Coexistence

Dogs

Animal Assistance to Humans
Assistance and Therapy Animals

Dolphins

Donkeys

Earthworms

Elephants

Emus

Falcons

Fish

Foxes

Geckos

Giant Pouched Rats (*See also* Rats)

Gorillas (*See also* Great Apes; Monkeys)

Great Apes (including Chimpanzees and Gorillas)

Honeybees (*See also* Insects)

Horses

Hyenas

Insects (*See also* Honeybees)

Kiwis

Lizards. *See* Geckos; Reptiles

Mice (*See also* Rats)

Mollusks

Conservation and Environment
Mollusk Extinction

Monkeys (*See also* Great Apes)

Animal Assistance to Humans
Assistance and Therapy Animals
Culture, Religion, and Belief Systems
Monkeys and Humans in India
The Monkey God Hanuman [sidebar]
Living with Animals
Rhesus Monkeys and Humans
Research
Animals in Space

Mountain Lions (*See also* Cats, Large)

Living with Animals
Mountain Lions and Humans

Octopus

Living with Animals
Giant Octopuses and Squid

Penguins (*See also* Birds)

Bonding
Discrimination between Humans by Gentoo Penguins (Pygosceli papua papua/ellsworthii)

Parrots (*See also* Birds)

Communication and Language
Interspecies Communication—N'kisi the Parrot: A Personal Essay
Culture, Religion, and Belief Systems
The Parrot Book

Pigs

Literature
Animals in Literature
George Eliot and a Pig [sidebar]
Living with Animals
Pigs and Humans

Pikas

Conservation and Environment
Ecosystems and a Keystone Species, the Plateau Pika

Preface

This is the third encyclopedia I have edited for the Greenwood Publishing Group, and it will be my last. Of course, that is what I said before tackling the *Encyclopedia of Animal Behavior* (2004), six years after editing the *Encyclopedia of Animal Rights and Animal Welfare* (1998). However, having the opportunity to compile a collection of essays concerned with the broad topic of human-animal relationships, a rapidly growing field worldwide also called *anthrozoology*, was just too inviting. There also are a rapidly growing number of national and international meetings devoted to human-animal interactions, and numerous people from all over the globe want to know more about this topic. Nonetheless, I truly had no idea what this journey would look like. But in the end it was a rewarding effort, and I learned an incredible amount about the nature of human-animal relationships from people who live all over the world. In the past two years I have been privileged to visit researchers, writers, and animal advocates at work in England, Scotland, Norway, Germany, Spain, Portugal, Brazil, Australia, China, India, Tasmania, Kenya, and Tanzania, and some of them have contributed to this encyclopedia. These visits not only reinforced my impression that there is worldwide interest in animals but also illuminated the cultural differences, about which you will read in these volumes.

Immediately after I solicited entries for this encyclopedia I was astounded by the response. People from different countries and cultures offered to write original essays on a staggering array of topics, some of which I had never imagined: *entomophagy* (the eating of insects), contributed by Juanita Choo, who has lived with an indigenous tribe in the Venezuelan Amazon, studying the use and cultivation of palm weevil grubs for food; biomimicry (research devoted to using models from nature for solving human problems); the ways in which giant African pouched rats help to save human lives; "man the hunted," which Donna Hart and Robert W. Sussman have studied, discovering the history of humans as prey for prehistoric predators; the role of animal monsters in film; the use of honeybees to study alcohol consumption by humans (in fact, honeybees play a surprisingly large role in different aspects of human-animal relationships); and the role of animals in the settlement of colonial North America, to name but a few. Many of my busy and over-committed colleagues even offered to write upwards of three or four essays because they, too, were excited that at last, an encyclopedia like this was in the works. Finally, it bears mentioning that this encyclopedia and its companions, the *Encyclopedia of Animal Rights and Animal Welfare* and the *Encyclopedia of Animal Behavior*, comprise the largest compilation of human-animal studies and writing to date. As my colleague and contributor William Lynn remarked, this is a significant achievement. It is also a testament to the growing interest in human-animal studies worldwide.

What You Will Find in This Encyclopedia

We encounter a wide array of animals in diverse circumstances worldwide, ranging from educational and research settings to dog parks, animal shelters, farms, ranches, zoos, rodeos, clothing stores, houses of worship, and factory farms. We study them, eat them, wear them, bless them, help them when they are in need, and train them to help us when we are in need. The essays in these volumes show how our interactions with animals are numerous, diverse, challenging, frustrating, and fascinating.

The *Encyclopedia of Human-Animal Relationships: A Global Exploration of Our Connections with Other Animals* is the most complete and comprehensive collection of original essays on the broad topic of human-animal relationships. It covers hundreds of topics from A to Z, from animals in art to animals in the zodiac. With more than 350 entries, this encyclopedia presents essays from these disciplines and others: anthropology; archeology; art; biology; cultural studies; ecology, engineering; environmental concerns and conservation; economics; education; film; history; law; literature, language, and mythology; mass media; medicine; military; music; philosophy; psychology and therapy; recreation and sports; religion and theology; sociology; and veterinary medicine. There simply are no rivals in breadth, depth, or scope. So please read on and enjoy the collection of exciting and remarkable essays that await you. It is a wonderful and comprehensive resource for school assignments as well as for general knowledge.

Essays range in length from about 400 to 7,000 words. Some are more technical than others because of the topic at hand, but they all offer important information. When measurements are given, the metric system is generally used, which is the standard for worldwide science writing.

Newly emerging, exciting, and sobering fields are represented, including the effects of climate change and global warming on animals, patterns of animal extinction, conservation medicine, conservation psychology, and interspecies communication. Also among the numerous topics considered are animism, cartoons, children and animals, feral children, ecofeminism, ecotourism, extinction, veganism and vegetarianism, the use of animals in education and research, zoos, animal protection, slang involving animals, activism, rescue and service animals, animals in disasters and war, animal rehabilitation, hunting, domestication, attitudes toward animals, animal therapy, companion animals, animal models of human trauma, and animals and human disease (including bird flu).

In these volumes, contributors' views of animals from different countries and cultures are considered, including North America, Latin America, China, Taiwan, India, Mongolia, Israel, Finland, Australia, New Zealand, the UK, and indigenous cultures (including Native Americans and others). Religions and their varying perspectives on animals are also explored, including Buddhism, Christianity, Islam, Judaism, and others. For those interested in how we use knowledge of human-animal interactions to make life better for both humans and animals, there are essays discussing "applied" aspects of anthrozoology, showing how people work with animals for the betterment of both. Finally, these essays will appeal to many readers because of the array of animals they feature, including bats, cats, dogs, mice, rats, rabbits, horses, bees and other insects, elephants, spiders, sharks, fish, turkeys and other birds, prairie dogs, pikas, wolves, hyenas, snakes, whales, dolphins, octopuses, chimpanzees and other great apes, and more.

You will discover, as you read the excellent, "user friendly," and up-to-date entries written by an international group of well-respected scholars from many different disciplines, that there are many people who devote their lives to study human-animal relationships and who took the time to write their essays because they want to share their zeal and knowledge with you. There also are essays by people who actively work with animals as aids to humans and in therapy. Although the vast percentage of essays are neutral and do not promote a point of view, a small number of provocative personal essays have been included, such as one by Jane Goodall, titled "Animal Beings I Have Known," in which she discusses the close relationships she has had with chimpanzees and dogs; a discussion of the plight of moon bears in China by Jill Robinson, called "The Bear Bile Industry in China"; a poignant piece by Dawn Prince-Hughes, "Gorillas and Me: A Personal Journey," which describes how forming a close relationship with a captive gorilla, Congo, helped her cope with autism; and my own essay on the many ways in which human activities influence the lives of animals (known as anthropogenic effects). There are a few more personal essays with a point of view in these volumes because people feel so strongly about their relationships with other animals, including the companions with whom they share their homes. Many people find it impossible to be neutral about how they interact with and treat other animals.

Much information is provided in the numerous references that are provided at the end of most entries. There also is a comprehensive list of general sources at the end of Volume 4 (books, journals, magazines, organizations, and Web sites) to help you find your way through the fascinating topic of human-animal interactions. I have also included a chronology of legislation, initiatives, and other topics related to the use of animals and to animal welfare and animal protection.

How to Use This Encyclopedia

The essays in this encyclopedia are organized alphabetically under broad topics. This arrangement will help students in researching specific areas of human-animal relationships. Two lists of all the entries are featured at the beginning of each volume. One alphabetically lists the broad topics of the encyclopedia and, following them, their entries. A second list of entries lists those essays that specifically focus on a particular species.

There are long essays that stand alone and short sidebars that are more focused on a specific topic or species. For some entries there are cross-references to other related essays. These cross-references, along with the comprehensive index and the two lists of entry names at the front of the volumes, will be very helpful to you for finding your way around the voluminous information in this encyclopedia.

Giving Thanks

A work such as this requires a lot of cooperation among different people. As she did for the *Encyclopedia of Animal Behavior,* Anne Thompson provided essential editorial support and encouragement throughout the process of soliciting, receiving, editing, and getting authors' responses to their essays. Anne also helped to organize the authors and topics into a variety of spreadsheets, a task that would have driven me crazy. Anne was always there to chat, and it is safe to say that without her unwavering help this encyclopedia would never have seen the light of day. Kevin Downing also fully supported this effort and, indeed, this encyclopedia was his idea. I can blame him for many sleepless nights and thousands of e-mails over the past two years! I also thank Liz Kincaid, our photo editor; Pete Feely, our copyeditor; and Kaitlin Ciarmiello and Ashley Bonora, who helped in organizing some of the entries. I would also like to thank Michael O'Connor and Susan Yates for overseeing the production of the final product.

Last, but surely not least, I want to thank all of the people who took time out of their busy lives to write exemplary essays. These people share our enthusiasm for learning as much as we can about human-animal relationships, and we all should be grateful for their efforts.

Marc Bekoff
Boulder, Colorado

Introduction

Anthrozoology: The Study of Human-Animal Relationships

The relationship between human animals and nonhuman animals ("animals") is a complex, ambiguous, challenging, and frustrating affair. The field of research that focuses on human-animal relationships, *anthrozoology,* is rapidly growing as humans interact with a wide variety of different animals in countless venues worldwide, ranging from educational and research settings to dog parks, animal shelters, farms, ranches, zoos, rodeos, clothing stores, houses of worship, and industrialized slaughterhouses. A search on the Internet search service Google for the phrase "human-animal relationships" generated more than 52 million hits. There are numerous journals, books, and societies that focus on our relationships with animals, and the number seems to grow daily. A recent cover story about our complex relationship with elephants titled "Are We Driving Elephants Crazy?" appeared in the magazine section of the *New York Times* (October 2, 2006). That's how much interest there is in anthrozoology.

As we come to understand the myriad interrelationships in which humans and animals participate, we begin to appreciate and respect the complexity and necessity for these encounters. Let me try to capture some of the important reasons why we form close connections between humans and animals. For example, many essays in this encyclopedia show that animals are important in diverse cultures and religions (see for example, in the section on Culture, Religion, and Belief Systems, *Islam and Animals; Religion's Origins and Animals; The Blessing of the Animals; Indian and Nepali Mahouts and Their Relationships with Elephants;* and *Birds as Symbols in Human Culture*). Animals also help us along when we need emotional and physical support (see for example, Animal Assistance to Humans—*Assistance and Therapy Animals;* Bonding—*Behavior of Animals and the Human-Animal Bond;* Health—*Benefits of Animal Contact; and Human Emotional Trauma and Animal Models;* Living with Animals—*Dogs as Social Catalysts;* and, Play—*Human and Animal Play*).

The essays in this encyclopedia show that we are only doing what comes naturally when we seek out the company of animals and try to learn as much as we can about them. Animals of all stripes—big and small, tall and short, beautiful and not so beautiful, with and without backbones—have played significant roles in the evolution of human societies, cultures, and religions. And there are always surprises. Recently, a group of researchers at the Santa Barbara campus of the University of California discovered that a brain parasite (*Toxoplasmosis goodie*) that comes from cats might actually cause personality changes that result in cultural changes if enough individuals in a given locale are infected (http://www.sciencedaily.com/releases/2006/08/060804085444.htm). People who are infected with this parasite are more prone to feel guilty than those who are not infected. Furthermore, since this parasite is affected by climate, there might be variations among groups of people inhabiting different geographical regions, with people living in more humid and low-altitude areas suffering more than people living elsewhere.

I mention this fascinating study as an example of the numerous surprises that are forthcoming concerning how animals influence humans. Who would have ever thought that there could be a relationship between a parasite found in cats and human cultural evolution? As additional research is conducted in anthrozoology, more and more mind-opening findings may occur. Animals also play a major role in human endeavors such as art, music, cooking, and literature to name but a few. When I attended the Asia for Animals meeting in Chennai, India, in January 2007, I heard a fascinating lecture on animals in Chinese literature by Professor Song Wei, director of the Law Institute at the University of Science and Technology of China in Hefley. He mentioned that 80 percent of the total number of poems in the *Complete Poetry of the Tang Dynasty* made reference to animals, including pigs, piglets, cattle, sheep, horses, and dogs. Song Wei also noted that there are more than 700 idioms in the *Chinese Idioms Dictionary* and that more than fifty species of animals are included among these idioms.

Animals can also be important in the political arena. In Australia, politicians' attitudes toward sharks played a role in political campaigns ("Sharks Swim into Political Waters," http://news.bbc.co.uk/2/hi/science/nature/5414410.stm) because of the voting power of the recreational fishing lobby. And politicians in the United States can be evaluated by their attitudes toward animals. The Society for Animal Protective Legislation recently developed a compassion index (http://www.compassionindex.org/) to assess how public representatives voted on legislation involving animal welfare and conservation. Animals also can be politically controversial when they are used by large employers as commodities, such as cattle on feedlots or chickens on factory farms, or when the land around factory farms becomes polluted because of animal waste. Politicians are often called upon both to support and to criticize large cattle and chicken businesses; for example, when they employ many people in a region and support it economically and when they treat animals inhumanely, destroy surrounding land, or foul the air.

I have always been interested in human-animal relationships. I decided to compile the essays in this encyclopedia for the same reason that I worked on my other two award-winning encyclopedias (*Encyclopedia of Animal Rights and Animal Welfare*, Greenwood, 1998; *Encyclopedia of Animal Behavior*, Greenwood, 2004): I love people and I love animals. And loving animals does not mean that I love people less. My parents have told me that I always "minded animals." I would always ask them what a dog was thinking and feeling. I still ask such questions, and I plan to forever. This encyclopedia, like my two previous collections, has allowed me to expand my horizons in anthrozoology because the essays in each are so broad and interdisciplinary. I cannot really put in words how much I have learned from my esteemed colleagues. Often, when I read these essays as they were submitted, I would sit back and smile and think I am so lucky to be able to benefit from these experts' and writers' knowledge and wisdom. Just when I thought I knew it all I realized how little I knew. I hope that you share this feeling as well.

Old Big Brains in New Bottlenecks: Why We Seek Nature's Wisdom

I have always wondered why it feels so good to interact with animals, why it feels comforting to know that they are waiting for us at home or that they are "out there" in nature. In 2002, as I was preparing an essay for a meeting called The Path to Nature's Wisdom, convened by His Holiness the Dalai Lama as part of his Kalachakra for World Peace 2002, I discovered the following quotation by the renowned author Henry Miller in his book *Big Sure and the Oranges of Hieronymus Bosch* (New Directions, 1957): "If we don't always start from Nature we certainly come to her in our hour of need." I was asked to write and talk about animals as a path to wisdom, highlighting my own experiences and outlining the gifts that animals bring to us when we open our hearts to their presence and essence. I realized that we could look to evolution to understand why we do seek out the company of animals, and why we feel good when we do so.

I find I am never alone, and neither do I feel lonely when I am out in nature and among animals. Nature's wisdom easily captures me, and I feel safe and calm wrapped in her welcoming arms. Why do we go to animals and to nature for guidance? Why do we feel so good, so much at peace, when we see, hear, and smell other animals, when we look at trees and smell the fragrance of flowers, when we watch water in a stream, a lake, or an ocean? We often cannot articulate why, when we are immersed in nature, there are such penetrating calming effects, why we often become breathless, why we sigh, why we place a hand on our heart as we sense and feel the beauty, awe, and mystery of animals.

Numerous studies, some of which are mentioned in this encyclopedia, show that when we pet a dog our heart rate goes down and so does the heart rate of our friend. Just seeing animals can also be soothing. Empathy for and feeling the presence of other animals and their emotions are critical to human well-being. Veterinarian Marty Becker wrote a wonderful book titled *The Healing Power of Pets* (Hyperion, 2002) and showed how pets can keep people healthy and happy—they help to heal lonely people in nursing homes, hospitals, and schools. In *Kindred Spirits* (Broadway Books, 2001), the holistic veterinarian Allen Schoen points to fourteen concrete ways in which a relationship between animal companions and humans has been shown to reduce stress. These include reductions in blood pressure, increases in self-esteem in children and adolescents, increases in the survival of victims of heart attacks, improvement in the life of senior citizens, aiding in the development of humane attitudes in children, providing a sense of emotional stability for foster children, reductions in the demand for physician's services for no serious problems among Medicare enrollees, and reductions in the feeling of loneliness in preadolescents. And Michelle Rivera, in her book *Hospice Hounds* (Lantern Books, 2001), tells numerous stories of how dogs and cats can help people as they near death. Furthermore, a recent study, "Cardio, the Canine Heart Dog, Is a Friend Indeed" (http://www.medpagetoday.com/Cardiology/2005AHAMeeting/tb/2166), showed that a visit from a friendly pup also might be good medicine for an ailing heart. In a randomized study of seventy-six hospitalized heart-failure patients, UCLA researchers found that anxiety scores dropped an average of 24 percent after one arm of the study interacted with pairs of human volunteers and cardio canines of a dozen different breeds. The dogs would lie for twelve minutes on the patient's bed, where patients could pat and scratch their ears. "This study demonstrates that even a short-term exposure to dogs has beneficial physiological and psychosocial effects on patients who want it," said Kathie Cole, a clinical nurse at the UCLA Medical Center.

Numerous people also work for the betterment of animals, and their efforts have improved the well-being of animals around the world. Many essays in this encyclopedia clearly show this. Here is a poignant story of just how far some people will go to help animals in stress. Recently, Suma, a 45-year-old elephant in the Zagreb Zoo in Croatia, was extremely upset after her partner of ten years, Patna, died of cancer. Suma's keepers discovered that she enjoyed listening to music by Mozart, and that she relaxed and "leaned against the fence, closed her eyes and listened without moving the entire concert." The music helped Suma cope with her grief, and zoo authorities bought a stereo and provided music therapy for her.

Perhaps the feelings that are evoked in our relationships with animals are so very deep—primal—that there are no words that are deep or rich enough to convey just what we feel: joy when we know that animals are doing well and deep sorrow and pain when we feel that they are being destroyed, exploited, and devastated.

Now, what about our ancestors? Surely there must have been more significant consequences for them if they interfered with the animals with whom they shared their homes. They were more exposed than modern humans and did not have all the mechanical and intellectual know-how to undo their intrusions into natural processes. Indeed, early humans were probably so busy just trying to survive that they could not have had the opportunities to wreak the havoc that we have brought to the world of animals. And the price of early humans' injurious intrusions would likely have been much more serious for them than they are for us because of their intimate interrelations with, and dependence on, animals.

Nonetheless, I imagine that our psyches, like theirs, suffer when animals are harmed, and there is a loss of balance. Modern sociocultural milieus, technology, and nature have changed significantly, and we face new and challenging bottlenecks and barriers. Cycles of nature are still with us and also within us, although we might not be aware of their presence because we can so easily override just about anything "natural." Much technology and "busy-ness" cause alienation from nature. This breach, in turn, leads to a wanton abuse of nature. It is all too easy to harm environs to which we are not attached or to abuse other beings to whom we are not bonded, to whom we do not feel close.

Our brains can distance us from nature, but they can also lead us back to her. Perhaps there is an instinctive drive to have close ties with nature—*biophilia*, if you will—a term coined by the renowned biologist Edward O. Wilson to refer to an innate connection between humans and nature. So when these reciprocal interconnections are threatened or ruptured, we seek other animals and nature to soothe us because our old brains still remember the importance of being an integral and cardinal part of innumerable natural processes and how good these deep interconnections felt.

Perhaps our close ancestral ties with nature offer reasons for hope for the future, reasons for being optimistic about how to reduce the impact we have on animals. Their losses are also our losses. It does not feel good to cause harm to animals and nature. And the essays in this encyclopedia reinforce why this is so. It's simple; animals play significant roles in innumerable human activities around the world.

The Roots of Human-Animal Interrelationships

In addition to there being evolutionary reasons for bonding with animals and for wanting to learn about the roles they play in our lives, it turns out that many of our closest connections with animals are formed during very early life, as mine were. It has been said that children have an instinct, an intrinsic connectedness or a feeling of biophilia, to be caring and mindful of animals. Children's attachment to pets, for example, is widespread and strong. Gail Melson, who studies child development at Purdue University, notes that more than 75 percent of children in the United States live with pets and are more likely to grow up with a pet than with both parents (see entries in the "Children" section of this book and Melson's book *Why the Wild Things Are: Animals in the Lives of Children* and also Education—*Nurturing Empathy in Children*). Furthermore, American boys are more likely to care for pets than for older relatives or younger siblings. A vast majority of children refer to their pets as "family" or "special friends" and confidants, and more than 80 percent refer to themselves as their pet's mother or father. More than half would prefer their pet rather than family members if stranded on a desert island. Numerous people, especially children, refuse to leave their pets in the midst of natural disasters. Recall how many people refused to leave flooded areas of New Orleans without their pets when Hurricane Katrina struck during the summer of 2005. They risked their own lives for their animal companions.

Animals Are Our Consummate Companions

Our close relationship with other animals brings us many gifts. Not only do we share our homes and other spaces with members of a vast number of species who enrich our lives, but we also have learned much about the significant role that animals play in numerous human endeavors, ranging from everyday activities to more scholarly research in anthrozoology. I believe that animals are our consummate companions because they complete us. Animals make us whole when we come to appreciate how much they have influenced humans worldwide.

Revising Stereotypes

This state-of-the-art collection highlights the integral and essential role that animals play in our lives and illustrates how we define and redefine ourselves in light of what we learn about their lives. It sets a standard for future research and provides an excellent introduction to the multifaceted field of anthrozoology. These essays also show how much we already know, and they surely will motivate you to seek out more information in your areas of interest. We must continually reassess our relationship with animals and revise stereotypes as we learn more and more about their lives and the role that they play in ours. In the future, we can all look forward to much exciting research that shows just how much we depend on other animals and how much they rely on us. By reading these essays and incorporating the knowledge that you gain into your own life and by sharing it with others, you will become part of the process of increasing our understanding of the nature of human-animal relationships worldwide. Please read on.

Animal Assistance to Humans
Animal-Assisted Interventions

Animal-assisted interventions (AAIs) are programs in which animals are employed to help a person or group of people benefit in a predetermined way. Though the terms people use to describe programs with animals vary considerably, there are generally considered to be three types of AAIs: animal-assisted therapies, animal-assisted activities, and animal-assisted education. In animal-assisted *therapy*, a person trained in that therapy incorporates an animal into a carefully prescribed therapeutic treatment, there are well-defined goals for the therapy, and progress is documented. An example of animal-assisted therapy is a dog assisting a physical therapist working on a patient's muscle tone. In an animal-assisted *activity*, the goals of the activity are less specific, and the role of the animal can be more easily adapted to changing situations. Animal-assisted activity practitioners may have received some training about how to conduct the activity, and have been evaluated by a national organization, such as Therapy Dogs International or the Delta Society. Visiting a retirement center to brighten the day of the residents is an example of an animal-assisted activity. Animal-assisted *education* includes any program that increases student learning of specific, predefined information or skills through animal involvement. An example is a certified teacher teaching children on a field trip to a dog park how to be safer with dogs by having them learn how dogs show their feelings.

It is likely that throughout history, animals—especially companion and domestic animals—have helped people cope and heal. AAIs, however, remained informal and did not generally inform each other until they were scrutinized and advocated by scientists and national organizations. Since about 1980, AAIs have grown and matured. Nonetheless, much work needs to be done to understand and regulate these programs.

Although other societies employed animals to help the sick, the ancient Greeks were the first to leave us documented accounts of it. At times, the terminally ill in Greece had their spirits lifted by horseback rides through the country. Associating dogs with the demigod of medicine, Asklepios, the Greeks also let dogs help the sick in a curious manner. In *Greek Hero Cults and Ideas of Immortality* (1921), Lewis Farnell reports that some temples let sacred dogs wander between the convalescing worshipers to lick their wounds. Although canine (and some other) saliva contains small amounts of bactericide, the dogs were probably most effective simply by giving comfort.

Anne and Alan Bowd suggest that the first organized use of AAI was in the ninth century in Geel, Belgium. The families of Geel extended their compassion to nonrelatives with mental and/or physical challenges. Among the ways the residents helped those in their care was with *therapie naturelle,* part of which involved learning to care for domestic animals. However, their practice did not spread.

Medical books in the seventeenth century mention using horseback riding to treat low morale, nervous disorders, and even gout. Small, domestic animals were incorporated into some of the treatments conducted at the Retreat, a revolutionarily humane psychiatric facility in York, England, founded in 1796 by Quaker William Turk. Even Florence Nightingale suggested letting small pets comfort the sick. Levinson (1972) reported that the American Red Cross developed an animal-assistance program to help console emotionally traumatized airmen at the American Air Force Convalescent Center in New York in 1944. However, there was no real attempt to spread this type of therapy.

The current interest in animal-assisted interventions began when Boris Levinson, at Yeshiva University in New York, brought his dog, Jingles, to work. With Jingles in the room, Levinson was finally able to communicate with patients who had previously been unreachable. Levinson was not the first person to incorporate animals into therapy, but he was the first to provide a thorough and compelling account of it by publishing *Pet-Oriented Child Psychotherapy* in 1969.

The now-documented successes of using animals to help people heal inspired more programs and research. Dr. Leo Bustad, one of the founders of the Delta Society and its first president, was also instrumental in bringing attention to this research and promoting the status of animals. Certainly more research needs to be done, but findings typically confirm what people suspect: animals are excellent "social lubricants." They make it easier for people to talk to each other. Companion animals can also be a person's most intimate confidant and most stable source of emotional support; as Levinson said, a "pet is an island of sanity in what appears to be an insane world." Professors Alan Beck and Aaron Katcher also posit that the physical contact animals allow—something normally missing in other types of psychological therapy—is quite important. Animals can help people overcome depression and loneliness. They can help people stay healthy and survive illness.

The number and types of animal-assisted interventions have grown tremendously. AAIs—therapies, activities, and educations—are implemented at libraries, schools, farms, hospitals and hospices, retirement centers, community centers, and other locations. The animals involved include companion animals (dogs, cats, fish, rabbits, guinea pigs), domestic animals (horses, cows, pigs, goats), or others, including dolphins and silk worms. Those incorporating animals include educators; museum curators; occupational, physical, and clinical therapists; counselors; psychologists and psychiatrists; and negotiators. Nearly all AAIs are volunteer driven.

A recent census of care facilities in Brabant, Belgium, found that 86 percent of AAI programs began without initial planning or formal preparation. Although practitioners should know their animals quite well and be adept at training and handling them, most practitioners haven't received training in incorporating their animals into their programs. It is important to note that many exceptions do exist and that some AAIs—especially some animal-assisted activities—do not require extensive planning. Nonetheless, there is general agreement among AAI organizations and experts that the field would benefit from better oversight and training. High quality AAI training programs do exist, but they are few.

AAI is still a field in its infancy, and much work remains to be done. But if implemented correctly, AAIs can provide low-cost, complementary treatments and activities for a wide range of conditions.

See Also

Applied Anthrozoology and Veterinary Practice—*Leo Kenneth Bustad (1920–1998): Pioneer in Human-Animal Interactions*

Further Resources

Adams, T. *Anthrozoology: A brief history*. Retrieved January 28, 2006, from http://www.anthrozoology.org/media/anthrohistory.htm/

Arkow, P. (2004). *Animal-assisted therapy and activities: A study, resource guide and bibliography for the use of companion animals in selected therapies* (9th ed.). Stratford, NJ: Author.

Beck, A. M. (1996). *Between pets and people: The importance of animal companionship—revised edition.* West Lafayette, IN: Purdue University Press.

Beck, A. M., & Katcher, A. H. (1984). A new look at pet facilitated therapy. *Journal of the American Veterinary Medical Association, 184*, 414–21.

Beck, A. M., & Rowan, A. (1994). The health benefits of human-animal interactions. *Anthrozoös, 7,* 85–89.

Bustad, L. K. (1988). Living together: People, animals, environment—A personal historical perspective. *Perspectives in Biology and Medicine, 31,* 171–84.

Corson, S. A., Corson, E. O., Gwynne, P. H., & Arnold, L. E. (1977). Pet dogs as nonverbal communication links in hospital psychiatry. *Comprehensive Psychiatry, 18*(1), 61–72.

Delta Society. http://www.deltasociety.org/

Farnell, L. R. (1921). *Greek hero cults and ideas of immortality.* Gloucestershire, UK: Clarendon Press.

Hines, L. M., & Bustad, L. K. (1986). Historical perspectives on human-animal interactions. *National Forum, 66*(1), 4–6.

Levinson, B. M. (1972). *Pets and human development.* Springfield, IL: Thomas.

Therapy Dogs International, Inc. http://www.tdi-dog.org/

Lieve Meers, Debbie Coultis, and William Ellery Samuels

Animal Assistance to Humans
Assistance and Therapy Animals

Assistance Roles Improving the Lives of People and Animals

From horses carting loads to cats killing mice and dogs sniffing out bombs, animals have provided varied forms of assistance to people throughout the ages. The term *assistance animals,* however, defines a subgroup of working animals. These animals, sometimes called "service animals," are trained to provide medical care or support to a person with physical, mental, or emotional disabilities, and typically reside with that person. *Therapy animals* are another subclass of working animals that also provide specific medical care or support to people with physical, mental, or emotional disabilities, but they differ from assistance animals by working with more than one person and residing with a handler other than the client or at the facility where they work. The relationships of assistance and therapy animals can and should benefit humans and, ideally, should benefit the animals as well. Also, there should not be any potential animal welfare concerns or unnecessary risk for the involved animal.

Guide Dogs

Perhaps the best known of all assistance animals is the guide dog. Guide dogs were introduced in the United States after the First World War by the organization The Seeing Eye, which is why guide dogs are often called "seeing eye dogs" by people today. Guide dogs were developed to assist people with varying levels of blindness by safely maneuvering them through their daily tasks. Many people find guide dogs to provide better assistance than such nonliving assistive devices as canes because they can detect and protect from danger (e.g., an oncoming car) while canes only allow the user to feel out objects. The dogs chosen to work as professional guides must not only be highly trainable but must also be intelligent enough to ignore their handler's command, such as to cross a street, if it is unsafe. Being a guide dog, therefore, is a very demanding job and can be quite strenuous, but most dogs appear eager to accompany their handler everywhere they go, as long as they are given appropriate time for rest and play. Overworking can be a great risk to guide dogs because, unlike humans, they cannot speak up if they are unhappy, although they may act out. The greatest risk is potentially fatal to dog, handler, or both, when a dog makes a poor decision in an unsafe situation. A concern that has only been anecdotally observed but not scientifically validated is that "guide dogs often go blind and hearing dogs often go deaf" at some point in their career. If guide or hearing dogs really do overuse the senses they "replace" for their handlers, resulting in decreased quality of life, early retirement, and possibly euthanasia, this is indeed an animal welfare concern.

Service Animals

The success of guide dogs led to new applications of assistance animals beginning in the 1970s, starting with the service dog. Like guide dogs, service dogs are trained to provide people with improved mobility and independence; however, they work with people who have mobility impairments ranging from quadriplegics to those with some ability to walk and/or use their arms but who need additional support for specific tasks. These tasks may include picking up dropped items, opening doors, helping someone when they have fallen down, and dialing the phone in an emergency. Specific tasks are often developed for specific disabilities, such as Parkinson's disease. Service animals are considered to be "adaptive technology"; that is, the animal can be trained to adapt to a client's needs as his abilities change—which can be particularly useful for people with degenerative diseases, such as Multiple Sclerosis. Service dogs do not need the energy of guide dogs because much of their working day may be spent lying by their handler's side waiting to be given a task. A somewhat mellow dog that is motivated to serve a handler may be very happy acting as a service dog because it gets to accompany its handler almost everywhere. There may be some animal welfare concerns though, because some service animals are trained to bear the weight of their handler when they are pulling themselves off the ground or to pull a wheelchair when the handler can no longer push.

Service Monkeys

Although not commonly used, service monkeys—a classification that includes monkeys as well as other nonhuman primates—may be familiar to the general public because they are frequently seen on television and in the movies. The idea of a primate that can provide 24-hour care to people with very limited mobility, particularly quadriplegics, may sound like an ideal solution to a difficult problem. The benefits to the primates are less evident. It is known that most nonhuman primates are highly intelligent, and some may enjoy having such a demanding job. However, it must be considered whether the animal would not be happier in a wild environment or at least not placed in a position of servitude. In addition, both the human and monkey are at risk of transmittable diseases. Humans and nonhuman primates are closely linked genetically, and they can easily infect each other with mild-to-severe diseases.

Hearing Dogs

Hearing dogs assist the hard of hearing by alerting them to such things as fire alarms, phone calls, baby cries, and car horns. These active dogs also benefit by having a job where they are always alert and get to use their excitability in a positive manner. Many dogs abandoned by owners who could not keep them active enough are rescued from animal shelters or pounds for just this reason. The hard of hearing handler not only profits from her dog's direct service, but also benefits by having the disability made visible to the world: sometimes it is useful for people to know why a handler is acting a certain way.

Guide Ponies

Guide dogs provide many benefits to their human partner by increasing their mobility and their independence, yet their working careers are often short, averaging eight to ten years if not shortened by illness or injury. For this reason some training programs have begun to explore the use of miniature ponies as substitutes for guide dogs. Ponies not only have longer working careers (20–30 years), but also have much better vision and may be more trainable and intelligent than dogs. These guide ponies can live in their handler's home and should receive the same right to enter commercial buildings as guide dogs do in the United States under the Americans with Disabilities Act. Despite these protections, it is not difficult to imagine that many buildings are not laid out for easy access to a person and his miniature pony, despite outfitting the pony with special tennis shoes to minimize damage to the facility.

Seizure Response and Seizure Alert Dogs

A type of assistance dog that has recently attracted much attention is the seizure alert dog. Seizure *response* dogs are trained to respond to a person during a seizure. This response may entail seeking help, protecting the person, or getting medication for the person after they have the seizure. What dogs have not yet successfully been trained to do, but some seem to take on of their own volition—even if they are the handler's untrained companion animal—is to be a seizure *alert* dog. These dogs let their handler know a seizure is coming on *before* it happens. This can allow the handler to get to a safer place, take medication, or call for help before the seizure occurs. Since the frequency of seizures often increases with stress, many people find that knowing they have an animal who will alert or respond to their seizure reduces their anxiety, decreasing their likelihood of having a seizure in the first place. Seizure response and alert animals may enjoy the benefits of accompanying their handlers in their daily lives, especially since many of them may go long periods of time between needing to perform a task. Regarding the dog's welfare, it is not known whether waiting for a seizure or seeing one's handler in a compromised position causes added stress, but it would not be surprising that these are some of the pitfalls of this job.

Similar to seizure response or alert animals are specially trained dogs who respond to such occurrences as diabetic emergencies or other body-chemical-related disorders. These animals typically provide similar benefits to their handlers, receive similar benefits as seizure alert/response dogs, and likely face the same potential animal welfare concerns.

Therapy Animals

Therapy animals, those who work with a handler or a facility to provide service to more than one client, lead very different lives than those of assistance animals. These animals may engage in either animal-assisted therapy (AAT) or animal-assisted activities (AAA). Per the Delta Society:

> AAT is a goal-directed intervention in which an animal that meets specific criteria is an integral part of the treatment process. AAT is directed and/or delivered by a health/human service professional with specialized expertise, and within the scope of practice of his/her profession. AAT is designed to promote improvement in human physical, social, emotional, and/or cognitive functioning [cognitive functioning refers to thinking and intellectual skills]. AAT is provided in a variety of settings and may be group or individual in nature. This process is documented and evaluated. . . . AAA provides opportunities for motivational, educational, recreational, and/or therapeutic benefits to enhance quality of life. AAA are delivered in a variety of environments by specially trained professionals, paraprofessionals, and/or volunteers, in association with animals that meet specific criteria.

Animals involved in either AAT or AAA are typically called therapy animals. Dogs, cats, birds, and other animals involved in either AAT or AAA are often someone's pet or companion, and that person chooses to volunteer his or her animal for a specific program. When treated properly, these animals get adequate downtime as a companion animal while being able to enrich the life of several clients by increasing their mobility, acting as "nonjudgmental" listeners to clients' traumas or psychoses, giving children the confidence to read aloud, or simply teaching kindness. These animals also act as advocates for their species by showing how well trained and intelligent they may be. The number of clients and staff they interact with, coupled with the sheer variety of species of therapy animals, may put them in a better position to be advocates. In countries where companion animals are feared or eaten, this is particularly important. These animals often lead people to a greater concern for all animals, both domestic and wild, while fostering a greater respect for the environment.

Although valuable, volunteer-handled animals may not provide results as beneficial to clients as those that are handled by professionals, depending on the dedication of the handler and the training of both handler and animal. These animals may also be at greater risk of stress, because volunteer handlers are not always sufficiently trained to identify those signs or may not always address problems sufficiently. An additional concern is that volunteer handlers may not remove their animal from a potentially dangerous situation early enough, putting the animal, clients, and staff at risk.

Horseback Riding (Hippotherapy)

Many therapy animals are also handled by professional handlers or institutions. Besides dogs, cats, and birds, they include horses, other livestock, rabbits, and in some areas amphibians and reptiles. Hippotherapy, or therapeutic horseback riding, is popular throughout the world. This particular type of human-animal relationship benefits clients by helping people with physical disabilities learn to balance themselves or those with emotional disorders, particularly anger disorders, learn to act without upsetting the animal. Hippotherapy is also helpful to autistic people, who may have a difficult time "reading" the social cues that others use to convey thoughts and ideas. It is hypothesized that it is sometimes easier for autistic people to learn cues from animals because they have fewer, more exaggerated, yet simple cues than humans.

In regards to the benefits to horses, many of them are retired to therapeutic riding programs at the end of their lives. Ideally, this provides these programs with calmer, more mature animals. Many of these animals enjoy the quiet days interrupted by short, simple lessons, but concerns can arise if animals are inappropriately matched with clients (e.g., putting a high-strung horse with an emotionally unstable client). Signs of stress, both short and long term, should always be monitored in therapeutic riding programs—as well as any other therapy program.

Psychiatric Service Dogs

A number of new classes of assistance animals that appear to bridge the gap between therapy animals and assistance animals have arisen recently. These animals are available for therapy work whenever it is needed. For example, psychiatric service dogs and social dogs help most by providing a permanent companion for a handler to rely on in times of need. Psychiatric service dogs go beyond that by addressing specific psychiatric needs, such as providing medications or interrupting obsessive-compulsive routines or manic behaviors. Having an animal that depends on a handler for its care may be enough for people to stay on medications or to leave the house, which they may have otherwise avoided. Psychiatric service dogs also improve the visibility of a handler's disorder, which some find helpful. For some people their psychiatric service dog may be their only friend, and this alone may help them cope with their disorder. The psychiatric service dogs benefit by the constant companionship of their handler, but some may clearly be at risk if their handler cannot provide such adequate care as feeding, walking, or emotional stability.

Social Dogs

Social dogs truly bridge the gap between therapy animals and assistance animals. These dogs primarily provide companionship for people who may otherwise have none. Social dogs are particularly useful to children with mental or emotional disabilities, particularly autism or Asperger's disorder. These children have difficulty understanding social cues and can therefore have a difficult time making friends. Ideally, social dogs can teach the children cues and also serve as a playmate. They can also teach autistic children to enjoy touch—something that many autistic people do not inherently appreciate. An added benefit of social dogs is that their good behavior and ability to perform unusual tasks may help some clients by attracting human friends. They also enhance the visibility of a disorder, which parents find beneficial when out in public with a child who would otherwise be seen as behaving inappropriately. Social dogs may lead wonderful lives if their work is minimal and they get to accompany their handler to enriching activities outside of the home. However, their welfare may be at risk if their handler's disability is so severe that he or she cannot learn to handle animals properly, causing them physical pain or emotional discomfort.

Assistance and therapy animals provide a wealth of benefits to their human handlers, including increased mobility, increased independence, visibility of a disorder, and simple companionship. The animals themselves may also benefit by the companionship, provision of an enriching job, or simply through advocating for their species. However, it is clear that there are animal welfare considerations to be made for all assistance and therapy animals. These should be given with equal if not more weight when evaluating the potential benefits; after all, this particular type of human-animal relationship should be beneficial to all involved parties.

See also

Animal Assistance to Humans—*Assistance Dogs*
Animal Assistance to Humans—*Giant African Pouched Rats Saving Human Lives*
Animal Assistance to Humans—*Horse-Assisted Therapy*
Animal Assistance to Humans—*Psychiatric Service Dogs*

Further Resources

Allen, K., & Blascovich, J. (1996). The value of service dogs for people with severe ambulatory disabilities: A randomized controlled trial. *Journal of the American Medical Association, 275*(13), 1001–06.

Coppinger, R., & Coppinger, L. (2001). *Dogs: A startling new understanding of canine origin, behavior & evolution.* New York: Scribner.

Davis, B. W., Nattrass, K., O'Brien, S., Patronek, G., & MacCollin, M. (2004). Assistance dog placement in the pediatric population: Benefits, risks, and recommendations for future application. *Anthrozoos, 17*(2), 130–145.

Delta Society. http://www.deltasociety.org/TextOnly/AnimalsFAQFAQ.htm

Hart, L. A., Zasloff, R. L., & Benfatto, A. M. (1996). The socializing role of hearing dogs. *Applied Animal Behaviour Science, 47*(1/2), 7–15.

Mader, B., Hart, L. A., & Bergin, B. (1989). Social acknowledgments for children with disabilities: Effects of service dogs. *Child Development, 60*(6), 1529–1534.

Salotto, P. (2001). *Pet assisted therapy: A loving intervention and an emerging profession: Leading to a friendlier, healthier, and more peaceful world.* Norton, MA: D.J. Publications.

Valentine, D. P., Kiddoo, M., & LaFleur, B. (1993). Psychosocial implications of service dog ownership for people who have mobility or hearing impairments. *Social Work in Health Care, 19*(1), 109–125.

B. Witkind Davis

Animal Assistance to Humans
Assistance Dogs

On a warm Friday evening near the turn of the millennium, a tall, trim woman in her sixties took hold of the handle on the leather harness of large young golden retriever, and with an exclamation of pure delight she set off at a brisk pace along a crowded pedestrian mall on Miami Beach. She had never met the dog—her eighth guide dog in nearly fifty years—but without hesitation she entrusted herself to him. Feeling her calm certainty, the young male expertly navigated, for block after block, a mass of pedestrians, dogs, and rollerbladers, moving in all directions at different speeds. The dog stopped at cross streets to look for cars, bicycles, motor scooters, or other obstacles before proceeding. For safety, the dog's trainer followed the pair's maiden voyage, but he had nothing to do.

Guide dogs for the visually impaired are the best known of the assistance dogs—animals trained to help people with disabilities overcome their physical or psychological limitations and lead more independent lives. Terminology in this field is fluid, but Assistance Dogs International, comprising nonprofit organizations providing these specially trained animals, identifies three main types of assistance dogs—guide dogs for the blind and visually impaired; hearing dogs for the deaf and hearing impaired; and service dogs for people with other disabilities. Service dogs retrieve objects, pull wheelchairs, open doors, anticipate and warn about epileptic seizures, turn on lights, locate people, provide balance and stability, dial 911 on specially equipped phones, alert to fire alarms and other specific sounds, and perform other tasks for which they are trained.

Studies have shown that disabled people with assistance dogs generally have higher self-esteem, lower blood-pressure, less anxiety and stress, and are less isolated than those without dogs. Trained service dogs are no panacea; sometimes the human and dog simply do not get along temperamentally or physically. But people with dogs say that when they do meld as a team the result can be nothing short of magical.

In ancient mythologies from around the world, dogs appear as guides for the dead. Although when they began serving as guides for the living is unknown, what records do exist indicate an association of long duration, which is not surprising, given the dog's talents. In the early ninth century, town elders in Geel, Belgium, reportedly urged their fellow citizens to give shelter and assistance to disabled people and their dogs. In the sixteenth century, French essayist Michel de Montaigne praised the guide dogs of Paris and neighboring villages for their ability to avoid human and natural hazards. "I have seen them, along the trench of a town, forsake a plain and even path, and take a worse, only to keep their masters further from the ditch," he wrote.

After that, references to guide dogs and, by the nineteenth century in France, dogs to alert the hearing impaired to visitors at their door became more common. But training the dogs was apparently an individual or small-scale enterprise until World War I, when Germans began systematically training their shepherd dogs to guide soldiers blinded by mustard gas. The French followed suit, and the modern guide dog soon spread to other industrialized nations. The feats of some of these animals became legendary.

The first of the German-trained dogs, Lux, was brought to the United States in 1925, a gift to Minnesota Senator Thomas D. Schall. Helen Keller was given a privately trained guide dog around the same time, and in 1929 Dorothy Harrison Eustis established the first guide dog program in the country, the Seeing Eye Foundation. Believing the American stock of poor quality, Eustis bred German shepherd dogs in Switzerland to serve as guide dogs in America.

Following World War II, a growing body of research documenting the value of dogs—and pets in general—in therapy, combined with the beginnings of a shift away from compulsory training methods to humane techniques emphasizing positive reinforcement, led to a blossoming in the 1970s of new ways to employ dogs' special talents and unique

Frank Morris and Buddy, the first seeing eye dog in America, at the Missouri Blind School in the early 1930s. Courtesy of Animal Image Photography.

relationship with humans. Therapy or facility dogs were trained to go with their handlers into hospitals and nursing homes, where they were known to be effective in connecting to patients isolated by mental illness, Alzheimer's, and other debilitating diseases. Service and hearing dogs, like guide dogs, served the needs of individuals with disabilities that limited their mobility and independence.

Founded in 1975 in Santa Rosa, California, Canine Companions for Independence was the first of these new service dog training programs in the United States, and it remains a leader in the field. The National Education for Assistance Dog Service (NEADS) was incorporated the following year in Princeton, Massachusetts. In 1978 the Delta Society was established in Oregon—it is now headquartered in Seattle, Washington—to promote research into the dog-human bond and the health benefits of companion animals, as well as to sponsor education and advocacy programs. Its National Service Dog Center provides dogs from its own breeding lines, from other breeders, and occasionally from animal shelters.

The American with Disabilities Act (ADA) of 1990 mandates that dogs trained to help people with disabilities overcome their limitations and to behave in public with proper decorum be allowed to go wherever the public is admitted. No certification papers or proof of training are required, nor does the act specify standards for training and performance. The ADA is widely credited with providing legal protections for the estimated 52 million disabled Americans.

It is estimated that only about 1 percent of disabled Americans use assistance dogs. Yet demand for trained dogs has risen, putting pressure on providers, many of them nonprofit organizations, to provide dogs, which should pass rigorous training and temperament standards. Local entrepreneurs have moved to help fill the need with varying degrees of competency and success.

The situation has led to calls in some quarters for establishment of national standards for certification, but at this writing no consensus has emerged on that subject. Assistance Dogs International has developed what it calls "standards of excellence in all areas of Assistance Dog acquisition, training and partnership." Guide dog schools have their own organization, as do assistance dog partners and trainers. All of these groups have their own agendas, a well as a desire to promote the use of assistance dogs.

The German shepherd, the original modern guide dog of choice, has been supplanted by the Labrador retriever, golden retriever, and crosses between, partly because of the difficulty in obtaining high quality German shepherd dogs and partly because the retrievers are believed physically and temperamentally better suited for assistance work. But other types of dogs, including mixed breeds, can be equally successful.

Most of the top guide dog programs and large service and hearing dog programs breed—and sometimes employ select individuals to breed—train, certify, and place their own dogs. Some organizations, such as NEADS, use rescued and donated dogs. At the best programs, puppies are farmed out to carefully selected foster homes for generally eighteen months to two years of intensive socialization before they are brought to the organization's training center. There they undergo extensive temperament testing and six months or more of training before they are certified and matched with a handler. Training is rigorous and highly specialized; the failure rate is high. In addition to learning to navigate the obstacles, dangers, and helpful devices of the machine age, for example, guide dogs must learn "intelligent disobedience"—that is, when to ignore an explicit command from their handler in order to prevent an accident.

Dogs that fail in training are generally placed in homes as pets or employed as detector dogs. Those who make the grade have a working life of six to eight years before their retirement. During that time, they have the potential to transform their handlers' lives.

Further Resources

Assistance Dogs International, Inc. http://www.adionline.org/
Canine Companions for Independence. http://www.caninecompanions.org/
Delta Society. http://www.deltasociety.org/
Guide Dog Users, Inc. http://www.gdui.org/
Montaigne, M. (1892). Apology for Raimond de Sebonde. In W. C. Hazlitt (Ed.), *The essays of Michel de Montaigne.* (C. Cotton, Trans.). New York: A.L. Burt.
National Education for Assistance Dog Services. http://www.neads.org.

Mark Derr

Animal Assistance to Humans
Giant African Pouched Rats Saving Human Lives

Giant African pouched rats (*Cricetomys gambianus*) save human lives by detecting buried landmines in suspected minefields; by detecting earthquake victims under the rubble of collapsed buildings; and, in some cases, by detecting pulmonary tuberculosis in possibly afflicted patients.

While landmine detection is currently deployed on minefields in Mozambique, and has been accredited according to International Mine Action Standards (2006), the two latter applications (use in earthquakes and for tuberculosis detection) are still in research and development.

Use of Pouched Rats for Humanitarian Tasks

Giant African pouched rats are an African indigenous rodent species, prevalent throughout the tropical zone of the continent. Until the mid-1990s, very little was known about these giant rodents, apart from being used as a protein food source in several African cultures and as a curiosity in some Western zoos.

Giant African pouched rats have an extremely developed olfactory organ. They have been domesticated, and they live much longer (6–8 years in captivity) than most rodents, which results in a good return on the initial training investment.

It is well known that rats are suitable for repetitive tasks in experimental settings. Less known is that they are very social and intelligent animals. Though pouched rats are bigger and calmer then most rodents, they are too light to set off landmines. Giant African pouched rats have many logistic advantages: they are easy to train, convenient to breed and maintain, and easy to transport in large numbers to an operational area. Moreover, they are cheap, and available in large numbers, and resistant to most tropical diseases (including tuberculosis).

How Mine Detection Rats Are Trained

Giant pouched rats are trained in a step-by-step process, through positive behavior reinforcement. It takes 10–12 months to train a landmine detection rat. At the age of four weeks, when babies start opening their eyes and developing fur, socialization begins. Babies are weaned from the nest at the age of five weeks, and classical conditioning begins. A click sound is associated with such food rewards as peanuts, bananas, and avocados.

A giant African pouched rat in training in an evaluation cage at APOPO, an organization that trains rats to find land mines in Africa, February 2005. Here the rat is pointing to a positive sample by keeping its nose in the sniffer hole. Photo by Christophe Cox. Used by permission.

The clicker is used as a conditioned reinforcer to shape the desired behavior. Rats learn to discriminate the scent of explosives set in a line, and pinpoint the source of the target scent with their noses. Once the imprint is well established, the animals learn to walk in a harness on a leash and search for the target scent while under the surface of a small sandbox. The pouched rats learn to scratch the topsoil above the target to indicate its position.

In subsequent steps, they gradually learn to cover longer distances and more open spaces to indicate buried landmines and explosive devices. A fully trained rat can scan 100 square meters of suspected area in less than a half hour, and three rats are used in a row to maximize detection probability.

Where This Training Is Done

The training of the giant African pouched rats is the result of the work of the nongovernmental organization (NGO) APOPO, a cooperation between Antwerp University (Belgium) and Sokoine University of Agriculture (Tanzania). APOPO has programs in Tanzania and Mozambique, involving over 100 staff and 250 rats.

Further Resources

APOPO. http://www.apopo.org/

Cox, C., Weetjens, B., Machangu, R., Billet, M., & Verhagen, R. (2004). Rats for demining: An overview of the APOPO program. Proceedings of the Eudem Conference on humanitarian landmine detection technologies. VUB, Brussels.

HeroRat. http://www.herorat.org. [This user-friendly and animated website contains very useful information on the effects of landmines, why rats are important in landmine detection, and how to adopt a rat.]

Verhagen, R., Cox, C., Machangu, R., Weetjens, B., & Billet, M. (2003). Preliminary results on the use of *Cricetomys* rats as indicators of buried explosives in field conditions. Geneva International Centre for Humanitarian Demining, Mine Detection Dogs, Training, Operations and Odour Detection, 175–93.

Verhagen, R., Weetjens, F., Cox, C., Weetjens, B., & Billet, M. (2006). Rats to the rescue: Results of the first tests on a real minefield. *Journal of Mine Action, 9*(2). http://maic.jmu.edu/JOURNAL/9.2/RD/verhagen/verhagen.htm

Bart Weetjens

Animal Assistance to Humans
Horse-Assisted Therapy: Psychotherapy with Horses

Equine-facilitated psychotherapy (EFP) is the use of the horse in mental health therapy. Horses are used in equine-assisted activities for improvements in physical, psychological, emotional, and social well-being. Over 40,000 people currently participate in some form of equine-assisted activity in the United States. Although much of the terminology and research in the area is relatively new, horses have been used for therapeutic purposes for centuries; the ancient Greeks noticed how much riding lifted the spirit and prescribed it for the terminally ill.

The formal use of horses for therapy became prominent in the 1950s after polio survivor Lis Hartel of Denmark, having rehabilitated with the aid of horse riding, won an equestrian Olympic silver medal in 1952. Interest has grown over the years, resulting in hundreds of national and international membership organizations, which promote ethical and safety standards, provide training, and encourage sound research within the growing field. There are now many subtypes of equine-assisted activities, including hippotherapy (physical therapy with the aid of the horse's movement), equine-experiential learning, vocational rehabilitation, therapeutic driving, therapeutic riding, and equine-facilitated mental health therapy or psychotherapy.

Equine-facilitated psychotherapy became a distinct specialty in the 1990s as anecdotal evidence suggested there were multiple psychosocial benefits in addition to the physical improvements from equine-assisted activities. Some of the psychosocial benefits include enhanced self-confidence, trust, self-acceptance, impulse control, boundary maintenance, nonverbal and verbal communication, empathy, compassion, mutual trust, and a reduction in anxiety, fear, and aggression. Psychotherapists see the horses as their partners, cotherapists, or tools. A typical session includes the participant, the horse, the horse professional, and the licensed clinical mental health professional. The collaboration of both professionals is important for safety and ethical reasons.

The experiential nature of EFP allows people to connect to aspects of the natural world, which may help them to reconnect with their inner selves. Unlike conventional talk therapists, horses do not listen to the words people say; they simply observe and respond to the subtleties of human behavior. Horses are dynamic, large, powerful, social animals. In domesticated settings, horses are dependent upon humans for their existence, yet they retain their distinct and sometimes complicated personalities. In order to form an effective partnership with horses, people need to develop trust and respect for the horses and for their own emotions. They cannot mask their emotions, as the horses will rapidly mirror their repressed feelings. With the aid of the therapist, participants can work through their emotions and patterns of behavior within the relationship that they have formed with the horse.

A wide variety of psychosocial issues can be addressed by EFP, ranging from severe mental illness, substance abuse, eating disorders, domestic violence, and child abuse to relationship and communication problems. EFP can take place on trail rides, in an arena, while riding, driving, vaulting, or performing ground work with the horse, or simply while grooming or being in the field or stable with the horse. Equine-facilitated therapy involves all the senses, can promote teamwork, problem solving, creative thinking, communication, leadership, and encourage overall emotional growth and learning.

Further Resources

Equine Assisted Growth and Learning Association (EAGALA). http://www.eagala.org/

Equine Facilitated Mental Health Association (EFMHA). http://www.narha.org/

Federation of Riding for the Disabled International (FRDI). http://www.frdi.net/

Fitzpatrick, J. C., & Tebay, J. M. (1998). Hippotherapy and therapeutic riding. In C. C. Wilson and D. C. Turner (Eds.), *Companion animals in human health* (pp. 41–58). Thousand Oaks, CA: Sage.

Horses and Humans Foundation (HHF). http://www.horsesandhumans.org/

North American Riding for the Handicapped Association, Inc. (NARHA). http://www.narha.org/

Thomas, C. E. (Ed.). (2006). *Equine facilitated mental health and learning: An annotated bibliography.* Denver, CO: EFMHA.

Clare Thomas

Animal Assistance to Humans
Horse-Assisted Therapy: Reaching Troubled Young Women

Makita (not her real name), a young girl who has not spoken for months, grazes a horse by a riding field in San Antonio. She is a patient in an equine-facilitated psychotherapy (EFP) program for at-risk juveniles. EFP instructor Leslie Moreau had Makita partly in her line of vision from the riding ring when she heard a voice coming from the girl's direction.

"You're really pretty," said the voice to the horse. Moreau realized it was the once-silent girl. Moreau said she had witnessed a "profoundly powerful" catharsis.

The horse-as-healing agent in the mental health field often produces results more effectively and quickly than conventional talk therapy. The client works with an animal partner that provides unconditional acceptance. Interaction with the horse provides a safe psychological place to process feelings and create a healing bond. "One of the most compelling aspects of equine-facilitated psychotherapy is the opportunity to form positive bonds and attachments," says Moreau, well established in the field. Significant bonds lacking love or consistency are a place of core psychic injury in EFP patients. And riding provides an occupational therapy vehicle for trauma survivors, who disengage physically or psychically. Those with PTSD from abuse often don't allow themselves to feel because that can bring back sensations related to abuse. Having worked eighteen years in EFP, Moreau has seen clinical improvement in clients with conduct disorder and post traumatic stress disorder, in victims of sexual and psychic abuse, and with capital and sexual offenders ("Outlaw Riders," 2001).

Some of Moreau's population have learned to bamboozle the system by saying what they think the therapist wants to hear. The equine advantage, quips Moreau, is, "You can't snow a horse." The interaction of horse and handler is a bridge to human relations: in various situations with the horse, the client is asked 'where does this situation happen in other aspects of your life?' For instance, a client becomes bonded with the animal and wants to treat it kindly. "I wonder if you could do this with your brother?" prompts the therapist. In this session the client processes questions about behavior and is then assigned them as her homework. If the client is angry at the horse for some reason, Moreau confronts her in colloquial terms (note some kids are more apt to relate when the experience doesn't sound like "therapy"): "Let's talk about what's pissing you off. What are you doing that makes you angry?"

Another opportunity for exploration is when the horse exhibits behavior the clients identify with: if a horse becomes unruly or is docile when treated well, Moreau probes, urging answers the horse has personified with a nonthreatening presence: "Who does this remind you of?" "And why did you continue to behave that way?"

EFP, a long-term (three months or more) intervention, is one of several mental health approaches using horses. It is a dynamic of understanding between horse and human, supported through a horse-savvy psychologist and an EFP-certified instructor. At Horsepower in Temple, New Hampshire, director Boo McDaniels, a pioneer in the equine–facilitated mental health profession, recounts a riding lesson with Jennifer, a girl once chemically addicted and repeatedly raped by a family member.

"Honey, you've got to stop her," McDaniels implored to the rider on her pony. "You've got to take charge. You have the power to make it happen." Upon learning her sense of mastery and discovering new boundaries, Jennifer stopped the pony many times. The session applied positive psychology to build up strength and confidence.

Short-term therapy with horses, in a different format, can also serve as a preventive measure: "We help with kids who are just starting to slip through the cracks," says Priscilla Marden, director of Horse Warriors in Jackson, Wyoming. In the equine-assisted experiential program (a form of equine-facilitated therapy) there, kids ride through the sagebrush mountains, rain or shine. The curriculum of Horse Warriors is designed to build character and camaraderie. The goal is to restore kids before they are sent to expensive residential centers and juvenile courts.

In the intensive work of EFP, time with the horse can be cathartic. Cassandra was in her mid-twenties, afflicted with the effects of domestic violence and sexual abuse: the emotionally harmful edges of critically low self-esteem and fear. In the EFP program, HorseMpower, in Ocala, Florida, Cassandra grooms a bay Morgan horse in a stall. The horse, a calm nineteen-year-old gentle enough to be a handicap-therapy horse, had unwittingly crowded her. "Just move him over by pushing on his haunches and shoulder," cofounder and psychologist Marilyn Sokolof instructed. But the girl was paralyzed by a fear of asserting herself. In that moment, she was afraid the horse would strike at her in some form (the deeper significance stemmed from her history that if she asserted herself with her abusive partner or family member, she'd be harmed).

"The therapist was available to process the deeper issues with the client," explained Sokolof. "As the client gained an understanding of her fears, she was able, with support, to begin asserting herself with the horse (and in her life)," she said.

In another case, a girl named Mariah had tried to stab her mother with a knife. She came to Moreau after trying myriad therapies. Her lessons in Moreau's program were an experiential metaphor: she refused to assert herself as a rider—slouching, becoming a dead weight. To not ride as a passenger, Mariah had to learn to be an assertive communicator. Moreau urged her with the question: if the horse isn't moving, whose responsibility is it?

The match of horse and rider is important. Moreau described the horse Mariah was teamed with, Caesar, as one who would give her what she needs: a partner who asked that he be handled right—one who could "outstubborn" her.

Therapy with horses is fun. Most adolescents are excited about working with a horse, and they see therapy as "I get to ride a horse!" A client commented on the dynamic relationship created with the animal: "I feel really relaxed around horses."

Further Resources

Bowers, M. J., & MacDonald, P. M. (2001). The effectiveness of equine-facilitated psychotherapy with at-risk adolescents: A pilot study. *Journal of Psychology and the Behavioral Sciences, 15*, 62–76.

Macauley, B. (2006). *The resources for research and education in equine assisted activities and therapy*. Denver, CO: North American Riding for the Handicapped Association.

McCormick, A., & McCormick, M. (1997). *Horse sense and the human heart*. Deerfield Beach, FL: Health Communications, Inc.

Moreau, L. *Outlaw riders: equine facilitated psychotherapy with juvenile capital offenders*. http://www.cyc-net.org/cyc-online/cycol-0405-outlawriders.html

———. (Winter 2001). Outlaw riders: Equine facilitated psychotherapy with juvenile capital offenders. *Reaching Today's Youth, 5*(2).

North American Riding for the Handicapped Association; Equine Facilitated Mental Health Association. http://www.narha.org/SecEFMHA/WhatisEFMHA.asp

Rector, B. (2001). *The handbook of equine experiential learning: Through the lens of adventures in awareness*. Self-published. [This text is available from: http://www.adventuresinawareness.net/products.]

Andrea Reynes

Animal Assistance to Humans
India and Animal Therapy

While the beneficial aspects of companion animals has been generally well known in India for many years, there was no formal pet therapy program until 2001. It is thought that the first animal used for organized pet therapy in India was Cleopatra, a friendly dachshund that worked with autistic children at the Saraswathi Kendra Learning Centre for Children in the late nineties.

In 2000 Jill Robinson, founder and CEO of Animals Asia Foundation (AAF) of Hong Kong, offered to help assess the suitability of the first group of dogs in India to start a pet therapy program with the Blue Cross of India in Chennai, India. AAF had helped start a similar program in the Philippines at Manila, with the Philippines Animal Welfare Society. The assessment of the dogs is very important because even one incident of hostility could jeopardize the entire program.

Twenty-one prescreened dogs were assessed by Robinson and the staff of AAF, and seventeen were certified to be "Dr. Dogs." The program aims to partner happy dogs with happy people. Jumble, an eight-year-old mixed breed, was the first such dog in the program. (Most of the other dogs were Labrador and golden retrievers.) All dogs used in the program must be at least one year old; must obey certain basic commands; must be spayed or neutered; and must be accompanied by their caregiver on all visits to the designated institution—be it a school for special children or a senior citizen's home.

Since 2001, these and other Dr. Dogs have been helping hundreds of children in many schools for special children in Chennai with spectacular results, especially with autistic children. There is clear evidence that this is more effective if there is a one-to-one session between the dog and the child. Bringing a dog to a large group of children is not only poor therapy; it could be stressful to the dog.

With almost a decade of experience, AAF has brought out a manual for the Dr. Dog program, which is followed by the Blue Cross of India for its Dr. Dog program. Without exception, every institution where this program has been in operation in Chennai has found it most beneficial.

The medical benefits of having a companion animal are not "proven" quantitatively. In the fifties, doctors in the Soviet Union used dolphins for pet therapy and showed that swimming with dolphins markedly reduced depression in humans. A review of recent studies published in the *British Medical Journal* (2005) disputed the claims that pet ownership was linked to a lower risk of heart disease, or that pet owners made fewer doctor visits, or that owning a pet boosted the mental and physical health of older adults. However, Dr. Leonard C. Marcus, a veterinarian and medical doctor in Newton, Massachusetts, who is also the founding Director of the Program for Health Care Negotiation and Conflict Resolution at the Harvard School of Public Health (HSPH) told me that, "Pets provide companionship, they are a source of comfort, and people who have difficulty communicating with other people can express themselves freely with pets" (personal communication, 2007).

It is this last aspect of people expressing themselves freely with dogs, who are nonjudgmental, that makes the use of dog therapy so valuable for children with autism and for stroke victims in speech therapy. But pets need not be kept only by those with a special need. There is sufficient evidence to show that just the mere action of stroking a dog or cat has the effect of calming a person down and reducing blood pressure.

The two programs of Animals Asia Foundation, Hong Kong—Professor Paws and Dr. Dog—have been widely replicated with amazing results in many Asian countries. As mentioned above, the Blue Cross of India started their Dr. Dog program in Chennai with help from AAF in 2001. Several cities in India—notably Bangalore, Vishakapatnam, and Ludhiana—have pet therapy programs, and AAF has several hundred canines in their Dr. Dog programs in Honk Kong, China, Vietnam, and Thailand.

Further Resources

Animals Asia Foundation, Doctor Dog. http://www.animalsasia.org/index.php?module=4&menupos=3&lg=en

McNicholas, J., Gilbey, A., Rennie, A., Ahmedzai, S., Dono, A., & Ormerod, E. (2005). Pet ownership and human health: A brief review of evidence and issues. *British Medical Journal, 331*, 1252–54. http://www.bmj.com/cgi/content/full/331/7527/1252

S. Chinny Krishna

Animal Assistance to Humans
Psychiatric Service Dogs

Three Stories

Mary is a gifted PhD student in sociology. She sits in her living room typing away on her laptop. A crescendo of noise disrupts her concentration. "It sounds like a crowd of people in a heated argument, but where are they?" Mary wonders. Then she hears her name called loudly. In moments such as these, Mary has learned to rely on her dog's natural response to its immediate environment to discern what is real. Paxil, her dog, is sound asleep on the couch next to her. Mary breathes a sigh of relief. "It is just another hallucination," she concludes. "The people aren't really there, and I should probably take some more medication." Mary has schizophrenia, and Paxil is her psychiatric service dog.

> Bill is driving to work, and today's commute is especially bad. Suddenly Bill slams on his brakes and sounds his car horn. An "idiot" just pulled in front of him. Bill can't wait to get around the car to leer at its incompetent driver. A few minutes later, a yellow stoplight turns red, and Bill screeches to a halt. "Pedestrians are so annoying," he grumbles to himself. At that moment, Bill catches a glimpse of his dog, Barney, who is wearing a seatbelt in the backseat of the car. Barney is wide eyed and looking at Bill as if he does not recognize him. Bill registers Barney's fear, which jolts Bill back to a more reasonable consciousness. "I am scaring my dog. I must be irritable hypomanic," he reflects. Bill pulls over to the curb and parks his car. He takes a few deep breaths and begins stroking Barney's fur. He thanks Barney for the alert and decides to walk with him the rest of the way to work. Bill has bipolar disorder, and Barney is his psychiatric service dog.

Colleen passes through a metal detector at the local courthouse with her dog, Rachel. The machine beeps loudly, and Colleen is directed towards secondary screening. She thinks to herself, "I should never have left home today. Police make me so nervous." Colleen waits in line for secondary screening. Meanwhile, Rachel begins to lean into Colleen's legs with all her weight. The sensation of Rachel's body against her own provides Colleen with somatic awareness. Colleen realizes that she is beginning to hyperventilate. "This always happens right before I'm going to have a panic attack," she thinks. Colleen kneels on the floor next to Rachel and takes deep conscious breaths while stroking Rachel's fur and gazing into her serene eyes. Another panic attack is averted. Colleen has panic disorder, and Rachel is her psychiatric service dog.

Canine-Human Partnerships to Combat Mental Illness

The common theme among these real-life vignettes is a critical moment of therapeutic insight that is facilitated by canine-human partnership. In a psychological context, insight may be described as knowing what is happening to you, as it is happening, and knowing what to do about it, and then doing it. For persons disabled by severe mental illness, insight is vital to mental health disability management. With insight, a person with panic disorder can remember to take deep breaths to prevent a panic attack. A person with bipolar disorder is better able to appreciate that he is driving aggressively and should get off the road. Episodes of mental illness compromise insight, namely, one's ability to figuratively and mentally "connect the dots." Canine-human partnerships facilitate therapeutic insight among human handlers debilitated by episodes of mental illness.

The Americans with Disabilities Act of 1990 defines a service animal as one that is "individually trained to do work or perform tasks for the benefit of an individual with a disability."[1] Psychiatric service dogs are trained to accompany and interact with their mentally ill handlers in structured and therapeutic ways on a 24-hour basis. Some therapeutic interactions take the form of trained tasks that a dog may perform on command. Table 1 provides an abbreviated list of trainable tasks organized by diagnostic category.

Not all forms of psychiatric service dog assistance may be construed as tasks, however. In many cases, the dog assists by providing a therapeutic function for its human handler, such as helping an agoraphobic person leave his or her home safely for the first time in years. Hallucination discernment is another example in which a handler leverages her

Table 1 Trainable Tasks by Diagnostic Category

Disorder	Symptom	Trainable Task
Major depression	Apathy Hypersomnia Feelings of isolation Sadness or tearfulness Memory loss	Tactile stimulation Wake up handler Cuddle and kiss "Hug" Lick tears away Remind to take medication Help to find keys or telephone
Bipolar (manic phase)	Aggressive driving Racing thoughts or distractibility Insomnia	Alert to aggressive driving Tactile stimulation Alert to insomnia
Schizophrenia	Hallucinations Forgotten personal identity Confusion or disorientation Social withdrawal Feeling overwhelmed	Hallucination discernment Carry handler identification documents Take handler home Facilitate social interactions Buffer handler in crowded situations
Panic disorder	Pounding heartbeat, nausea, or sweating Dizziness Fear or fight/flight response	Staying with and focusing on handler Brace or lean against the handler Lead handler to a safe place
Post-traumatic stress disorder	Hypervigilance Fear or startled response Fear or anxiety Nightmares	Alert to presence of other people Threat assessment Turn on lights and safety check a room Interrupt by waking up handler Turn on lights for calming and reorienting

An expanded version of this table is available at www.psychdog.org.

dog's natural response to its immediate environment to attain insight into what is real and what is not. Through the mechanism of constant partnership, psychiatric service dog handlers learn to trust the feedback they receive from their dog. Such feedback may take the form of a characteristic "look," or facial expression, as in the case of Bill's aggressive driving; or it may take the form of an alert that has been shaped behaviorally, such as nudging a handler's elbow when the dog perceives an incipient neurochemical event in its human handler (such as "identity switching" in persons with dissociative identity disorder). Research has not yet resolved whether such alerts are merely behavioral observations made by the dog or they are in response to unidentified olfactory cues.

Further Resources

Fields-Meyer, T., & Mandel, S. (2006). Healing hounds: Can dogs help people with mental-health problems get better? *People*, *66*(3), 101–02.

Hendrickson, E. (2005). Well-heeled: Psychiatric service dogs are a lifeline to those with mental illnesses. *AKC Family Dog*, May/June, 44–45.

Johnson, A. (2005). Guard dogs of mental health. *Bark Magazine, 31,* 41–42.

Psychiatric Service Dog Society; 1911 Key Blvd #568 Arlington, VA 22201; (571) 216-1589; http://www.psychdog.org

Smith, M. J., Esnayra, J., & Love, C. (2003). Use of a psychiatric service dog. *Psychiatric Services, 54*(1), 110–11.

Joan Esnayra

Animal Assistance to Humans
Service Dogs: A Personal Essay

My Freedom Journey

To live with a disability is a challenge. To endure domestic violence is horrifying. To experience them together is overwhelming. Yet, despite the odds, my service dog and I overcame them both.

My journey began when I lay sprawled on the kitchen floor, punished for serving a TV dinner. That I'd worked a twelve-hour day, had a rod holding my spine together, and was his wife didn't matter. My husband had to "discipline" me. He had picked me up and thrown me six feet across the kitchen.

I tried to get up, but my legs were useless. Something was very wrong, and I could only lie there reliving the seventeen years of abuse from this man. Why couldn't I escape? The obvious answer was scoliosis. At the age of twelve I'd had spinal surgery for severe curvature of the spine. My spine was deformed into an excruciating S-shape, making daily activities difficult or impossible to accomplish. Equally crippling was the fear that I couldn't make it in the world by myself. My husband took full advantage of my fear: winning an argument merely took a well-placed kick to my back.

While I lay there assessing the pain that was burning my legs, my husband fixed a meal to replace the offending TV dinner. His boots clomped from refrigerator to stove; cans opened, pots clanged, and steak spit in the broiler. Not once did he turn in my direction. "Why doesn't he help me?" I thought. He sat and ate, turning pages of the newspaper. The boots that kicked me looked oddly benign in repose.

Finally I called, "Please take me to the emergency room. I think you've really hurt me." His chair scraped backward. Two dogs scurried like frightened mice, but the third, Penny, bared her teeth, growling in my defense. My husband bent down, his blue eyes glaring into mine. If I could have moved, I would have recoiled as he spat, "Get up and take yourself!" Laughing, he stomped out of the room. I vowed that even if it killed me, I'd find a way to get myself out of the marriage.

I had no clue how. Despairing, I watched the circling paws of my Shelties. In the midst of my suffering, they cared and would have helped if they could. I thought of all the dogs I'd loved and been loved by, their constant support and acceptance. Now, my Shelties' concern reminded me of a guiding eye dog I'd seen when I was seven. If a dog could be trained to help the blind, surely one could help me. I smiled and thought; "I will walk out of here, but I won't walk alone."

That night resulted in a year in bed and did permanent damage to my spine, but I continued my journey to freedom. I devoured books on domestic violence, learned helplessness, positive-reinforcement training, dog breeds, and service dog laws, and I studied for a degree with books propped on my chest. When my husband was at work, I studied and planned; when he was home, I remained still and silent.

After a final spinal surgery, the day came to act. I called my friend, Nancy, who taught puppy classes based on positive reinforcement. A member of the local kennel club, she pointed me toward breeders of golden retrievers, a dog big enough to help me. I visited breeder after breeder. My friend, Roy, wanted to puppy test the dogs, and we did, but none of them were quite right. Irrationally, I knew I would "just know" her and she me when the time came.

I found her in Frederick County, Maryland. The breeder said, "You don't want her. I call her Houdini because she escapes every pen I put her in." I walked over and looked into her eyes while she leapt up scratching to reach me. I smiled, held out my arms and said, "Journey! I've been looking for you!"

Our flight began with tiny steps, secret five-minute training sessions that ballooned into forty-five minutes. My husband wondered about the dog, but ignored her. Responding to the positive-reinforcement techniques learned in Nancy's puppy classes, Journey learned at an astounding speed. I learned how to teach her without using fear or punishment, horrors I knew too well. By six months of age, her repertoire included a hundred tasks, including doing the laundry, helping me dress and undress, picking up dropped items, grocery shopping, going to college with me to carry my books, and finding the phone if I fell.

I had spent most of my life confined by emotional and physical limitations. Journey hadn't. She was not bound by impossibilities. While my spirit had been imprisoned by fear, Journey's was free. She wasn't afraid of failure, because failure incurred no reprisals. When presented with a puzzle, she wouldn't give up. I'd say "try again," and she did.

If ever two species became one, it was Journey and me. It seemed we communicated on a spiritual level: I would think of something I needed help with, and she would

Journey doing the laundry, just one of hundreds of tasks she can perform. Courtesy of Karen Kraft.

anticipate it. During her training she never faltered or let me down, and I vowed never to let anyone hurt her.

My vow to protect her was unexpectedly called upon. I found that despite the Americans with Disabilities Act, we were not welcome everywhere. People stepped in our path to bar us from public places. We were taunted by teenagers, complained about by customers, and sneered at by employees. A neighbor threatened to shoot Journey, and we were removed from an airplane. In the face of discrimination, I defended her, and in so doing, learned to defend myself.

I was no longer an injured player benched on the sidelines. When someone threatened Journey; I stormed off the bench and fought. Because I loved Journey, I set boundaries. I said "no, you can't do that to her;" "no, you can't do that to us;" and finally, "no, you can't do that to me."

Still, I wondered if we could survive on our own. I hesitated to leave until the night my husband threatened to kill her. He was in another rage: he turned over tables, smashed chairs, and hurled dishes to the floor. I shook in fear; a rabbit afraid to act. Journey walked up to him with wagging tail, as if to assuage his anger. He lifted his fist, glared at me, and screamed, "Get her away from me or I'll kill her."

Every protective instinct I had was aroused. I thought, "You may hit me, but nobody hits Journey." I grabbed her and pulled her away. He hit the wall instead and stormed down the hall. Quietly, I noted where Journey's orange cape and leash were. I waited for what seemed eternity. When he answered the phone, I grabbed my keys and ran with Journey into the night.

I never went back. I didn't have to go back. Journey was truly my ticket to Freedom. She didn't let me down. Together we got our own apartment and went back to school. Together we trained other dogs for people with disabilities, and together we were in the newspapers and on television. Today, ten years later, together we write this story. Even now, she lies with her head on my feet. Today, we are free.

She is my hero, my Freedom Journey.

Journey fetches a library book from the shelf. Courtesy of Karen Kraft.

Further Resources

Bancroft, L. (2002). *Why does he do that?* New York: G. P. Putnam's Sons.

Barnett, O. W., & LaViolette, A. D. (1993). *It could happen to anyone: Why battered women stay.* Thousand Oaks, CA: Sage Publications.

Donaldson, J. (1996). *The culture clash.* Berkeley, CA: James & Kenneth.

Nosek, M. A., & Howland, C. (1998). Abuse and women with disabilities. *Violence against women, online resources,* from http://www.vaw.umn.edu.

Pryor, K. (1984). *Don't shoot the dog.* New York: Bantam.

Sidman, M. (2001). *Coercion and its fallout.* Boston: Authors Cooperative, Inc.

Walker, L. E. (1979). *The battered woman.* New York: Harper Perennial.

Karen Kraft

Animal Rights.

See Ethics and Animal Protection

Animals as Food

Dog Eating in the Philippines

Dog eating in the Philippines has been the subject of much controversy both in the Philippines and overseas, with animal welfare groups trying to end the practice.

The practice of eating dogs is found in many Asian countries, including China, Korea, Vietnam, and Thailand, to name a few. But people often associate dog eating with the Philippines partly because there has been much publicity around the practice there. During the 1904 St. Louis Exposition in the United States, there were living exhibits of various ethnic groups from the Philippines, including one where *Igorots*—the generic term for people living in the northern Cordillera region—would slaughter a dog each day. Pilapil (1994) speculates that the publicity around this exhibition may have left long-lasting impressions on Americans about dog eating and Filipinos.

While dog eating is found throughout the Philippines, its prevalence varies. It is most often associated with the Cordillera region, where the relatively cold weather is used to justify the eating of dogs, because the practice supposedly generates body heat. There is no scientific basis to this, considering that dog-eating Koreans say that the practice cools the body and is especially suited to hot summers.

The anthropologist Marvin Harris (1986, p.179) suggests that "[d]og-eating cultures generally lack an abundance of alternative sources of animal foods, and the services which dogs can render alive far outweigh the value of their flesh and carcass. In China, where perennial shortages of meat and the absence of dairying have produced a long-standing pattern of involuntary vegetarianism, dogflesh eating is the rule, not the exception."

Harris' hypothesis does not quite fit the Philippine case, with the dogs serving many functions besides that of food. Among the people of the Cordillera, the dog is a valuable companion animal for hunting and is a watchdog for agricultural fields and the home.

Dawang (2006), who is himself from the Cordillera region, argues that the slaughter of dogs used to be part of ritual sacrifice, done, for example, in times of illness or death. Emphasizing the gravity of the situation, the family dog has to be the one sacrificed. Dawang emphasizes that "[d]ogs were not butchered as drunkards' fare, nor as a daily or regular part of the Igorot diet. Igorot families much preferred to avoid the circumstances which might lead them to sacrifice their dog."

The idea that dogs are eaten because of protein shortage does not quite hold either for the Philippines, where there are different sources of animal protein, including wild animals such as boar and deer as well as domesticated pigs and chickens. There may be an element of convenience in the use of dogs—they are easily available and are free or of minimal cost compared to other animals used as food. Stray dogs are common and become easy prey for men looking for dog meat to accompany a drinking session. In urban areas, there are known criminal syndicates that specialize in the kidnapping of dogs, which they then sell to particular restaurants.

In 1998 the Philippine Congress passed Republic Act 8284, or the Animal Welfare Act, which prohibits cruel treatment of animals. The law has been used to crack down on dog eating, but the practice continues quite openly.

A growing Filipino middle class that views dogs as companion animals rather than just as watchdogs or as food has become more conscious about animal welfare issues and will probably become more vocal in opposing dog eating as well as other forms of maltreatment of animals.

See also

Animals as Food—*Dogs and Cats as Food in Asia*
Culture, Religion, and Belief Systems

Further Resources

Animal Welfare Act of 1998. Retrieved March 21, 2006, from Chan Robles Virtual Law Library Web site: http://www.chanrobles.com/republicactno8485.htm.

Dawang, B. A. (2006). Dog-eating and my Culture. Retrieved March 21, 2006, from The Igorot Web site: http://shaleys.tripod.com/id2.html.

Harris, M. (1986). *Good to eat: Riddles of food and culture*. New York: Simon and Schuster.

Pilapil, V. (1994). Dogtown U.S.A.: An Igorot legacy in the Midwest. *Heritage*, 8(2), 15.

Michael L. Tan

Animals as Food
Dogs and Cats as Food in Asia

The practice of eating dog meat has occurred in various parts of the world since ancient times. Places where there are records of dog eating include southeast Asia and Indochina, North and Central America, parts of Africa, and the islands of the Pacific. Archaeological evidence suggests that during the Neolithic and Bronze Ages, the practice of dog eating was widespread in Europe too. Less is known about cat eating, and although there is no evidence that it has ever occurred to the degree that dog eating has, it does have a long history and has occurred for ceremonial, nutritional, gastronomic, and medicinal purposes.

Today the eating of dog and cat meat has largely been extinguished; however, it still continues in a number of countries. This essay will focus on three of these: China, South Korea, and Vietnam.

China

Dogs as Food

In early North China (c. 6500–3000 BCE) and South China (c. 5000–3000 BCE) the dog (*Canis familiaris*) was one of the most important domesticated animals kept by farmers. Apart from being used in hunting, as guards, and in ritual and sacrifice, dogs were also a major source of meat; dog flesh was served at ceremonial dinners and was eaten by kings, and a special breed, the chow, was developed for eating.

Apart from being eaten for its taste and as a source of protein, dog meat was also eaten by the Chinese because it was considered to be good for the *yang*—the male, hot, extroverted part of human nature (as opposed to the female, cool, introverted *yin*). It was believed to "warm" the blood and so was eaten most commonly during winter. Indeed, as early as the fourth century BCE, Mencius, a Chinese philosopher, praised dog meat for its medicinal properties. He recommended it to treat liver ailments, malaria, and jaundice and to enhance virility.

Dog eating has occurred throughout China's history, even after it was officially banned by the first emperor of the Manchu (Qing) Dynasty (1644–1911), Nurhachi (because his life had been saved by a dog). Southern Chinese ignored this ban, and dog eating developed into a symbol of the anti-Manchu revolution. Dog meat was codenamed "three-six meat"—a play on the Chinese word for the number nine, which rhymed with the word for dog. Apparently, this code name is still used today in places such as Hong Kong, where dog meat has been illegal since 1950.

After the fall of the Qing Dynasty in 1911, the prevalence of dog eating fluctuated. The importance of this practice, however, was highlighted in 1983, when municipal authorities ordered the destruction of all dogs in the city, some 400,000. The only dogs exempt from this were police dogs, laboratory dogs, and those licensed for the meat trade.

During the 1990s, a number of mining firms and farmers faced economic ruin and so turned to new markets to survive. The intensive production of dog meat was seen as a profitable enterprise with much room for expansion. The Chinese government was enthusiastic and gave significant financial support to those undertaking this new endeavor.

Today dog meat is produced and consumed in many parts of China, but most commonly in southern China (especially the province of Guangdong) and the provinces of northeast China (Jilin, Heilongjiang, and Liaoning). Some dog meat is exported to North and South Korea and Russia. The Chinese prefer the meat of puppies to adults and dogs on sale are usually only six to eight months old. Puppy hams are a specialty.

In general, consumption is seasonal, with most dog meat being eaten in the winter months, as it is meant to be good as a warming food. Increasingly, however, it is being eaten throughout the year. The price of dog meat is comparable to the cost of beef, the most expensive meat in China, and it has been calculated that dog-meat farming is four times more profitable than pig farming and two times more profitable than chicken farming.

The sale of dog meat is not uncommon in southern China. Courtesy of Shutterstock.

Dogs used for their meat are sourced from commercial dog-breeding farms (those that produce pups for meat and those that breed dogs for both fur and meat) and small-scale family farms and some are purchased or even stolen (alive or poisoned) from people's homes.

Historically, the Chow Chow was the main breed of dog especially bred for food, although other breeds and cross-breeds were eaten and made available at markets, as well. In the 1990s, people became interested in producing a better and more economical breed of meat dog, and so a variety of breeds were imported to see which would be most suitable for crossing with their native dogs. Crossing indigenous, female dogs (e.g., Mongolian dogs) with breeds such as the St. Bernard, Tibetan Mastiff, Great Dane, Newfoundland, Leonberger, or Dalmatian proved successful, producing large, fast-growing dogs. The St. Bernard, in particular was considered the best breed to work with because it has a gentle nature, is fast maturing, and produces large-yielding litters of puppies. The Chinese wanted to introduce these qualities to a new breed of meat dog.

The St. Bernard cross-bred hybrid dogs are crossed again with a local, female dog, and it is the resulting puppies that are eaten at around six to eight months of age; in effect, it is the St. Bernard's grandchildren that are eaten. Purebred St. Bernards are not eaten, as the meat is considered to be too bland.

But while St. Bernard cross-bred puppies were developed as *the* meat dog for the expanding commercial farming of dogs, small-scale farmers and individuals continue to supply meat dogs through the breeding of mongrel/pariah types of dog.

No accurate figures are available on the amount of dog meat that is eaten in China, but it has been reported in the popular press that the demand for dog meat in 2000 exceeded 100,000 tons but that only 40,000 tons were supplied. If one takes as an average that each dog slaughtered has a carcass weight of approximately 20 kilograms, it would mean that about 2 million dogs were slaughtered for food in 2000 and that 5 million dogs would have been needed to meet demand. And with a population of 1.3 billion people, 100,000 tons of dog meat equates to only approximately 70 grams (140 ounces) per person per year. Compared with the amount of lamb (4.6 kg), beef (7.6 kg), chicken (22.2 kg), fish (34.9 kg), and pork (47.3 kg) eaten per person per year, the amount of dog meat eaten is very low.

A study commissioned by the International Fund for Animal Welfare (IFAW) of 1,738 city residents (aged 18 to 65 years) in China found that 43 percent had eaten dog meat at some stage in their lives, while a survey conducted by Li and coworkers (2004) of 1,300 Chinese university students revealed that 53 percent had eaten dog meat before. However, neither study explored how frequently people consumed this meat.

A man sits beside caged dogs in a market selling wild animals for dishes in Guangzhou, China, 2003. © AP Photo/str.

Cats as Food

When cat eating began is unclear, but there is evidence of "wildcats" being a novelty dish of wealthy urbanites in Sung times (960–1279 CE), and there is reference to cat eating occurring in Guangdong in the fourteenth century. There are also accounts from the eighteenth, nineteenth, and twentieth centuries of domestic cats being eaten and/or live cats being sold along the streets for eating, and cats' eyes were apparently offered in Cantonese food shops in the nineteenth century.

Today, like dog meat consumption, cat meat consumption is also seasonal and regional. It is mainly eaten in the winter and is most popular in southern China, particularly the province of Guangdong. However, cat is becoming more common in some cities in northeast China, especially where Cantonese have settled. The most popular and well-known cat dish is "dragon, phoenix, and tiger" (Lung fung foo), a soup that consists of snake, chicken, and cat.

Cats for meat come from a variety of sources: cat-fur farms, village farms, families/individuals, and the streets (as strays or ferals). Farms are the main source of cats. With family cats, the animals are taken by their owners to the markets and sold to butchers, purchased by traveling vendors, or more commonly, are stolen; family cats fetch a higher price than farmed cats, as they are thought to have had better lives and so, as a consequence, taste better.

No specific breed of cat is eaten; cats in markets are basically of the domestic shorthair variety. Long-haired cats are not killed for food or fur; instead, these tend to be kept as pets. And while the Chinese prefer puppy flesh to the meat of adult dogs, with cats it is the adult flesh that is preferred; eating kittens is considered by many Chinese people to lead to bad luck.

No figures are available on the total number of cats bred/supplied for their meat or on the amount of cat meat that is sold each year nationally, but they would be much lower than for the dog-meat trade—there are no commercial, intensive, cat-meat farms, and cat eating has historically been much less common than dog eating. The supply of meat does, however, also come from cats bred for their fur, and it has been estimated that about 500,000 cats are killed for their fur each season (between October and February).

Pet Ownership

Not only have dogs been eaten in China since ancient times, they have also been kept as pets. During the Qin Dynasty (221–07 BCE), aristocrats and people from wealthy and influential clans began to keep dogs, rabbits, doves, and caged birds as pets, and by the early Han Dynasty (206 BCE–220 CE), a detailed account of dog keeping, *The Classic on Dogs,* was produced.

It is not known exactly when domestic cats (*Felis silvestris catus*) first arrived in China, but the first convincing evidence dates from Han times (206 BCE–220 CE). Most probably the cat arrived from India, where it had been domesticated for some time, along the same trade routes that brought Buddhism in first century CE. Cats would have played a major role in vermin control, which was essential as the safe storage of grain was crucial to Chinese life. From about the sixth century, the cat was also linked to the supernatural, but a general fear of cats does not seem to have lasted beyond the ninth century, when there is evidence that these animals became more than vermin killers—they became pets.

Toward the end of the Han Dynasty, the Chinese Emperor Ling showed how much he regarded his dogs by investing them with the rank of senior court officials. This entitled them to the best food available, plush oriental rugs to sleep on, and personal body guards. Successive emperors and foreign dynasties showed similar fondness for dogs, especially little, pug-faced ones, ancestors of the modern Pekingese. Only the Ming (1268–1644) removed dogs from court and replaced them with cats. However, during the Manchu (Qing) Dynasty, the Pekingese was reinstated at court, and the breed enjoyed a privileged status until 1911, when Communism took control. In their heyday, Pekingese puppies were suckled at the breasts of imperial wet-nurses, and adults were attended to by servants. To supervise their care and husbandry, the emperors created a special elite corps of royal eunuchs.

Before the founding of the People's Republic of China in 1949, keeping dogs in households was quite common in rural villages and cities. Free-ranging and ownerless dogs, usually of the mongrel variety, were very common, and a small proportion of people enjoyed eating dog meat. However, during the 1950s the dog population declined dramatically, especially in the cities. This can be attributed to a government prohibition on keeping dogs in the city and the elimination of dogs to prevent rabies. Social and economic conditions also contributed, including limited monthly quotas of food grain, extremely crowded housing, and persistent propaganda about dogs being pets of the exploiting classes. Also in 1952 there was a campaign to kill dogs—officially, this was to protect the public by thwarting an American germ-warfare scheme during the Korean War involving the use of dogs to transmit disease organisms to the Chinese people. It is more likely, though, that the government was trying to get rid of dogs because they ate food that was needed by the public.

During the nationwide famine of 1959–61, dogs were almost wiped out, and in 1963 another dog-killing campaign began, the main reason being that as these animals provided nothing useful to humans and ate valuable food, they should be eliminated. After the famine, dog numbers began to increase but fell away again during the Cultural Revolution (1966–76), when pet-keeping was banned as "bourgeois"; given its long-standing links with imperialism, pet-keeping acquired a tarnished image in post-revolutionary China. By the late 1970s, however, dogs were visible again.

Major social change has taken place since China began to carry out economic reforms in the 1980s: incomes have increased, there is improved housing and living conditions in cities, and there is more tolerance of individual diversity. This has led to an increase in the keeping of pets.

South Korea

Dogs as Food

The eating of dog meat has a long history in Korea, originating during the era of Samkug (Three Kingdoms, 57 BCE–676 CE). It was not very common after this period, though, as Buddhism grew in popularity and became the state religion during the Koryo Dynasty (918–1392). However, during the Choson Dynasty (1392–1910), Confucianism became the state ideology, paving the way for the return of dog meat as food. Indeed, Confucians enjoyed the meat so much it was, according to oral tradition, nicknamed "Confucians' meat." During this period, dog meat was served in many ways, including *gaejangguk* (original name for dog soup), *sukyuk* (meat boiled in water), *sundae* (a sausage), *kui* (roasted meat), and *gaesoju* (extract).

The consumption of dog meat has mainly been associated with farmers trying to maintain their stamina during the oppressive heat of summer. However, exceptions to this have been found. For example, in 1534 there is evidence that during the reign of King Chungjong dog meat was offered to a high official as a bribe, and in 1777 reference is made to government officials going out to eat dog meat soup.

It is important to note that, as in China, dog meat has always been a medicine as well as a food. This is not surprising, as in East Asia there has always been a lot of interest in the medicinal qualities of food. In "Precious Mirror of Korean Medicine," written by royal physician Hoh Jun (1546–1615) and first published in 1613, the medical qualities of different parts of the dog are given; dog meat is said to be "warming" (*yang*).

Although there were criticisms about dog eating in the 1940s, these disappeared during the Korean War (1950–53). During this time, people were faced with severe food shortages and so dogs became a valuable source of protein. However, criticisms resurfaced in the 1980s in the form of international condemnation. This campaign was led by former French actress Brigitte Bardot. During preparations for the Seoul Olympic Games in 1988, protests were made to the South Korean government by individuals, animal welfare charities (local and international), foreign governments, and the world mass media about the slaughtering and eating of cats and dogs. The government reacted by banning the sale of dog meat at markets, moving restaurants serving dog meat to places where foreigners would be less likely to see them, and changing the name of dog meat soup (*boshintang*) to a variety of "more appealing" ones: *yongyangtang* ("nourishing soup"), *kyejoltang* ("seasonal soup"), and *sagyetang* ("soup for all seasons"). In the lead-up to the 2002 World Cup (football/soccer), which South Korea was co-hosting with Japan, international and national pressure again was put on the South Korean government to ban the consumption of dogs and cats.

It is clear that if dog eating had remained a purely rural phenomenon, it would have disappeared in modern times, as 80 percent of Koreans now live in cities. Instead, dog eating seems to be increasing. For example, in April 1997, a chain of dog-meat restaurants was launched by entrepreneur Yong-sup Cho, and in 2002, in cooperation with a cosmetics company, Dr. Yong-Geun Ann, Department of Food Nutrition, Chungcheong College, South Korea, released a range of cosmetics based on dog meat, including, dog oil cream, dog oil essence, and dog oil emulsion. In addition, he also produced a range of foods, including dog meat soy sauce, dog meat kimchi (kimchi is a traditional Korean food made of fermented vegetables), dog meat mayonnaise, canned dog meat, and dog meat candy. Also, *Seoul Searching Magazine* reported on a dog meat festival held on October 3, 2003, in Seocheon on the west coast of South Korea, where dog meat and a variety of products derived from dog parts, including dog wine and dog oil, were promoted and sold.

Today, dog meat is eaten nationwide all year round, although it is most commonly eaten during summer, especially on the three hottest days (bok days) between July and August. Eating dog during summer is believed to fight the debilitating effects of the heat and humidity; it is meant to energize people. Although this meat is most often eaten in the form of a stew or soup (tang), it can also be taken in the form of a liquor, *gaesoju*, which is actually an extract derived by boiling parts of the dog until they liquefy and then adding special herbs.

Dogs for food are sourced from farms and individuals who breed a few litters each year. Some dogs will also have been collected from the streets as strays. The type of dog most commonly farmed for food is known as *nureongi* (yellow dog), which is a midsized, yellow-furred dog. However, other types of dog may appear at markets, for example, pointers, mastiffs, and terriers, but these are less common. *Nureongi* are not normally kept as pets.

It was reported by the Ministry of Agriculture and Forestry (MAF) in 1997 that approximately 2,250,000 dogs were bred on farms and that 958,000 (43 percent) were used for human consumption; 702,000 were used at 6,484 *boshintang* restaurants and 256,000 were turned into *gaesoju*, which was sold at 10,689 *Youngyangso* or *Boshinwon* (nutritional or body health) stores. This resulted in approximately 11,500 tons of dog meat being consumed (either directly as meat or in the form of *gaesoju*), which with a population of 45 million, equated to 256 grams per person. This is most likely an underestimate, as not all dog meat is sold through restaurants; some is sold directly to consumers at markets, and undoubtedly some breeders/farmers eat their own product. Whatever the true figure, it is unlikely that dog meat consumption was greater than that of pork (700,000 tons), beef (370,000 tons), chicken (280,000 tons), or duck (40,000 tons) that year.

A recent survey of 1,502 South Korean adults by Yong-Geun Ann (2000a) showed that 83 percent (92 percent of men, 68 percent of women) had eaten dog meat at some stage in their lives and that 86 percent (92 percent of men, 72 percent of women) were in favor of dog meat as a food. Most commonly, these respondents ate dog meat only two to three times per year and believed that it was good for their health and that it gave them energy. Of those objecting to the eating of dog meat, most (69 percent) thought that it was barbaric or inhumane. A survey of 1,000 South Koreans commissioned by the World Society for the Protection of Animals (WSPA) showed that 40 percent of respondents ate dog meat at least occasionally and that young people (18–24 years) were less likely to eat it than older people (45–54 years). Overall, 70 percent of the respondents thought there was nothing wrong with eating dog.

Cats as Food

There is no evidence of cats ever having been a regular source of food for South Koreans; traditionally, cats have been kept to catch rats and mice. However, in the 1980s the consumption of cat appeared in the form of a cat liquor/extract, *goyangi soju*, made by boiling cats in a pressure cooker until they liquefied and then adding special herbs. It is claimed that this "extract of cat" is good for treating rheumatism and arthritis.

The number of cats at markets is much lower than the number of dogs. They are of the domestic shorthair variety, usually gray and ginger tabbies and black and whites. They are not farmed as such, but are bred by individuals who may just have a few cats to breed from. Stray, feral, and stolen pet cats also find their way to markets.

Pet Ownership

While historically dogs have commonly been kept to guard property and cats have been kept to catch mice and rats, the keeping of animals as companions, or pets, did not really take off until the 1990s, when the economy rapidly improved, standards of living rose, and people had more disposable income. In addition, the government encouraged pet-keeping in a bid to shake off the negative image that the country acquired because of negative media reporting of dog eating before and during the Seoul Olympics in 1988. According to the Ministry of Agriculture and Forestry, at the end of March 2004, there were 2.23 million dogs, cats, and other pets being raised in 758,000 households. The most popular breeds of dog were, in descending order of popularity, Maltese, Shih Tzu, Yorkshire Terrier, and Poodle.

With pet-keeping becoming more popular, the related industries of pet food and pet services have developed, too; recently, the pet industry was estimated to be worth 1.5 trillion won (US$1.3 billion) and rising rapidly. This interest in pets is perhaps best personified by the opening in 2003 of Asia's largest pet department store, Mega Pet, in Ilsan, one of Seoul's satellite cities. In this eleven-story building, one can purchase pets, pet food, and accessories and visit pet beauty salons, restaurants, a hotel, and a gym.

In addition, over the past ten years, the electronics giant Samsung has moved to develop and promote animal-assisted therapies and activities (AAT/AAA) in South Korea. These services included guide dogs for people with visual impairments, hearing dogs for deaf people, therapeutic horseback-riding programs and dog-handling and training programs in juvenile detention centers. The breeds of dog used are Toy Poodles, Pomeranians, Cavalier King Charles Spaniels, Labradors, and Border Collies. Currently, there are fifty-one guide dogs, fifteen hearing dogs, and one service dog in South Korea.

Vietnam

Very little has been written on the history of peoples' relationships with animals in Vietnam. What we do know is that dog eating has a long history there, especially in north Vietnam. Much of north Vietnam was ruled by the Chinese for around 1,000 years until 938, so many Chinese customs have been assimilated, including dog eating. Also, the north has a long history of poverty; straying, scavenging dogs were a cheap and ready source of protein.

Dogs as Food

In Vietnam today, *Thit cho* (meat dog) and *Cay to 7 mon* (young dog made into seven different dishes) are the names for restaurants that only serve dog cuisine. The dishes served are expensive and are mainly eaten by middle-class and wealthy people, particularly men having drinking sessions with their male friends and colleagues. In the late 1990s, these restaurants experienced a surge in popularity, especially in Hanoi. This has been attributed to an improved economy, with more people being able to afford luxury cuisines.

Dog meat is thought to enhance health and longevity. It can be cooked in a number of ways; one of the most popular dishes is dog sausage. Leftovers such as bones are sold to factories to be ground up as feed for chickens and pigs.

Unlike other meats, dog meat cannot be eaten every day, as tradition prohibits consumption during the first week of the lunar month; supposedly, bad luck will befall anyone who breaks this rule. And while most dog eaters start eating this meat after the tenth of the month, the best time to eat dog meat is on the last day of any month. The most important time of the year to eat dog meat is during the festival of Tet, which marks the beginning of a new lunar year, as it is meant to wipe away bad luck. As in China, there is also a seasonal peak in demand for dog meat—it is mainly eaten during winter, as it is considered warming.

Although less popular than in the north, dog meat is also eaten in the south. Its lesser popularity is thought to be because, historically, there has always been more food available in the south and Buddhism had a stronger influence there. Also, the warmer climate in the south means there is less of a demand for "warming" foods than in the colder north.

As in China, the preference in Vietnam is for young dogs, as the flesh is tenderer. These range in age from eight to ten months, and those weighing around 10 kilograms are thought to be ideal for restaurants.

Apparently, only "Vietnamese" dogs, as opposed to nonlocal, "Western" dogs such as Pekingese and poodles (which are kept as pets), are eaten. These "Vietnamese" dogs are all of a similar type: pariah-type, with a coat color ranging from very light (almost white) through yellow to red, although occasionally, these dogs are black/dark gray in color. "Vietnamese" dogs may also be kept as pets, but these are not eaten; the Vietnamese make a distinction between dogs they eat and those they raise as pets.

In 1998, the Ministry of Agriculture and Rural Development reported there were at least 14 million dogs in Vietnam and that their numbers were increasing as more and more farmers turned to raising dogs instead of pigs. It was not clear, though, just how many of these dogs were being eaten. One estimate is that between 4 and 5 million dogs are eaten each year in Vietnam.

Cats as Food

Cat meat cuisine does not have a long history; it is believed that it began to appear on restaurant menus in the north in the 1990s. This appears to have been in response to the economy doing well, with people having more money to spend on food; exotic, unusual foods such as field mice and cats (thought of as "small tigers") became fashionable. And apart from being eaten just for its taste, cat meat was also thought to cure asthma, and it was believed that sexual arousal could occur or sexual prowess be enhanced in men if they ate four raw cat galls pickled in rice wine.

Although cat meat became very popular—in one district of Hanoi a dozen restaurants specializing in cat meat opened and about 1,800 cats were eaten each year in each restaurant—in 1998 the government ordered the closure of all restaurants serving cat meat and stopped exports of cats to China. This was in response to rats devastating approximately 30 percent of the grain being produced in Vietnam; apparently, there had not been enough cats around to keep the rat population under control.

Pet Ownership

Dogs and cats are also commonly kept in people's households. Because the crime rate is very high in Vietnam and it is becoming a more affluent society, many people purchase dogs to guard their property. However, some dogs are now being kept purely for companionship; these tend to be small pure breeds such as Pekingese and toy poodles, and are popular amongst wealthy citizens. Cats are needed to catch mice, rats, and insects, as they are all plentiful in the hot, humid climate of Vietnam and so are a threat to crops and grain supplies.

Animal Welfare

There are no regulations concerning the humane rearing, transport, and slaughter of cats and dogs in any of the countries covered in this essay. And although in China dogs and cats should be inspected by government veterinarians before and after slaughter (for public health reasons), this rarely happens as there are no designated slaughterhouses (and none in South Korea or Vietnam either); animals are killed on farms, in markets or restaurants, or at the backs of buildings, etc., with farmers having little fear of being inspected. And even when inspectors know where to look, a shortage of personnel means they cannot regularly check what is going on. Another problem is that, apart from South Korea, these countries do not have any animal welfare legislation, so dogs and cats, along with all the other species in those countries, are not protected.

Campaigns to Ban

A number of animal charities currently campaign to have the consumption of cats and dogs banned. This is not surprising when you consider that the campaigns are largely driven by Westerners—people who do not eat dogs and cats, but who keep them as pets or companions. Indeed, the dog and cat are the two most popular pets worldwide, with owners commonly stating that companionship is the most valuable benefit derived from them. In addition, research over the past few decades has indicated that pets may be good for people's health too. However, in the West, millions of pets are abandoned and needlessly euthanized each year, and tens of thousands of dogs and cats are used in medical research. Are these fates really any better or more morally acceptable than the cooking pot? Why are cats and dogs exempt from the food table when the majority of us happily barbecue a steak or literally fish for our supper? Why is it okay to kill and eat cattle and to catch and fry fish? Are these animals not worthy of our concern too? This lack of coherence in the West's criticisms of Vietnam, China, and South Korea, understandably causes consternation amongst their peoples.

See also

Animals as Food—*Dog Eating in the Philippines*
Animals as Food—*Dogs in China May Be Food or Friend*

Further Resources

Ann, Y.-G. (2000a). Korean's recognition on edibility of dog meat. *Korean Journal of Food and Nutrition, 13*(4), 365–71.

———. (2000b). The Korean's recognition of dog meat food. *Korean Journal of Food and Nutrition, 13*(4), 372–78.

Barrett, T. H. (1998). The religious affiliations of the Chinese cat: An essay towards an anthropo-zoological approach to comparative religion. In Louis Jordan (Ed.), Occasional Papers in Comparative Religion No.2. SOAS: University of London.

Harris, M. (1985). *Good to eat: Riddles of food and culture*. Prospect Heights, IL: Waveland Press.

Hopkins, J. (1999). *Strange foods: Bush meat, bats, and butterflies*. Singapore: Periplus Editions (HK) Ltd.

International Fund for Animal Welfare. (1998). *Public opinion survey on animal welfare*. Yarmouth Port, MA: Author.

Li, P., Zu, S., & Su, P-F. (2004). Animal welfare awareness of Chinese youth. *Animal People*, May 2004. Retrieved February 11, 2007, from http://www.animalpeoplenews.org/04/5/animalAwareofChinese5.04.html.

Schwabe, C. W. (1979). *Unmentionable cuisine*. Charlottesville: University Press of Virginia.

Serpell, J. (1996). *In the company of animals: A study of human–animal relationships*. Cambridge: Cambridge University Press.

Simoons, F. J. (1994). *Eat not this flesh: Food avoidances from prehistory to the present* (2nd ed.). Madison: University of Wisconsin Press.

Walraven, B. (2001). Bardot soup and Confucians' meat: Food and Korean identity in global context. In K. Cwiertka & B. Walraven (Eds.), *Asian food: The global and the local* (pp. 95–115). Honolulu: University of Hawaii Press.

World Society for the Protection of Animals. (2004). Dogs for food: Public opinion survey—South Korea. Paper presented at the 10th International Association of Human–Animal Interaction Organisations Conference "People and animals: A timeless relationship," Glasgow, Scotland, October 6–9.

Wu, F. H. (2002). The best "Chink" food. Dog eating and the dilemma of diversity. *Gastronomica*, 2(2), 38–45.

Anthony L. Podberscek

Animals as Food
Dogs in China May Be Food or Friend

China is a vast country with a complex history. It is a place with people of many different cultural backgrounds and one where many civilizations have risen and fallen. The world today has China to thank for numerous significant ideas, innovations, inventions, and foods. For instance, what would today's world be without noodles, rice, and dogs?

DNA research strongly indicates that dogs were first fully domesticated in East Asia, probably in part of what is now China. Exactly when this occurred is the subject of much debate, but certainly dogs were domesticated by 15,000 years ago, and there are now a number of early dog remains from Europe, Southwest Asia, North Asia, East Asia, and even North America dated to between 8,000 and 15,000 years old. The DNA evidence suggests that dogs were domesticated from wolves on more than one occasion and that this could have occurred as early as 40,000 years ago. Other research indicates a widespread and close association between wolves and humans even earlier, with wolves-dogs playing a key role in the development of modern humans in various parts of the world.

Dogs and humans have had an incredibly large range of relationships since the time that dogs were first domesticated, from competitor to companion, from food to friend. There are hundreds of ways in which dogs have helped humans, with many breeds developed for particular tasks. They have saved human lives more than any other species and today play vital roles for those with a range of disabilities. They are a unique species in this regard, and dogs are also commonly found throughout the world as loving companions for people. In many cases they are much more than pets and have become full members of human families. This is particularly the case in Western cosmopolitan areas but also is true of Asian cities and among many people in rural areas globally. However, in various parts of the world, especially China and some nearby Asian countries, dogs also have a widespread role as food. It is this relationship that is most contentious for people outside East Asia, but it is also emerging as an issue within contemporary East Asian societies.

In China, dogs were likely initially domesticated for a range of reasons, including food. Until recently, the raising of dogs for food has been a cottage industry. Traditionally, it was a specialty meat in various parts of China, reputed to give strength and warmth during winter. In summer it is usually not on the market. However, there also are some parts of China where dog is not regularly or even seasonally consumed. Legend has it that dog eating really took off about 2,000 years ago, in what is today Jiangsu Province. The first emperor of the Han Dynasty, Liu Bang, apparently was particularly fond of the taste of dog, and there is an elaborate story involving a giant turtle that led to a local specialty that is sought after today, turtle-flavored dog meat. It can be purchased in restaurants, on street corners, in vacuum-sealed plastic bags from shops, and even in brightly-colored gift boxes at airports.

Dog eating was curtailed during the 1966–76 Cultural Revolution, when Red Guards rampaged through the country killing dogs, even those raised for food, because they were regarded as a bourgeois extravagance. Since the late 1980s dog meat has again grown in popularity. This is partly because its relative rarity meant that it was expensive, and therefore dog eating for some became a status symbol of prosperity. As more and more people became wealthy in China, demand for dog meat increased so that today a major emerging industry is being fuelled. But new found wealth is also leading to a big increase in people living with companion animals, with many people now able to afford dogs as pampered pets. This seemingly contradictory use of dogs is leading to new breeds being introduced to China both as companions and as livestock for the food industry.

In 1998, ten Saint Bernard dogs from Europe and the United States were exported to China for dog meat trials, along with Great Danes and Tibetan Mastiffs. As a result, the Chinese government declared the Saint Bernard to be "The Meat Dog of Choice," advertising its aphrodisiac qualities in business brochures designed to encourage the farming of these highly intelligent animals. Large breeding farms were set up, supported by Web sites detailing rearing and housing requirements and spelling out the financial benefits of farming Saint Bernards for table meat.

Big dogs, such as Newfoundlands and the Saint Bernards, loved as loyal companions or human life-savers in other parts of the world, are particularly sought after because they have lots of meat on them. Today dog farms are emerging across China, and many large foreign breeds are either being bred for meat or crossbred with local dog varieties, such as Mongolian dogs, traditionally used for meat. Crossbreeds are reputed to grow fast, and raising dogs for food is much more profitable than farming any other animal. Buyers of large foreign dog breeds include farmers and small businessmen, who come from all over China hoping to make some money on the side by breeding dogs. They are drawn by ads boasting of a high rate of return, three times as profitable as raising poultry and four times as profitable as raising pigs.

Conditions on dog farms are usually cruel and crude, more so than chicken and pig battery farms found in Western countries. Dogs are often bled to death and skinned, although some are barely alive when butchered. As in South Korea, dogs are tortured before death in some parts of China as it is claimed that this makes the meat taste better, possibly because of an increase in adrenaline. There are no animal welfare organizations in China, and the dogs are subjected to the most inhumane treatment imaginable. Methods include the following:

- Pouring boiling water over the live animal to increase the adrenaline production. Their throat is then cut and the meat left to dry.
- Holes are cut in the paws. The animal is then left to bleed to death. This takes ten minutes or so but is believed to make the meat taste better.
- Legs are broken the night before slaughter, and then the dog is skinned alive the next morning.
- Beating with sticks and slow strangulation/blow torching.

Dog meat is stewed, roasted, or otherwise cooked in a range of ways, with some recipes, such as the turtle flavored meat, a specialty of particular regions. Hides are sold to factories for processing into clothing, hats, blankets, and other products.

Although China is the biggest dog-eating country in the world, an adequate account of the scope of the food-dog industry in China is not available. Some organizations estimate that 10 to 20 million dogs are processed each year in China, but it is predicted this will soon grow dramatically. Today some Chinese people passionately and vocally express their distaste for eating dogs, while others advocate just how delicious they consider dog meat to be. New ways of dogs helping Chinese people are also being developed or introduced into China. For instance, in late 2004 a program using pet companions to improve mental and physical health was initiated by the Animals Asia Foundation. As China engages the wider world more and more, it is coming under increasing pressure to curtail the dog meat industry or to strictly regulate it. There are also calls for the dog meat industry to be made illegal given just how important dogs are to humans, but widespread and longstanding cultural traditions to do with food are usually very difficult to change.

See also

Animals as Food—*Dogs and Cats as Food in Asia*
Culture, Religion, and Belief Systems—*Chinese Youth Attitudes toward Animals*

Further Resources

Maynard, E. (2003). *The China Dogs*. Retrieved on February 11, 2007, from http://www.labouranimalwelfaresociety.org/articles/chinadogs.htm.

———. (2004). *Introduction: Sirius Global Animal Organization*. Retrieved on February 11, 2007, from http://www.2kat.net/sirius.

McLaughlin, K. E. (2005). With woofs and wet noses, dogs help heal in China. *The Christian Science Monitor*. Retrieved in June 1, 2005, from http:/www.csmonitor.com/2005/0601/p07/s02-woap.htm.

Savolainen, P., Zhang, Y.-P., Luo, J., Lundeberg, J., & Leitner, T. (2002). Genetic evidence for an East Asian origin of domestic dogs. *Science, 298*, 1610–13.

Serpell, J. A. (Ed.) (1995). *The domestic dog: Its evolution, behaviour and interactions with people*. Cambridge: Cambridge University Press.

Taçon, P. S. C., & Pardoe, C. (2002). Dogs make us human. *Nature Australia, 27*(4), 53–61.

Vilá, C., Savolainen, P., Maldonado, J. E., Amorim, I. R., Rice, J. E., Honeycutt, R. L., Crandall, K. A., Lundeberg, J., & Wayne, R. K. (1997). Multiple and ancient origins of the domestic dog. *Science, 276*, 1687–89.

Paul S. C. Taçon and Elly Maynard

Animals as Food
Entomophagy (Eating Insects)

The Beginnings of Entomophagy

Insects have served as food to humankind and our primate relatives for thousands of years and as far back as early proto-hominids. While we can only speculate about the beginnings of entomophagy in human societies, humans may have first been drawn to insects that produce sugars and subsequently to insects rich in fats and lipids. Based on linguistic studies, the origins and spread of entomophagy are postulated to have begun in southern India and Southeast Asia. Today, entomophagy is actively practiced in Latin America, Africa, Asia, and the Pacific and exists to a lesser degree in North America and Europe.

Social Patterns of Insect Consumption

Entomophagy is practiced by people from every socioeconomic status. Ranging from the high-priced insect foods sold in European and Japanese gourmet restaurants to those collected in the forest by rural peasants and indigenous tribes, insects are highly sought after by many for their taste, flavor, and texture. In rural settings, however, entomophagy can provide important supplements to nonmeat protein, fat, vitamins, and minerals. The amount of energy, protein, and fat derived from insect diets varies depending on the insect and the stage at which the insect is eaten. Caterpillars, beetle larvae, and adult and immature ants have higher caloric value by weight than soy beans and pork. Dried insects sold in village markets in developing nations usually possess more than 60 percent crude protein. In some Amazonian tribes, insect protein can constitute a significant portion (up to 26 percent) of their daily protein intake. Insect-derived vitamins and minerals are also important supplements, particularly for women in developing nations, where mineral deficiency, such as iron, is particularly problematic. It is important to emphasize that while insects provide valuable dietary supplements to many indigenous and rural communities, insect food is frequently sought after solely for its gastronomic value rather than as an emergency meat-protein supplement.

Main Taxonomic Groups of Insects Eaten

Food insects span a range of groups, including species belonging to at least ten orders and representing at least 90 families and 370 genera. Globally, an estimated 1,400 species are recorded as insect foods. The primary food insects are beetles (Order Coleoptera), larval butterflies and moths (Order Lepidoptera), bees and ants (Order Hymenoptera), and grasshoppers (Order Orthoptera). Beetle larvae from over 100 genera and seventeen families are generally collected for food. Within Lepidoptera, over eighty genera in twenty families are collected for food, while over fifty genera in seven families of Orthoptera are consumed. With the exception of Orthoptera, humans generally collect and eat insects in immature stages rather than the relatively hard adults. The larvae of coleopteran insects such as the palm weevil larvae of the genera *Rhynchophorus* and *Rhinostomus* are efficient feeders of palm tissues and effectively convert plant tissue into a rich and palm-flavored insect fat, giving it its popularity among several indigenous groups of Latin America, Africa, and the Pacific. In a few cases, even newly enclosed beetle adults, with their soft exoskeletons, are collected for food.

Preferred and common insect foods vary from region to region and across cultures, although a number of cultures have converged on similar entomophagy favorites. In Africa, termites, palm weevil larvae, and caterpillars are three common insect foods. Moth caterpillars belonging to Family Saturniidae are the most frequently collected and eaten in Africa. The moth caterpillar's popularity amongst locals is evidenced by the market value of the caterpillars and their importance as sources of income for local harvesters. Caterpillar food is a particularly important protein source for women and children during times of low meat procurement.

In Asia, silkworms are the popular insect foods, especially in India, China, Japan, and Korea. More commonly known for their silk production, silkworm pupae are highly regarded delicacies, and their popularity as insect foods provides an energetically effective form of recycling the silkworm that would otherwise be discarded once the silk cocoon is removed. Fried grasshoppers, locusts, and cicadas are amongst the insect-food favorites in Thailand and Japan, while in Latin America and Papua New Guinea palm weevil larvae are highly valued as important insect food sources. In Australia, indigenous aborigines commonly consume insects, most famously the witchetty grubs. The witchetty grubs are larvae of cossid moths, named after the witchetty plant, where larvae develop in the roots. While it is not clear why particular insect foods are preferred, and sometimes why insects are eaten at all, the combination of cultural traditions and beliefs and the availability of palatable insect species are certainly two important determinants of the popularity of insect food within a culture or region.

Entomphagy has yet to enter the restaurant or market scene in the United States. Yet unbeknownst to most consumers, many processed foods contain insect parts that were intentionally added. A ubiquitously eaten insect is the cochineal bug (*Dactylopius coccus*). The cochineal is used to color foods produced industrially. A red dye from the pulverized female bodies of this cactus herbivore is frequently used to color candies, drinks, and meat products.

Food Insects Collected and Cultivated

In more traditional settings, insect foods are collected opportunistically in forest habitats, but they may also be managed and semicultivated so that greater numbers can be harvested for food. In several Amazonian tribes, knowledge of the ecology and habitats of various insect species enables insect-gatherers to seek out termites (*Syntermes* sp.), leaf-cutter ants (*Atta* sp.), and caterpillars during times of the year when these insects are in abundance. Termites are drawn out using sticks or other plant materials. Certain caterpillars are collected from underneath *Bactris* palm leaves, and these are roasted before being eaten. Soon after the beginning of the first rains of the rainy season, indigenous tribes collect reproductive male and female ants exiting by the thousands from their nests for their mating flights. To facilitate the capture of the ants, enclosed structures made from Musaceae leaves are sometimes built over the ant nests so that ants can be easily captured as they escape through a limited number of exits. Similar strategies are used by termite-collecting humans in Africa.

Insect foods may also be semicultivated by providing conditions ideal for mass production. The Hoti people of the Venezuelan Amazon, for instance, cultivate palm weevil larvae by cutting down old palm trees, which provide ideal habitats for breeding large numbers of the larvae of *Rhinostomus barbirostris*. The larvae are harvested approximately three months after the palms are felled and are eaten live, roasted, or cooked in soups.

An extreme form of human-managed insect food is the silk worm. Silkworm pupae have been cultivated in China and India for at least 5,000 years, resulting in flightless silkworm moths that depend on their human cultivators for food. In industrialized regions such as China and South Korea, silkworm pupae are reared and mass-produced in factory settings and then canned and sold in local as well as international markets.

Entomophagy with Medicinal Function

Insects are eaten not only for their nutritional and culinary value, but also for their medicinal value. While many medicines remain scientifically untested, chemical analyses of a number of medicinally ingested insects indicate the presence of bioactive compounds with antipathogen and anticancer properties. In China, ants, locusts (*Oxya* and *Locusta*), termites (*Macrotermes barnyi*), and even the feces of a noctuid moth are eaten for medicinal purposes. Insect bodies are usually ground up and taken as a tea, powder, or paste. In Brazil, insects are eaten for medicinal purposes by both indigenous and nonindigenous people. Insect medicines are often ingested as teas made from body parts or entire insects such as grasshoppers, tenebrionid beetles (*Palembus dermestiodes*), and pompilid wasps. At least fifty insect species are eaten as medicines in Brazil, with Hymenoptera (ants, bees, and wasps) forming the largest group of insects recorded as insect medicines.

Entomophagy as Pest Control

A number of authorities have suggested promoting entomophagy as an ecologically friendly use of insects as meat protein supplements and nonchemical pest control.

A cook dresses up a plate of deep-fried worms served with a dollop of guacamole at the Hosteria Santo Domingo in Mexico City, 2005. © AP Photo/Dario Lopez-Mills.

Children sell insects as snacks opposite the Royal Palace of Phnom Penh, Cambodia, 2006. The snacks include crickets, wild birds, river mussels, and silkworms. © AP Photo/Heng Sinith.

During periods of grasshopper or locust outbreaks, such insects could be collected and eaten to not only provide supplementary protein, but also a nontoxic form of pest control. In Mexico more than twenty species of grasshoppers and locusts are already sold in markets. Entomophagy could not only provide a viable option for pest control but also a nutritional and economic return to the human predators involved.

Further Resources

DeFoliart, G. R. (1999). Insects as food: Why the Western attitude is important. *Annual Review of Entomology, 44*, 21–50.

———. (2002). *The human use of insects as a food resource*. Retrieved October 13, 2006, from http://www.food-insects.com.

Menzel, P. D., & D'Aluisio, F. (1998). *Man eating bugs: The art and science of eating insects*. Berkeley, CA: Ten Speed Press.

Meyer-Rochow, V. B., & Changkija, S. (1997). Uses of insects as human food in Papua New Guinea, Australia, and North-east India: Cross-cultural considerations and cautious conclusions. *Ecology of Food and Nutrition, 36*, 159–85.

Ramos-Elorduy, J. (1997). Insects: A sustainable source of food? *Ecology of Food and Nutrition, 36*, 247–76.

Sutton, M. Q. (1990). Insect resources and Plio-Pleistocene hominid evolution. In D. A. Posey, W. L. Overal, C. R. Clement, M. J. Plotkin, E. Elisabetsky, N. de Mota, et al. (Eds.), Proceedings of the First International Congress of Ethnobiology, Museo Paraense Emílio Goeldi (pp. 195–207), Belém, Brazil.

Juanita Choo

Animals as Food
Global Diversity and Bushmeat

Biodiversity is the sum total of all life on Earth, that wealth of genes, species, ecosystems, and ecological processes that make our living planet what it is—the only place in the universe where we know with certainty that life exists. We are fortunate to be an integral part of such a rich and diverse planet, especially at a time when scientific understanding and systems of transport and communications enable us to see, to visit, to learn about, and to fully appreciate and understand the amazing range of life forms with which we share this Earth.

Unfortunately, in spite of our growing knowledge of biodiversity and our increasing appreciation of its complexity, its magnificence, and its value to human beings, some believe that we are in a crisis of epic proportions. Some scientists think that we stand at the threshold of one of the most overwhelming losses of life in Earth's history—a planet-wide series of extinctions, coming in great spasms unlike anything since the loss of the dinosaurs some 65 million years ago. This time the cause is not a giant meteorite crashing into the Earth or some other uncontrollable cosmic force. This crisis is caused by the inability of one species to control its consumption at the expense of the many millions of others.

Humans are a profound and powerful species. Whether envisioned as the end point of creation or as a product of ongoing evolution, humans have surpassed every other known form of life in our ability to transform the world to suit our needs and ambitions. We have made the whole Earth our dominion and all life our resource. Yet we also bring with us a voracious hunger, progressing as if we are destined to consume everything in our path.

Humans are the ultimate omnivore. On a planet fantastically rich with life, there is very little that we do not eat. From ants to elephants, everything that moves has turned up in the cooking pot. International traders travel the world in search of rhino horn, tiger penis, bear gall bladders, and whale meat. Exotic birds and rare turtles are worth more than their weight in gold. Monkey brains are an illegal gourmet delight in Asian restaurants. The consumption of wild species witnessed in Southeast Asia was catalyzed by the extension of global economics into the region. Similar trends are beginning to emerge in Central and South America and are rampant in much of Africa.

This broad capacity to consume has helped make human beings the most adaptive and the most deadly species on earth. Human adaptability in different situations once assured our survival in competition with other animals. Now our position at the head of the food chain, long considered a key to success, may be turning against us. Human population has grown exponentially, and since there is practically no natural limit on what we consume, the end result has become in some places the local and global extinction of other species and the massive destruction of natural habitat.

Threatened Forests and Wildlife in Africa

Nowhere are the destruction of wilderness and the consumption of threatened wildlife expanding faster than in equatorial Africa. The Congo River, its tributaries, and the other great rivers of the region pass through a myriad of swamp and forest stretching across the center of the continent to the Atlantic Ocean. A wonderland of flora and fauna wanders in the shadows of the great forest canopy. With luck a traveler may see the elusive forest elephant tramping across a grassy clearing, but for the most part the great mammals are as secretive as the smallest of jungle animals.

Why should an elephant hide? For millennia these powerful and sensitive beasts roamed freely, losing only the most lame and weak members of their communities to predators. Humans who dwelt in the forests rarely killed an elephant, and when they did it was cause for reverent celebration and gratitude. When outsiders came to Congo with guns and a lust for wealth, the balance changed. Humans now had the upper hand, and the gun hunters chased the great mammals deep into the forests. Elephants, apes, buffalos, and leopards survived the early invasion by learning to stay clear of humans. Today, there is no longer any place to hide.

Apes, elephants, indeed all the edible animals, have become consumer items as people with guns and a hunger for wild game scour the wilderness in search of commercial gain. Those who once considered the forest a place of danger now see it as a frontier for business, where money can be made without the controls and costs of urban political and legal structures. Like the gold rush in America in the nineteenth century, today's quest for natural resources in the remaining wilderness is fraught with problems of lawlessness, social disruption, and environmental destruction. Like every rush for quick riches, the plundering of rain forest resources moves at a pace and in a fashion that cannot be continued without severe consequences for nature and humanity as a whole.

To the outsider, the tropical forests are a mysterious Eden, the birthplace of our own species, a place of romance and adventure that captivated the imagination of the Victorian explorers and Hollywood film makers of the twentieth century. In some places, this is still the case. The immense rain forests of the Congo Basin are surpassed in size only by those of Amazonia, and more than 70 percent of them are still intact, placing this region among the three highest priority wilderness areas on Earth. Nearly 10,000 plant species are found in the Congo forests; at least 3,300 of them are found nowhere else. The diversity of animals is equally impressive. The Congo forests are home to more than 1,260 species of mammals, birds, reptiles, and amphibians; more than ninety of which are found only in these forests. Of particular note are the large mammals of the Congo wilderness—the forest elephant and buffalo, okapi, bongo, giant pangolin, the great apes, and a host of other large and spectacular species. This region is one of the last great strongholds of the great Pleistocene mega fauna that are already gone or rapidly disappearing elsewhere in the world.

The adjacent Guinean forests of Western Africa, which once formed a continuum with those of the Congo, are similarly diverse, but have already been much more adversely impacted. These highly endangered "biodiversity hotspots" have lost 90 percent of their original area and have very high concentrations of threatened species. The combined diversity of Guinean and Congo forests includes at least 12,000 plant species, 800 species of birds, 315 mammals, 190 reptiles, and 230 amphibians. Clearly, equatorial Africa is one of the most prolific and endangered cornucopias of biodiversity and wonder on this planet.

Why should people be so concerned about these forests and about their future? It is because the Congo Basin is now being subjected to unmanaged, uncontrolled, unsustainable forms of exploitation that have reduced forests over most of the tropical world to fragments of their former magnificence—to biodiversity hotspots in which only 5–10 percent of the original forest and wildlife remain. This devastation has already happened to a large extent to the forests of East Africa and in Western Africa and is taking place around the globe.

Contrary to the talk about sustainable development and sustainable utilization of forests, wildlife, and fisheries, it has become obvious that our present consumption of living creatures and other natural resources is anything but sustainable. Species and ecosystems that in theory should be useful to us on a renewable basis are being grossly overexploited to the point of elimination in many parts of the world, especially in the tropical countries where so many of Earth's unique life forms are found. To make matters worse, this is very rarely being done to alleviate poverty or to improve the quality of life of people living in the most desperate countries. This destruction is occurring for the profit of a small number of already wealthy companies or to meet the desires of urban elite. A relative handful of people are exploiting some of our planet's richest and most diverse ecosystems in a manner that will impoverish us all forevermore and in a way that is truly irreversible. This is happening in many parts of the world, but it is especially evident in the forested regions of Southeast Asia, Central Africa, and Western Africa, with similar patterns emerging in the Amazon rain forests in South America and the boreal forests of the far North.

Leading the international exploiters of the world's biodiversity-rich forests are the loggers. Modern technology has enabled the timber industry to reap high profits by extracting hard wood for sale worldwide. Rain forests in Africa and worldwide are often leased to foreign timber and agribusiness barons with local connections. International financiers and bilateral and multilateral donors are underwriting construction of roads that stretch from major cities to the core of the wilderness. Logging towns and monoculture farms are springing up in once impenetrable forests. Unsustainable overharvesting and farming is the standard practice. With no effective way to audit these businesses in far-flung wildernesses, some scientists believe their invasion marks the death knell for primary forests and their irreplaceable biodiversity.

As the commercial destruction of forests expands rapidly across most of the tropical world, timber companies are now cutting the last few large blocks of tropical forest, with the Congo Basin being an especially appealing target. Logging of these forests is a nineteenth-century, colonial-style form of resource exploitation that is no longer justifiable, necessary, or defensible in the twenty-first century. These forests are too rich in biodiversity and too important for the vast array of local, regional, and global ecosystem services that they provide to permit their widespread destruction. Often this activity is being carried out under the banner of "sustainable forest management," which refers to the ability to sustain timber harvesting for profit, not to sustain forest ecosystems for posterity. Claims by loggers that they are sustainably managing rain forests are rejected by knowledgeable observers who have seen such logging operations firsthand and have observed what is left behind.

The Bushmeat Crisis

One of the principal causes of what most consider a severe crisis is the strong connection between the timber industry and the commercial bushmeat trade. With every tree that is felled, thousands of forest animals are hunted to feed a burgeoning demand for game meat that starts with the logging workers and ends as an expensive meal for the elite in cities and towns. With increasing frequency, illegal bushmeat is being shipped and sold worldwide. Suitcases and cargo containers filled with the carcasses of a wide variety of forest animals are arriving in the ports and restaurants of Europe, Asia, and North America.

To be sure, subsistence hunting has been taking place since the dawn of humanity and is part of our biology and our culture. What is happening now, however, has little to do with subsistence and is contributing to a rapid decline in the quality of life of those people most dependent on the forest for their livelihood. Unfortunately, timber companies are opening roads deep into pristine areas, making them readily accessible to large numbers of invading bushmeat hunters who follow quickly in the loggers' tracks and who eradicate everything bigger than a rat for sale in distant markets.

Most timber companies do not import food to their logging crews. Instead, they support commercial hunters or provide guns and ammunition to their own workers so they can live off the local wildlife, thus further increasing their corporate profit margin. Hunters may use logging trucks, which are constantly moving back and forth between cities and the forest, to transport themselves and their prey to the most lucrative markets. Illegal transport of bushmeat is a bonus for logging truck drivers. In the urban areas, bushmeat goes not to starving people, but rather to elite who are willing to pay much more for elephant, gorilla, mandrill, or bongo than for chicken, goat, or beef.

Recent projections estimate the volume of bushmeat commerce in the Congo Basin at over 3 million metric tons per year—a multibillion dollar business that is largely illegal. A wide variety of species are affected, but perhaps the most dramatic example is the impact on our closest living relatives, gorillas, the chimpanzees, and the bonobos. These rare great apes are found only in the forests of equatorial Africa and are the animals with which we share the highest percentage of DNA and the most physical and social characteristics. All ape species are now in danger of extinction. Many less-known animal species are also being driven toward extinction; some of them hang on by the slimmest of threads.

Further complicating the situation is growing evidence that the spread of AIDS and other diseases to human populations is related to consumption of primates. American and French scientists have traced the evolution of HIV back to its origins in the forests of West and Central Africa. They are now convinced that slaughtering primates harboring Simian Immune Virus (SIV) set the stage for the most virulent plague in modern times. Researchers have found significant proportions of primate bushmeat to be infected with potential pathogens. It is also now clear that there is a connection between eating ape meat and outbreaks of Ebola virus, which has killed many hundreds of rural Africans in recent years. The large increase in direct contact with dead and infected forest animals is exposing people to diseases for which they have no immunities. It is becoming increasingly evident that eating our biological kin into extinction is risky business.

These adverse public health outcomes of bushmeat commerce are being recognized with increasing frequency. The international medical community is now on the alert. Although experts in conservation and public health have stepped up their dialogue, the centers of world power and influence have yet to recognize this problem. Many scientists and activists have urged the World Health Organization, national governments, and international agencies to address the issue. Conserving the wildlife of tropical areas and helping people to avoid new diseases and further epidemics will require extraordinary political will, great financial investment, and a host of new players from diverse fields.

Most bushmeat trade is illegal, but enforcement of laws is rare. Wild meat is preferred over domestic meat in many societies and cultures, and the increase in supply from logging concessions has catalyzed demand. While few will argue for an immediate cessation of all wildlife hunting, most people agree that the illegal slaughter of threatened species for gourmet meat is unnecessary and should be stopped. It is difficult to convince people not to poach and eat the rare animals from their forests when global consumers are paying a premium to extract rare trees and international exploiters are destroying the habitats of rare and endangered species.

Preliminary efforts are underway in some biodiversity hotspots to convince people to control bushmeat consumption. Most major conservation and wildlife protection organizations in the world have now focused at least some attention on bushmeat commerce, and some are investing significant time and talent to study and confront the crisis. Bushmeat committees and task forces are active in many countries in Europe, seeking to educate and influence the public and people in power. A consortium of U.S.-based nongovernmental organizations (NGOs) has formed a Bushmeat Crisis Task Force, which has compiled and disseminated information on the biodiversity crisis to key decision makers, the media, and the general public. New efforts by the United Nations Environmental Program, the

World Conservation Union (IUCN), and the Convention on International Trade in Endangered Species (CITES) are among the higher profile attempts at an international level to find solutions to this growing tragedy. So far the results are largely awareness, education, and problem analysis. Systems to monitor and control illegal bushmeat commerce on the supply, delivery, and demand sides have yet to be established.

In fact, the refusal of most people to eat certain species comes not from the animal's rarity in nature, but from its commonality with humankind. While relatively few people will fight to keep the endangered forest antelopes out of the cooking pot, millions will rally to stop the slaughter of great apes. Not only do the apes share over 97 percent of our DNA, but they also share our capacity to love friends, grieve loss, suffer failure, and celebrate success. Apes are the animals that evoke in us the deepest humane concern and whose killing elicits the most potent moral outrage. By emphasizing the plight of humankind's ape cousins and using them as "flagship species," conservationists hope the public will choose to protect the biodiversity-rich forests of equatorial regions of the world and all that lives there.

To reverse the enormous growth in illegal bushmeat and wildlife commerce demands humane concern, financial investment, and political will. It is apparent that wildlife and biodiversity conservation will become a theme and a central activity in human society only when people everywhere become aware of the importance of nature as a source of spiritual renewal, cultural identity, and national, economic, and ecological security. This awareness must lead to positive change both within and outside the biodiversity-rich nations. It is the privilege and obligation of the leaders of the global community to promote true conservation values, install good governance, and work for the good of all life on earth. It is every person's responsibility to protect and preserve what remains of our planet's amazing biodiversity.

Further Resources

Butynski, T. M. (2001). Africa's great apes. In B. B. Beck, T. S. Stoinski, M. Hutchins, T. L. Maple, B. Norton, A. Rowan, E. F. Stevens, & A. Arluke (Eds.), *Great apes and humans: The ethics of coexistence* (pp. 3–56). Washington, DC: Smithsonian Institution Press.

Mittermeier, R. A., Mittermeier, C. G., Pilgrim, J., Fonseca, G. A. B., Konstant, W. R., & Brooks, T. (Eds.). (2002). *Wilderness: earth's last wild places.* Washington, DC: Conservation International.

Mittermeier, R. A., Myers, N., Robles Gil, P., & Mittermeier, C. G. (1999). *Hotspots: Earth's biologically richest and most endangered terrestrial ecoregions.* Washington, DC: Conservation International.

Rose, A. L. (2001a). Bushmeat, primate kinship, and the global conservation movement. In B. M. F. Galdikas, N. E. Briggs, L. K. Sheeran, G. L. Shapiro, & J. Goodall (Eds.), *All apes great and small—Volume 1: African apes* (pp. 241–58). New York: Kluwer Academic/Plenum Publishers.

———. (2001b). Conservation must pursue human-nature biosynergy in the era of social chaos and bushmeat commerce. In A. Fuentes & L. Wolfe (Eds.), *Conservation implications of human and nonhuman primate interconnections* (pp. 158–184). Cambridge: Cambridge University Press.

———. (2001c). Social change and social values in mitigating bushmeat commerce. In M. I. Bakarr, G. A. B. Fonseca, R. A. Mittermeier, A. B. Rylands, & K. W. Walker (Eds.), *Hunting and bushmeat utilization in the African rain forest* (pp. 59–74). Washington, DC: Conservation International.

Rose, A. L., Mittermeier, R. A., Langrand, O., Ampadu-Agyei, O. & Butynski, T. (2003). *Consuming nature: A photo essay on African rainforest exploitation.* Photography by Karl Ammann. Los Angeles: Altisima Press. [Adapted from a volume on bushmeat and the biodiversity crisis, supported by Conservation International, Wasmoeth Wildlife Foundation, and The Biodiversity Institute.]

Anthony L. Rose and Russell A. Mittermeier

Animals as Food
Veganism

Vegans (pronounced VEE-guns) are people who choose not to eat any animal products, including meat, eggs, dairy, honey, and gelatin. Vegans do not wear fur, leather, wool, down, or silk or use cosmetics or household products that were tested on animals or contain ingredients that were derived from animals. Most vegans also do not support industries that feature captive and/or performing animals, including circuses, zoos, and aquariums.

The American Vegan Society (2006) defines veganism as "an advanced way of living in accordance with Reverence for Life, recognizing the rights of all living creatures, and extending to them the compassion, kindness, and justice exemplified in the Golden Rule."

The word *vegan* was derived from *vegetarian* in 1944 by Elsie Shrigley and Donald Watson, the founders of the U.K. Vegan Society. Shirgley and Watson were disillusioned that vegetarianism included the consumption of dairy products and eggs. They saw "vega" as "the beginning and end of vegetarian" and used the first three and last two letters of vegetarian to coin the new term.

There is no conclusive estimate of the current number of vegans in the United States. An American Dietetic Association (ADA) report indicated that in 2000 approximately 2.5 percent of the U.S. adult population (4.8 million people) consistently followed a *vegetarian* diet, meaning that they never ate meat, fish, or poultry. Slightly less than 1 percent of those surveyed were considered vegans (American Dietetic Association, 1993).

A poll published in *Time* magazine on July 7, 2002, showed that 4 percent of the 10,007 American adults surveyed consider themselves vegetarians, and 5 percent of those who described themselves as vegetarians also considered themselves vegans.

Retail sales of vegetarian and vegan food have more than doubled in the United States since the late 1990s. In 1998 the total sale of vegetarian and vegan foods was $729.6 million. In 2003 that amount increased to $1,558.9 million. (*USA Today,* 2004).

Several schools and universities have begun offering vegan meals in recent years. A 2005 nationwide survey conducted by ARAMARK, a company that provides food to universities and school districts, indicated that approximately 24 percent of college students say that vegan meals are important to them.

Although most people go vegan for ethical reasons, health and environmental concerns are also motivating factors.

Ethical Reasons for Veganism

People for the Ethical Treatment of Animals (PETA) noted that approximately 27 billion cows, pigs, chickens, turkeys, and other animals are killed for food each year in the United States. Our modern factory farming system strives to produce the most meat, milk, and eggs as quickly and cheaply as possible and in the smallest amount of space possible.

Some people, such as Jewish Nobel Prize–winning author Isaac Bashevis Singer, have equated the treatment of animals in slaughterhouses with the treatment of humans during the Holocaust. Having fled Nazi Europe in 1935, he took a room above a slaughterhouse and watched as cows were prodded, kicked, and sworn at as they were herded down a ramp to their deaths. He proclaimed that "as long as human beings go on shedding the blood of animals, there will never be any peace" (Dujack, 2003).

There is evidence of cows, chickens, pigs, and other meat animals being raised in poor conditions, where they may be fed high-bulk food, such as grains, or substandard or inappropriate food. They are sometimes kept in very small spaces in order to raise as many animals as possible. Most disturbing, there is evidence that at slaughterhouses, animals are not always humanely killed, such as when stun guns do not work. United States Department of Agriculture inspection records documented fourteen humane slaughter violations at one processing plant, including finding hogs that "were walking and squealing after being stunned [with a stun gun] as many as four times" (Warrick, 2001).

Some people believe that it is unethical to eat other animals and base their vegetarianism and veganism on these beliefs. Vegans also believe that it is wrong to use animals for their milk or eggs.

Health Reasons for Veganism

Animal products, particularly meat, eggs, and dairy, are generally high in saturated fat, cholesterol, and concentrated protein. Numerous studies have linked the consumption of certain animal products to serious illnesses, such as heart disease, strokes, diabetes, and breast, colon, prostate, stomach, esophageal, and pancreatic cancer.

Unlike animal products, plant-based foods are cholesterol free and generally low in fat and high in fiber, complex carbohydrates, and other vital nutrients. Researchers from the University of Toronto have found that a plant-based diet rich in soy and soluble fiber can reduce cholesterol levels by as much as one-third (Fauber, 2003). According to David Jenkins, professor of nutrition and metabolism at the University of Toronto, "the evidence is pretty strong that vegans, who eat no animal products, have the best cardiovascular health profile and the lowest cholesterol levels" (Callahan, 2003).

Studies have shown that, on average, vegetarians and vegans are at least 10 percent leaner and live six to ten years longer, than meat eaters. The ADA has reported that "vegetarians, especially vegans, often have weights that are closer to desirable weights than do non-vegetarians" (American Dietetic Association, 1993).

In *Dr. Spock's Baby and Child Care,* the late Dr. Benjamin Spock, an authority on child care, wrote, "Children who grow up getting their nutrition from plant foods rather than meats have a tremendous health advantage. They are less likely to develop weight problems, diabetes, high blood pressure, and some forms of cancer" (Spock, 1998).

According to the ADA and Dietitians of Canada, "well-planned vegan and other types of vegetarian diets are appropriate for all stages of the life cycle, including during pregnancy, lactation, infancy, childhood, and adolescence" (ADA Web site).

It is possible to get most vital nutrients from a vegan diet; however, because vitamin B_{12} is primarily found in animal sources, vegans need to take a multivitamin or B_{12} supplement to get ample B_{12}. Vitamin B_{12} is also found in nutritional yeast and many fortified cereals and soy milks.

Environmental Reasons for Veganism

The process of turning cows, pigs, chickens, and turkeys into "meat," "pork," and "poultry" takes a toll on the environment. According to *E: The Environmental Magazine,* almost every aspect of animal agriculture—from grazing-related loss of cropland and open space, to the inefficiencies of feeding vast quantities of water and grain to cattle in a hungry world, to pollution from factory farms—can cause an environmental disaster with wide and sometimes catastrophic consequences (Motavalli, 2002).

The Environmental Protection Agency has reported that factory farms pollute our waterways extensively. Animals raised for food produce approximately 130 times as much excrement as the entire human population—87,000 pounds per second (PETA Vegetarian Starter Kit).

Livestock waste emits ammonia, nitrous oxide, carbon dioxide, and other toxic chemicals into the atmosphere. A study by Duke University Medical Center showed that people living downwind of pig farms are more likely to suffer from tension, depression, fatigue, nausea, vomiting, headaches, shallow breathing, coughing, sleep disturbances, and loss of appetite (Schiffman et al., 1995).

Gidon Eshel and Pamela Martin of the University of Chicago compared the amount of fossil fuel needed to cultivate and process various foods, including running machinery, providing food for animals, and irrigating crops. The researchers found that the typical U.S. diet generates nearly 1.5 tons more carbon dioxide per person per year than a vegan diet with an equal number of calories (*New Scientist*, 2005).

Raising animals for food also requires massive amounts of water and land. It takes 2,500 gallons of water to produce a pound of meat, but only sixty gallons of water to produce a pound of wheat, and a meat-based diet requires more than 4,000 gallons of water per day, whereas a vegan diet requires only 300 gallons of water a day (Robbins, 1987).

In the United States, animals are fed more than 70 percent of the corn, wheat, and other grains we grow (PETA). The world's cattle consume a quantity of food approximately equal to the caloric needs of 8.7 billion people; indeed it has been found that even if Americans just reduced their intake of meat by 10 percent, it would free more than 12 million tons of grain annually for human consumption (Robbins, 1987).

See also

Ethics and Animal Protection—*Compassionate Shopping*

Further Resources

American Dietetic Association. (1993). *Position of the American Dietetic Association: Vegetarian Diets*. Retrieved March 13, 2003, from http://www.fatfree.com/FAQ/ada-paper.

———. *Vegetarian Diets*. Retrieved March 23, 2005, from http://www.eatright.org/cps/rde/xchg/ada/hs.xsl/advocacy_933_ENU_HTML.htm.

American Vegan Society. *What Is Vegan?* Retrieved March 27, 2006, from http://www.americanvegan.org/vegan.htm.

ARAMARK. (2005). Vegan options more popular than ever on college campuses: ARAMARK focuses on meeting consumer needs in honor of Vegan World Day, June 21. Retrieved on March 1, 2007, from http://www.aramark.com/PressReleaseDetailTemplate.aspx?PostingID=552&ChannelID=210.

Callahan, M. (2003, June). Inside veggie burgers. *Cooking Light*, 74.

City of Santa Monica. *Sustainable City Progress Report—Food Choices*. Retrieved on March 27, 2006, from http://santa-monica.org/epd/scpr/EnvironmentalPubllicHealth/EPH12_FoodChoices.htm.

Choosing a meat-free option. (2004, February 23). *USA Today*. [Available online at http://www.usatoday.com/educate/et/ET04.06.2004.pdf]

Do you consider yourself a vegetarian? (2002, July 15). *Time, 160*(3), 48.

Dujack, S. R. (2003, April 21). Animals suffer a perpetual 'Holocaust.' *Los Angeles Times*.

EG Smith Collective (2004). *Animal ingredients A to Z*. Oakland: AK Press.

Fauber, J. (2003, July 22). Ape diet shown to lower cholesterol. *Milwaukee Journal Sentinel*, p. A01. [Available online at http://www.jsonline.com/story/index.aspx?id=157004]

It's better to green your diet than your car. (2005, December 17). *New Scientist, 2530*, 19.

Marcus, E. (2000). *Vegan: The new ethics of eating*. Ithaca: McBooks Press.

Motavalli, J. (2002). The case against meat. *E: The Environmental Magazine, 13*(1), 26.

People for the Ethical Treatment of Animals. *PETA Media Center-Vegetarian Fact Sheets*. Retrieved on March 27, 2006, from http://www.peta.org/mc/factsheet_vegetarianism.asp.

———. *Vegetarian Starter Kit*. Retrieved on March 27, 2006, from http://www.petaliterature.com/VEG297.pdf.

———. *Chew on This*. Retrieved on March 13, 2007, from http://www.goveg.com/feat/chewonthis/

Robbins, J. (1987). *Diet for a new America.* Walpole: Stillpoint Publishing.
———. (2001). *The food revolution.* Berkeley: Conari Press.
Schiffman, S., Saitely Miller, E., Suggs, M., & Graham, B. (1995). The effect of environmental odors emanating from commercial swine operations on the mood of nearby residents. *Brain Research Bulletin, 37*(4), 360–75.
Scully, M. (2002). *Dominion: The power of man, the suffering of animals.* New York: St. Martin's Press.
Singer, P. (1975). *Animal liberation.* New York: Avon Books.
Spock, B. (1998). *Dr. Spock's baby and child care* (7th ed.). New York: Simon & Schuster, Inc.
Stepaniak, J. (1998). *The vegan sourcebook.* Lincolnwood: Lowell House.
———. (2000). *Being vegan.* Lincolnwood: Lowell House.
Warrick, J. (2001, April 10). They die piece by piece. In Overtaxed Plants, Humane Treatment of Cattle Is Often a Battle Lost. *The Washington Post,* p. A01.

Heather Moore

Animals at Work
Animals in the London Blitz, 1939–1945

The London Blitz clearly demonstrated the depth of relationships that can exist between people and animals. While some rescuers undertook the dangerous work of rescuing human victims of the bombing, others set out to assist injured or trapped animals. At the same time, dogs worked with their handlers in bombed and demolished buildings, seeking signs of human or animal life. In this way, the rescue service became a joint venture.

People Helping Animals

In July 1939, at the request of the Ministry of Home Security, the National Air Raid Precautions for Animals Committee (NARPA) was set up to make arrangements for animals that might be caught in air raids. NARPA was assisted by several animal welfare organizations, especially the Royal Society for the Prevention of Cruelty to Animals (RSPCA), the People's Dispensary for Sick Animals (PDSA), and the Canine Defence League, and veterinary surgeons freely volunteered to give advice. Rescue centers for animals were opened, staffed by volunteers and funded largely by money raised by the American Humane Association, the Canadian Society for the Prevention of Cruelty to Animals, and Australia and South Africa. In all, there were nearly 700 rescue centers by the end of the war. Where possible, these were placed near rescue centers for the human victims of the bombings, so that owners and their pets could be close to one another. Despite this, many owners refused to be separated from their animals.

The majority of horses in London were working horses, used for pulling milk floats and carts. A BBC appeal was made on the radio, asking people whose cars had been requisitioned to lend empty garages for horses whose stables had been destroyed by bombs. His Majesty King George VI offered part of the Royal Mews as temporary accommodation and large dray horses might be seen leaving the royal stables at Buckingham Palace as part of their daily routine.

Animals Helping People

Although there was occasional criticism that people were sharing their food-rations with their pets or growing weeds where vegetables could have been grown, there was general recognition that with husbands and sons away, animals played a big part in keeping up family morale. Few could doubt the depth of affection people had for their animals.

Evelyn Le Chêne's book *Silent Heroes* and Jilly Cooper's *Animals in War* both tell the story of Irma the "Blitz Dog," one of the dogs serving in London during the bombing. After—and often during—the raids, Civil Defence teams would work in the rubble with trained dogs, who, with their excellent sense of smell and enormous dedication, were especially good at locating life in the rubble. Because London had to observe strict blackout regulations, streets and houses were frequently in darkness. Irma, an Alsatian, saved many lives. Once she believed that someone was trapped under a collapsed building, she refused to leave the spot. On one occasion, she convinced the rescue squad that there was still someone below, and after more searching, the rescuers dug out two little girls, both alive. Another time, she signaled the position of four people trapped fourteen feet beneath, one of whom had died. Once, she located under the rubble a family of five and their cat. Irma was later awarded the highest British animal honor for valor, the Dickin Medal, instituted by Maria Dickin, founder of the People's Dispensary for Sick Animals. One side of Irma's medal bore the words "PDSA" and "For Gallantry," with "We Also Serve" below. Inscribed on the other side was "IRMA. For being responsible for the rescue of persons trapped under blitzed buildings while serving with the Civil Defence Services in London." Irma and seventeen other dogs were presented with a Dickin Medal. By the end of the war, fifty-three Dickin Medals had been awarded to animals, and the RSPCA reported that in all they had rescued 256,000 animals.

A rescue worker uses a sniffer dog to search for casualties in the ruins of Farringdon Market, London, after a V-2 rocket attack, 1945. ©Getty Images.

Remembering the Animals

In November 2004, the Princess Royal unveiled a memorial in London's Park Lane to commemorate the contribution made by animals in wartime. Its wall shows the medals given to animals, including the Dickin Medal that was given to Irma and has been likened to the Victoria Cross presented for outstanding bravery. We owe an enormous debt to both the people and the animals who served in World War II between 1939 and 1945.

Further Resources

Cooper, J. (1983). *Animals in war*. London: Heinemann.

Le Chêne, E. (1994). *Silent heroes: The bravery and devotion of animals in war.* London: Souvenir Press.

Moss, A. W., & Kirby, E. (1947). *Animals were there: A record of the work of the RSPCA during the war of 1939–1945*. London: Hutchinson.

Margaret Fidler

Animals at Work
Animals in War

From the earliest times, animals have been conscripted to serve in humanity's wars. All manner of creatures, from elephants to pigeons, have been drawn into our conflicts to be used as offensive and defensive forms of weaponry, to serve as couriers, or, more recently, to be used as disposable subjects for chemical and biological weapons experimentation.

Hannibal of Carthage used Indian Elephants in his ambitious plan to defeat the Roman army on their home soil via a journey to Italy over the Alps in 215 BCE. With 50,000 foot soldiers as reinforcement, the elephants plowed into the Roman ranks like modern-day tanks, trampling the enemy and causing general chaos.

Horses were perhaps the most commonly employed of wartime animals because of their agility, endurance, and speed. Among the first to launch a war using horses were the tribal Hyksos (from modern-day Turkey), who conquered Egypt around 2000 BCE with horse-drawn chariots from which their archers could deliver their load with deadly accuracy. In 450 BCE, Attila the Hun used horses with the addition of saddles and a new invention—the foot stirrup—which gave his warriors superior balance and leverage to more accurately fire an arrow, swing a sword, or throw a spear. Horses continued in this same capacity in the wars to follow, serving as the mobile foundation from which strategic assaults could be launched.

With their innate devotion to humans and superior physical senses, dogs have been one of the more easily exploited animals in military history. The Egyptians, Romans, and Greeks all depended on barking dogs to give early warning of approaching enemies. Also common were large, aggressive mastiffs trained to maim and kill. Cloaked in padded armor or wearing collars studded with metal spikes, these "Mollosus" would be unleashed on an enemy infantry to tear out the throats and bellies of soldiers and horses. The advantages of using dogs as weapons were not lost on later strategists, either; upon arriving in Jamaica in 1494, one of Christopher Columbus's first acts was to unleash a large hound on a reception party of ceremonially painted natives, killing six of them within minutes. Subsequent conquerors of the New World brought their own detachments of killer dogs and quickly routed every native community in Latin America.

Like dogs, carrier pigeons have played a recurring role through centuries of warfare. News of the conquest of Gaul (modern France and Belgium) in 56 BCE by Caius Julius Caesar was dispatched to Rome via a homing pigeon with a papyrus message tied to one of its legs. Similarly trained birds also were present at the battle of Waterloo in 1815, when Wellington used them to convey word of his overwhelming victory against Napoleon's forces. And during the siege of Paris in 1870, messages were reduced and copied onto a primitive version of microfilm, thereby allowing more information to be compacted into a portable size and sent into the city via pigeons. Over the course of four months, these birds transported 150,000 official memorandums and a million personal letters.

Commencement of the "war to end all wars" in 1914 saw the largest mobilization of animals in history. Three million horses, mules, and oxen; 50,000 dogs; and scores of other creatures were ensnared in this protracted and devastating conflict. World War I would prove fatal for most of these animals because for the first time they were being pitted against mechanized weaponry and lethal chemical agents.

A dashing cavalry charge typical of earlier wars was impossible given the nature of the new battlefield landscape. It was fraught with deep artillery craters, pits of sucking

The skeletal remains of horses litter this World War I battlefield. Courtesy of Animal Image Photography.

mud, and impossible tangles of barbed wire. Trapped in this quagmire, whole regiments could be mowed down with a machine gun. Eyewitness accounts describe pitiful scenes of horses that, upon hearing the retreat bugle, would struggle to return to the defensive line despite being horribly wounded. Additional horses and mules had to navigate these "no man's lands" under cover of darkness to replenish trench-bound troops with food and ammunition. The bodies of soldiers and horses killed during the day often had to be used as stepping-stones lest the entire team of pack animals and their human handlers be pulled under by the mud and smothered.

Some horses seemed to know when an attack was imminent. One British former polo pony would stamp her feet and neigh loudly a full five minutes before enemy planes appeared overhead. Others could hear the faint whistle of incoming mortars and, like their human comrades, would drop to their bellies and press their heads to the ground.

Dogs too played a key role in this war, although they were no longer used as attack animals because of advancements in other forms of weaponry. Swift canines were invaluable for relaying messages in the heat of battle, as again were the carrier pigeons, and the two often worked in tandem. Records attest to the efficiency of the messenger dog, which could have a message affixed to its collar before being unleashed to make a mad, zigzag dash across a battlefield. Of particular note was a black greyhound named Satan who turned the tide at the battle for Verdun. The town was being smashed by a German battery when the besieged French spotted a black dog racing toward them. A German bullet caught the animal and sent him crashing to the ground, but moments later, he staggered back to his feet. Despite one shattered hind leg, Satan pressed forward and limped the remaining yards to his friends. His collar contained a note that reinforcements were on the way, and in his saddle pack were two homing pigeons, which the soldiers used to dispatch back the precise location of the enemy so that artillery could knock out the German position. Thanks to the heroic actions of these animals, Verdun was saved.

Every country had its own Red Cross organization, and each group trained "mercy dogs" to locate wounded soldiers who were lost on the battlefields. When they found an injured man, these dogs collected his helmet or a piece of uniform and returned to the trench to alert the stretcher-bearers, who then followed the dog back to the wounded man. After a single battle, one Red Cross dog named Prusco located more than a hundred wounded soldiers and was strong enough to drag many unconscious men into sheltering craters before fetching the ambulance team.

The years leading up to World War II in 1939 saw vast improvements in planes, weaponry, and forms of wireless communication, reducing the conscription of so many animals, particularly horses. Even so, dogs were needed to support the soldiers in various capacities. A civilian organization called Dogs for Defense formed in 1942 and issued a public call for dogs. Americans donated 40,000 canines, many of them household pets. Those that made it through a basic doggie "boot camp" went on to be trained primarily as sentries, patrolling defensive perimeters of military facilities with an armed human escort. Others worked in the field with detachments of soldiers, where they alerted to potential ambushes and hidden explosives.

Chips was one of the most celebrated dogs of this war. The German shepherd first worked as a tank guard and marched with Patton's Seventh army through eight

The first batch of American canine inductees in 1942 was stationed for training at the Quartermaster Corps in Front Royal, Virginia. Courtesy of the National Archives.

campaigns in Africa, the Mediterranean, and Europe. Chips's true mettle under fire was tested on the coast of Sicily, where, against the commands of his handler, he bolted down the beach and leapt into what was thought to be an abandoned pillbox. In fact, it held six German soldiers poised to open fire with a machine gun. In spite of being wounded in the scuffle, the dog subdued the gunner and frightened the other soldiers into surrendering. For his actions, Chips received the Purple Heart and the Silver Star, honorary medals usually reserved for humans. He was the last American animal to be officially recognized by the government for wartime service.

At the end of World War II, the public was outraged when it learned that the army planned to euthanize or auction off the surviving war dogs rather than return them to their original owners. Yielding to the protests, the Department of Defense agreed to release the dogs to their original families following a brief retraining period to acclimatize the animals back to civilian life. Several hundred dogs went home, including Chips (his family reported that he did not seem much changed by his wartime experiences, although he acted a little more tired and less interested in chasing the garbage men when they rattled the cans).

Several thousand canines were again deployed in the Korean War (1950–53), and as in World War II, they primarily worked as sentries and scouts. "Whenever they pricked up their ears and whined a little we knew the enemy was within 150 yards and we made ourselves ready," recalled former Staff Sergeant Melvin Powell, who served a thirteen-month tour in Korea (Davis, 1970, p. 183). Strategists determined that whenever dogs were used in times of imminent contact with the enemy, they reduced casualties by more than 65 percent.

The War Dog Memorial, dedicated to the dogs of World War I, is in the Hartsdale Pet Cemetery in Hartsdale, New York. Robby the bomb detection dog is buried in its shadow. Courtesy of Animal Image Photography

During the Vietnam conflict (1965–72), scout dogs were particularly vital in helping soldiers avoid jungle ambushes and trip-wired explosives. A harmless-looking footpath could harbor spring-loaded, poisoned spikes and shrapnel-packed mines, and it was up to the scout dogs to identify these hazards in time to avoid disaster. Walking off-leash and about twenty yards in front of the unit, these canines worked in stealthy silence. They signaled when something was amiss by stopping, sitting down, or returning to the handler. By war's end, the dog teams were credited with discovering over a million pounds of enemy supplies, several tons of ammunition, and 4,000 enemy booby traps. By some estimates, they saved as many as 10,000 soldiers' lives.

Fearing another public protest over the treatment of decommissioned military dogs, the government had quietly classified all canines as equipment shortly after World War II, meaning they could be disposed of in any manner. And in March 1973, just as the

United States formally announced its withdrawal from Vietnam, orders were issued to leave the dogs behind. Most of them were given to the Army of the Republic of Vietnam, which had little interest or experience working with dogs. American GIs who credited the scout and sentry dogs with saving their lives were stunned and heartbroken. After all these animals had done for them, many thought it the height of betrayal not to take them home as well. To this day, some combat veterans speculate that their dogs perished from neglect or were killed and eaten, as was customary through much of Asia at this time.

Today there are approximately 4,000 trained American dogs patrolling airbases and military installations or working in explosives detection in places such as Afghanistan and Iraq. They never know civilian life but rather are purchased from breeders when they are just weaned and are then put directly into training facilities. Official policy still classifies all dogs as equipment rather than living personnel.

The government continued to routinely euthanize the animals when they were too old or infirmed to work until 1999, when news broke of the Air Force's intent to put down an aging bomb-detection dog named Robby. A passionate Internet campaign to end the blanket euthanasia policy was sparked. Thousands of citizens e-mailed their congressional representatives to protest for Robby. One year later, the House of Representatives and the Senate unanimously passed a canine retirement law stipulating an adoption alternative for decommissioned dogs. Robby died before the bill could be signed by President Clinton, but he was buried with full honors befitting a human soldier at the Hartsdale Pet Cemetery outside New York City. His epitaph credits him as the "inspiration for America's first war dog retirement law."

Other animals continue to play a role in the military, although they are not as well publicized as the working dogs. Each year, an estimated 300,000 primates, small dogs, pigs, goats, sheep, rabbits, mice, cats, and other animals are experimented on by the U.S. Department of Defense or contracted private entities. They are subjected to experimental chemical and biological weapons or are purposely shot and burned so that their wounds can be studied for weapons efficiency. Animal advocacy groups have demanded greater accountability from these research programs, and with increased media coverage of how the military exploits animals, the public again is becoming vocal in its disapproval.

There is not one member of the animal community that has not been adversely affected at one time or another by man's wars. The only way to repay them is to ensure that they are treated with greater respect and kindness in times of peace and that in the future they are involved in our conflicts as little as possible. Ultimately, the exploitation of animals for wars of our own making dehumanizes us all.

The bond between handlers and their dogs go deep. This sentry dog took a bullet in the head at Guam in WWII. Courtesy of the National Archives.

Further Resources

Ambrus, V. (1975). *Horses in battle.* London: Oxford University Press.

Brereton, J. M. (1976). *The horse in war.* Newton Abbott, England: David & Charles.

Cooper, J. (1983). *Animals in war.* London: William Heineman.

Davis, H. P. (Ed.). (1970). *The new dog encyclopedia.* Mechanicsburg, PA: Stackpole Books.

Greene, G. (1994). *A star for Buster.* Huntington, WV: University Editions.

Grier, J., & Varner, J. (1983). *Dogs of the conquest.* Oklahoma City: University of Oklahoma Press.

Hamer, B. (2001). *Dogs at war: True stories of canine courage under fire.* London: Carlton.

Lemish, M. (1996). *War dogs: Canines in combat.* McLean, VA: Brassey's.

Putney, W. W. (2001). *Always faithful: A memoir of the marine dogs of WWII.* New York: The Free Press.

Redmond, S. R. (2003). *Pigeon hero!* New York: Aladdin Paperbacks.

Silverstein, A. (2003). *Beautiful birds.* Brookfield, CT: Twenty-first Century Books.

Thurston, M. E. (1996). *The lost history of the canine race: Our fifteen-thousand-year love affair with dogs.* Kansas City: Andrews and McMeel.

Mary Thurston

Animals at Work
Dog and Human Cooperation in Hunting

The first animal that humans, *Homo sapiens,* domesticated was the wolf, *Canis lupus*—the ancestor of the dog, *Canis familiaris*—about 20,000 to 100,000 years ago. Today, dogs are among the most popular pets all over the world, and several strains are bred for different purposes, such as guiding, hunting, protection, or companionship. One of the suggested motivations for domestication in the first place was the benefit of having a dog present during hunting, such as for finding and tracking prey. Also, a mutual behavioral pattern between wolves and humans might have promoted domestication, given that both species hunt prey larger than themselves in social groups. Cooperation requires effective communication among group members, and during the domestication process, dogs have acquired social-communicative skills used with humans.

Moose, *Alces alces,* have a long evolutionary history with wolves in which they have been wolves' main prey, particularly in the northern hemisphere. For humans, moose have been one of the most important and valued game species ever since the Stone Age, providing meat, large skin for clothing, and bones for tools. We have studied the effect of dog assistance in hunting by comparing the per-hunter moose-hunting success for hunting groups in Finland, including noting the differences in size and uses of dogs.

Currently, in the Nordic countries, there are about half a dozen dog breeds from the spitz family specialized for moose hunting. In moose-hunting groups using a dog, hunters wait for a released dog to direct moose toward them, or if the dog has halted the moose by barking at it, one hunter approaches the moose until within shooting range. In groups without a dog, part of the hunting group works to direct or track the moose toward a line of armed hunters.

In moose-hunting groups of all sizes, hunting success was higher with a dog than without one. The difference was most pronounced in small groups with less than ten hunters. If the benefit is measured as an average carcass weight obtained per hunter, it equals 8.4 kilograms (18.48 lbs) and 13.1 kilograms (28.82 lbs) per hunter per day in groups without and with a dog, respectively. If the corresponding effect was the same in the early hunter-gatherer societies in which the wolf was first domesticated, hunters with dogs probably had a markedly improved protein acquisition with possible positive effects on survival as well. The number of dogs in a hunting group with more than ten persons also had a significant effect: the more dogs they had, the more moose they killed. In groups of less than ten hunters, the larger number of dogs did not increase hunting success. The benefit of using a dog was, interestingly, dependent on moose density. With the increasing number of moose in an area, the benefit of having a dog decreased—using a dog was most beneficial when moose density was low.

Although the prevailing conditions during domestication cannot be repeated to study the process accurately, our results support the idea that cooperation in hunting was an important factor in increasing hunting success and consequently advancing domestication of the wolf.

See also

Hunting, Fishing, and Trapping

Further Resources

Clutton-Brock, J. (1977). Man-made dogs. *Science, 197,* 1340–42.

———. (1992). The process of domestication. *Mammalian Review, 22,* 27–34.

Hare, B., Brown, M., Williamson, C., & Tomasello, M. (2002). The domestication of social cognition in dogs. *Science, 298,* 1634–36.

Ruusila, V., & Pesonen, M. (2004). Interspecific cooperation in human (*Homo sapiens*) hunting: The benefits of a barking dog (*Canis familiaris*). *Annales Zoologici Fennici, 41,* 545–49.

Vilá, C., Savolainen, P., Maldonado, J. E., Amorim, L. R., Rice, J. E., Honeycutt, R. L., Crandall, K. A., Lundeberg, J., and Wayne, R. K. (1997). Multiple and ancient origins of the domestic dog. *Science, 276*, 1687–89.

Vesa Ruusila and Mauri Pesonen

Animals at Work
Stock Dogs and Livestock

The Scots say, “There is no good flock without a good shepherd, and there is no good shepherd without a good dog.”

There are two classes of useful stock dogs: dogs that guard livestock (mostly sheep and goats) and dogs that work livestock (many species, including poultry).

Popular guarding breeds include the Anatolian shepherd, Maremma, Akbash, and Great Pyrenees. The guard dog's primary flock defense is salvos of warning barks, but he will attack predators that ignore his warning.

Future guard dogs may begin working on these predator deterrents with an experienced mentor dog while little more than puppies. Absent the mentor dog, the farmer or rancher will confine the pup with docile ewes or lambs for several months until the dog is bonded with its charges. Although the guard dog doesn't believe it is a sheep, it comes to prefer the company of sheep and takes responsibility for their safety. If the bonding is done properly, the guard dog will not be a house pet. Although the guard dog is minimally trained, its owner should be able to catch it for veterinary care or routine medications.

U.S. Department of Agriculture (USDA) studies have shown that livestock-guarding dogs are effective against predation by coyotes, bears, mountain lions, and wolves.

Guard dogs sometimes move their flocks to safety, and it is reasonable to surmise that guard dogs were the ancestors of modern stock dogs that work livestock but do not guard them. Eighteenth-century accounts describe stock dogs that guarded *and* worked stock, and some dog breeds today are "dual purpose": the English and German shepherds are the most numerous. Dual-purpose dogs are rarely seen in commercial livestock operations.

The second class of useful stock dog *works* livestock: the dog gathers, fetches, drives, pens, and helps the farmer sort cattle, sheep, goats, or hogs. Three breeds have economic importance: the kelpie, Border collie, and huntaway.

The kelpie is a "loose-eyed" Australian collie that specializes in pen and chute work. Because of his relaxed working style, the kelpie is more heat tolerant than the Border collie.

The huntaway—perhaps a retriever–collie cross—originated in New Zealand and is used there to drive enormous ewe flocks by controlled barking.

The most widely used stock dog in the world, the Border collie, is an amalgam of many now-extinct regional British collies. The modern breed was not created by noble sportsmen or dog fanciers, but by illiterate poor shepherds, who slept in chimneyless hovels (bothies) with their dogs and bred them as work tools. In the 1680s, Samuel Pepys wrote what may be the earliest account of a shepherd with his sheepdog:

> We took notice of his woolen knit stockings of two colours mixed, and of his shoes, shod with iron, both at the toe and the heels, and with great nails in the soles of his feet, which was mighty pretty; and taking notice of them, "why," says the poor man, "the downes, you see are full of stones, and we are faine to show ourselves thus, and these," says he, "will make the stones fly until they ring before me." I did give the poor man something, for which he was mightily thankful, and I tried to cast stones with his horne crook. He values his dog mightily, that would turn a sheep any way which he would have them when he goes to fold them.

These collies run to the heads of stock and fetch them to the shepherd. They can gather a thousand ewes, running a hundred miles a day over rough terrain. They work stock in a characteristic crouch. Their crouch and the powerful glare (known as "eye") imitate predator behaviors, and the dogs control livestock by moral authority without barking or biting.

Although crouch, eye, bidability (trainability), eagerness to work, and heading livestock define the breed, traits such as courage, power over livestock, wide outruns, and balance on sheep are also, though less reliably, heritable.

Interestingly, although almost all working Border collie pups will work stock, their behavioral traits appear to be separately heritable. There is no single "working stock dog" gene.

Until puberty, the stock-dog pup is nurtured and socialized like any other. One day, when the pup is five months to a year old, it "sees" sheep. Its tail goes down, it drops into the characteristic crouch, and it begins eyeing and stalking them.

Pet dogs have no genetic desire to sit, heel, or stay. Hence, they must be motivated with treats or checked by corrections. The stock dog desperately wants to do the work his genetics are summoning him toward, and the trainer's job is eliciting the proper expression of the dog's genetics.

Because the dog's reward is the work itself, food rewards and effusive praise are superfluous. The dog starts its training by circling sheep and holding them (balancing) them to the trainer. He proceeds to short fetches and small gathers. When he can outrun a few hundred feet and fetch a half dozen ewes to his handler, the dog is taught to drive them away. Often, the sheepdog-in-training will come with his owner and a more experienced stock dog on routine chores, and the young dog is encouraged to help. The more experience the dog gets, the less training is needed, and the better dogs will have learned a great deal about livestock simply by daily work.

Jack Knox, the dean of American stock-dog trainers, explains the stock-dog trainer's philosophy pithily: "Allow the right, correct the wrong."

The best correction is the mildest *effective* one. Shock collars are contraindicated. They have ruined stock dogs.

Most stock dogs will be useful chore dogs by age two, and by four, they should be able to do the hardest work on the farm.

Many stock dogs on farms and ranches compete in sheep and cattle trials, which are models of the dog's daily work. This trial model is much more difficult than the tasks the dog does at home.

The first official sheepdog trial was in Bala, Wales, in 1873, and trials are held regularly in the United Kingdom, North America, South Africa, the Falklands, and recently in Europe and Scandinavia. Although trials vary by terrain, sheep breed, and the trial host's intent, most feature an outrun (the dog runs out to the sheep and gets behind them without alarming them); the lift (the few seconds during which the dog and sheep evaluate each other and the sheep move off the dog); the fetch (the dog directs the sheep in a straight line to the handler); the drive (the dog drives his sheep through several freestanding gates); the pen (the dog and handler put the sheep in a small freestanding pen); and the shed (the dog and handler sort off one or more sheep and take them away. The handler may not touch the sheep, and the dog may not nip them.

The important trials are the most rigorous, and at the International (UK) and the National Finals (North America) the dogs may find themselves doing precision work half a mile from their handlers on sheep who have never seen a sheepdog or a man off horseback.

Cattle dog trials are much like sheepdog trials.

Although no more than 5 percent of working stock dogs ever see a trial field, almost all farm and ranch dogs trace their ancestry to trial dogs.

The principal registries—the International Sheepdog Society (ISDS) in the United Kingdom, the American Border Collie Association (ABCA), and the Canadian Border Collie Association (CBCA)—will register outstanding unpapered stock dogs on merit. Although most registered dogs are purebred Border collies, doubtless outcrosses have slipped into the gene pool. The arduous trials and the relentless selection for working abilities have proved a successful genetic strategy. The coefficient of inheritance (COI) is low, and the farmer and rancher need no special knowledge or connections to get a sound pup who will work stock.

The poor shepherd's dog, the Border collie, did not appeal to nineteenth-century dog fanciers. Before 1986 in the United Kingdom and 1994 in the United States, Border collies were never shown in dog shows. These Border collies exhibited great morphological and physiological variation. Like lawyers, they did not "look" alike, but they "worked" alike. Those who knew and used Border collies resisted the unsought "recognitions" on the grounds that breeding for dog shows would create a new morphologically uniform breed with the name "Border collie" without the behavioral genetics that make the dog useful.

The show breed is rare on farms and ranches and almost never competes in traditional sheepdog or cattle dog trials.

Border collies are unusually biddable, and this combined with the dogs' athleticism and eagerness to work makes them star performers in pet dog venues such as agility, obedience, and flyball. They make fine assistance dogs, and in the United Kingdom, Border collies dominate search-and-rescue canine operations.

That said, they make poor pets for owners who have no work (or sport) for their dogs. The Border collie's intelligence and desperate desire to work create a nuisance in most pet homes.

Further Resources

Livestock guarding dogs: Protecting sheep from predators. (n.d.). USDA, Agriculture Information Bulletin No. 588. http://www.nal.usda.gov/awic/companimals/guarddogs/guarddogs.htm

Parson, A. D. (1986). *The working kelpie*. Melbourne: Penguin.

Scrimgeour, D. (2002). *Talking sheepdogs*. Lancashire, UK: Farming Books & Videos Ltd.

United States Border Collie Club. (n.d.). http://bordercollie.org/

Donald McCaig

Anthropomorphism
Anthropomorphism

Anthropomorphism is, at its most general, the assignment of human characteristics to objects, events, or nonhuman animals. Notably, belying this neutral definition is a non-neutral connotation to the word and to the phenomenon it describes. Specifically, an "anthropomorphic" characterization is generally held to be an *erroneous* one—at best, premature or incomplete, and at worst, dangerously misleading. That an anthropomorphism is, further, *incorrect* as a description is often assumed.

Anthropomorphizing is a natural human tendency, thought to be the result of a perceptual system designed to find order in a complex world. Contemporary humans (perhaps as with our forebears) tend to interpret a landscape entirely free of human presence as thick with human faces: on a slab of rock, in the gnarl of a tree knot, in the waxing moon, in a pendulous flower. The lexicon used to describe the human body is pervasive in our descriptions of nature: the shoulder of a hillside, the arms of a tree, the fingers of a stream, the waist of a peninsula—all examples of what literary criticism calls "personification."

Of greatest import to the present study of human-animal interactions are anthropomorphisms of animals as having attributes and mental states (especially cognitive and emotional) similar to *human* attributes and mental states. Pets are regular subjects—a dog's low, rapid tail-wagging explained as guilt for eating a shoe, or a cat rubbing against its owner interpreted as an expression of fondness. Wild animals are no less immune: two red-tailed hawks who have established residency atop a building on Fifth Avenue in New York City have been called "in love"; their journey, an "odyssey"; and the male is often called "daring," "self-assured," and "an ambassador to the wild." Research in the recently developed field of cognitive ethology, in essence, accumulates empirical data on precisely the kinds of mental states that anthropomorphisms claim (without the backing of science): the purposes, feelings, motivations, and cognition of animals; thus, the science and the attributions are interwoven. This is the form of anthropomorphism with which we shall primarily concern ourselves in this essay.

Questions asked by scientists interested in the subject include: What is the history of our use of anthropomorphism? What does it mean—both originally and by implication? Is it a bane or a blessing? Why do we anthropomorphize at all?

A Brief History of Anthropomorphism

Anthropomorphic representation dates to (at least) Paleolithic art of forty thousand years ago, when some drawings of animals included characteristically human features. Anthropomorphisms have appeared in human writings for thousands of years; reproach for such projections for nearly as long. The term originally referred to the "blasphemous" descriptions of gods as having human forms. Indeed, religious scholars suggest that all religious systems include anthropomorphisms. Ancient societies similarly projected motives and emotions onto natural phenomena—angry winds, vengeful storms—and animals and natural events were often named and ascribed personalities. Later, even physics was to be influenced by an anthropomorphic teleology. Aristotle described a rock's downward tumble not as the result of a force between bodies, but as the rock acting to achieve the desired end of being on the ground. Xenophanes (sixth century BCE) is well documented as the first to give voice to the negative tone of anthropomorphism—he called it an error; modern critiques date to seventeenth-century philosophers Francis Bacon and Baruch Spinoza. In fact, the rise in modern science is matched by the diminishment and increasing censure of anthropomorphic descriptions of natural phenomena. Still, both ancient and modern literature and folk psychology are replete with anthropomorphic language. The characterizations of Aesop—the happy dog, the persistent tortoise, the industrious ant—resonate and endure to this day.

In its current usage, anthropomorphism is tinged with the bad flavor that the anecdotalism of late-nineteenth-century scientists such as Charles Darwin and George Romanes left in science's mouth. While on the one hand epitomizing "modern science," Darwin also embraced a classically anthropomorphic attitude toward animals. Based on anecdotes and personal experiences, Darwin and his followers ascribed everything from emotions to insight to animals with abandon—and the future sciences of zoology, biology, and ethology developed in reaction against this. A comparison of the languages of description makes the distinction clear: Darwin spoke of "ants chasing and pretending to bite each other, like so many puppies" (1871, p. 448). A century later, a more typical description of the study of ants (taken from a biological research group's Web site) investigates "the presence of neurochemical mechanisms underlying the phenomena of social reward and social cohesion in ant colonies," and "the role of homo- and heterospecific social context in the control of the expression/suppression of ant behaviour." Similarly, while Darwin noted that dogs could be variously magnanimous and sensible, shameful and modest, sensible and proud, these words are notably absent from contemporary ethological descriptions of dogs.

Further Conceptual Considerations

The historical result, as we shall see, is the often-presumptive dismissal of anthropomorphism. Recent writers have claimed it to be sentimental and sloppy, at once libertine and lazy. Before discussing the current debate about its use, a brief interlude to introduce some attempts to understand anthropomorphism as more of a rhetorical device than a metaphysical assertion.

It could be argued that anthropomorphism (of animals) is a particular kind of metaphorical description—and it would profit from such a claim, as metaphor is granted an immunity in application not extended to anthropomorphism. "My love is a red, red rose" may in fact be an odd or unhelpful description of one's love, to some audiences, but it would not typically be subject to complaints that it is *prima facie* inappropriate. "My dog loves that little poodle," however, is taken as a claim with a different level of standards for acceptance.

A look at how anthropomorphizers or metaphorists might respond to challenges and questions about their use of language makes this clear. With an anthropomorphism of this kind, if a user is asked "Is your dog *really, truly* in love with the poodle?", the anthropomorphizer might assent "Yes, he is," or she might clarify by saying "Well, I don't know that he *loves* her as much as just *lusts* after her." In other words, the anthropomorphizer ordinarily treats her claim as a literal claim, and addresses any challenge by maintaining the claim or by refining it to clarify her meaning. Any retreat from the literal claim ("Oh, I didn't mean he was *really, truly* in love with the poodle") withdraws the entire assertion. It does not merely refine the trope; it eliminates the anthropomorphism outright.

In contrast, the metaphor-maker *distinguishes* his usage from the literal meaning of the word. If asked "Is your love *really, truly* a red, red rose?", the metaphorist surely replies, "No, not *really*"—and then explains that he meant that his love is vibrant like a rose, delicate like a rose, et cetera, but that the turn of phrase was not meant to be taken literally. The metaphor may be judged by listeners as "better" or "worse"—more or less poetic or evocative—but not as a use of language to describe the world in a way that contemporary science might verify. Most anthropomorphizers of animals do not seem to be using words metaphorically, insofar as they are prepared to defend their language use as a strict literal use and as making a claim which would be verified (or refuted) by the methods of science. In this way, though anthropomorphism sometimes uses images shared by metaphor-makers, it is otherwise distinct from metaphor.

More powerful is the proposal that anthropomorphism is less a straightforward factual claim than a form of analogy. Structurally, the claim of the attribution of poodle-love could be described as equivalent to an inference of the presence of analogous emotions, given a myriad of other (physical and behavioral) similarities between dogs and humans. In other words, the speaker may clarify that it may not be "love," per se, but it is *like* love: he follows her around, he wags his tail uncontrollably when she appears, he persists in attempting to mount her . . . and so on, more or less just like human love. This is credible, although it does not exempt anthropomorphizers from criticism on factual grounds; even if the claim is more attenuated than originally thought, it is still (in most cases) without scientific support. And even if all anthropomorphisms are simply analogies relying on particular similarities between the target and the source, not all such analogies are anthropomorphisms; forming analogies between humans and other animals is regularly considered *non*anthropomorphic. For instance, dissection of a sheep's brain in a class on human cognition is not taken to be an anthropomorphic activity. On the other hand, the protest outside the classroom airing claims about the suffering of the sacrificed sheep may be.

Arguments Against and for Anthropomorphism

Even as analogy, anthropomorphism garners disapproval from most commentators. This is unsurprising given the negative light that has long been cast on such attributions. However, more recently, a new debate has emerged in ethology and psychology over the phenomenon, matching, at its extremes, those who think that it is irredeemably erroneous and an anathema to science against those who argue that anthropomorphizing is potentially useful. We will consider each position briefly.

The primary complaint heard extends the reaction to the anecdotalism of Darwin and others: anthropomorphism is not based in science. There is no objective theory formation or testing, no careful consideration of evidence; there is merely unreflective application of human descriptions to nonhumans. Anthropomorphism is a category error, some argue, the treatment of an entity (an animal) as a member of a class (things with minds) to which it does not belong, or the comparison of that entity to one (such as a human) belonging in a different category. Describing a dog as feeling guilt, they claim, is like saying that ideas are green. Those who assert that there are distinctively human traits might so argue: if the trait is, by definition, what separates humans from animals, then to treat an animal as possessing the trait is a logical error. If consciousness is a defining characteristic of humans, for instance, then to claim consciousness in non-humans is a category mistake.

Indeed, some anthropomorphisms are clearly wrong for just these reasons. Happiness is commonly attributed to an animal on the basis of an upturn of the corners of its mouth; such a "smile," however, may be a fixed physiological feature (as with dolphins) or a sign of fear or submission (as with chimpanzees)—not happiness. Similarly, a yawn is likely not a sign of boredom, as might be assumed by extrapolation from our own behaviors; instead, it denotes stress.

Still, the implied suggestion that any mental ability exhibited by human beings is necessarily exclusive to humans is itself presumptuous. A number of researchers are increasingly proposing a careful application of anthropomorphic terms to explain and predict animal behavior. Interestingly, it is the professional observers of animals who often become, with exposure and despite their training, more likely to anthropomorphize. These advocates suggest that anthropomorphisms are not necessarily incorrect. On the contrary, they say, anthropomorphisms are used in reliable ways and are useful. The comparative psychologist Donald Hebb (1946) discovered, for instance, that taking pains to eliminate anthropomorphic descriptions resulted in a *diminished* understanding of the behavior of his chimpanzees. Anthropomorphisms, carefully applied, may be coherent guides to predicting the future behaviors of animals.

The advocate suggests that to treat anthropomorphism as a category error is itself an error; its appropriateness relies on its correctness, and its correctness is an empirical question—not *a priori* determinable. The category-error claim's insistence on the wrongness of shared predicates between human and nonhuman animals is a vestige of the faulty notion that humans are separate from animals. Finally, some argue that anthropomorphism is inevitable, an unavoidable result of viewing objects and animals from a human perspective. (Other defenders believe this inappropriately downplays a real human ability to perceive and detect subtleties removed from our own experience.) Regardless, the endurance of anthropomorphism indicates that it is worth examining anew.

Explanations for Anthropomorphism

Why do we anthropomorphize? The question can be formulated in two ways: as a question of ultimate—evolutionary—causes, and as one of proximate—local—prompts.

Anthropomorphism's endurance marks it as likely useful—or at least not irreparably harmful—in explaining and predicting animal behavior. Just as the developing child uses animism—the attribution of life to the inanimate—to make sense of the sensory chaos of his environment, anthropomorphism may have arisen as a strategy to make familiar an uncertain world. In normally developing humans, our characteristic propensity to attribute agency to others will become a theory of mind and will find use in social interaction. In the development of the human species, anthropomorphism may have provided a means by which to anticipate and understand the behavior of other animals. With themselves as models, our human forebears could ascribe motivation, desire, and understanding to animals to determine with which ones they may want to cooperate or from which ones they should flee—as well as which ones they want to eat.

If there *is* an evolutionary explanation, we might expect other animals to engage in some version of the behavior. In fact, many do appear to attribute animal characteristics to inanimate objects or occurrences—what anthropologist Stuart Guthrie (1997) has called "zoomorphism." In *The Descent of Man,* Darwin described his own dog growling and barking at an open parasol moving in a breeze, as though in the presence of "some strange living agent" (1871, p. 67). Primatologist Jane Goodall observed chimps making threats toward thunderclouds. Other ethologists have noted animals shying from, stalking, or attempting to treat as prey or playmate a variety of natural objects. Nonhuman animals seem to be subject to a similar version of animistic perception as humans.

As with all stories of the evolution of a behavior, this one cannot be empirically tested. It is naught but an appealing story. A final observation asterisks the notion of the universality of anthropomorphism: what gets called an anthropomorphism varies by culture. Xenophanes observed a cultural difference in describing gods (snub noses and black hair in one region, gray eyes and redheads in another). The twentieth-century philosopher Bertrand Russell noted, only partly in jest, that the results of behavioral experiments seem to show that animals studied by Americans solve problems through an exhausting (if energetic) process of trial and error, while German animals come up with the answer through quiet contemplation. While in contemporary Western cultures the human-animal divide is marked by cognition and a sense of self, Japanese culture places emotional experience as central to identification as human. Scientists of both cultures might find emotional attributions anthropomorphic, but some Japanese primate researchers describe their animals as having personalities, motives, and rich inner lives. All are verboten in Western science.

Proximate Causes

Not all animals are anthropomorphized: gorillas and dogs regularly are, but worms and manta rays rarely are. Some have suggested that frogs' lack of anthropomorphizable characteristics led to their dismal fate at the dissecting table when dissection was becoming a mainstay of biology classes. Why? The question as to the proximate causes may be framed thusly: What are the behaviors and physical features of animals which prompt us to anthropomorphize them?

The answer no doubt has much to do with the ease with which the animal can be mapped to the human in terms of isomorphisms of features and similarities of movement. Aristotle noted the importance of self-locomotion to identification of an autonomous creature; in the last century, psychologists and ethologists have begun to investigate specifics. In 1944, psychologists Fritz Heider and Marianne Simmel published a now-classic paper showing that humans consistently told anthropomorphic stories to describe the behavior of geometric figures moving on a computer screen. They concluded that the timing of movements was integral to the humans' projective storytelling. More recent ethology has added contingent timing of behaviors, expressive facial and bodily reactions to others, and attention to gaze to the growing list of behavioral metrics.

Physically, phylogenetic relatedness accounts for some anthropomorphizing (e.g., of great apes and monkeys); simple ease of matching of parts may account for other differential treatments (an eel's lack of limbs, the facelessness of a limpet). In particular, discernable and flexuous facial features, the ability to form a mouth into a smile, and the ability to move the head expressively and reactively are reliable prompts to certain kinds of anthropomorphisms. Paleontologist Stephen Jay Gould and ethologist Konrad Lorenz both noted that animals with neotenized features—a large head and big eyes, for instance—may prompt affiliation and selection because these are features of human juveniles.

The Future of Anthropomorphism

Karl Popper proposed that hypotheses go through a process of Darwinian selection. Anthropomorphism is a prime candidate for consideration as one of these hypotheses that has survived a selection process for ideas, despite (or, especially, *considering*) scientists' struggle to replace it with behavioristic, nonattributional language. This claim does not imply that our survival as a species depends critically on the particulars of our anthropomorphisms—only that the particulars continue to beat out other explanatory theories.

The extended definition of anthropomorphism as erroneous is itself premature. What the claims of anthropomorphism *are,* often, is scientifically unproven; they are simply extrapolations from our own condition. This should not defame the claims on their face. The onus of science is to find means to confirm or refute these assertions. Hence the future treatment of anthropomorphism by science should include empirical testing of specific attributions. In the case of attributions of mental states, the process should include a deconstruction of the concepts attributed, and a determination of any behavioral correlates as well as what would count as confirming (or disconfirming) evidence of the presence of the attributional state.

A better understanding of what prompts anthropomorphism may yield other fruits. It may give us insight into what features are important to us in interacting with members of our own species. Further, we can look to anthropomorphisms that humans make to natural objects to design robots that look and interact in ways which prompt our anthropomorphizing of them. Instead of faithful reproduction of the form and perceptual and social skills of humans—an enormous, possibly insurmountable task, as the field of artificial intelligence has discovered—one might focus on just those components of physical objects which lead to our anthropomorphizing; to solely those behaviors that lend authenticity to a social interaction.

Finally, the status of anthropomorphism—and the content of the attributions—is relevant in the ongoing discussion of the role of animals in our society: their status as pets, their use as food and entertainment, and their treatment in medical and behavioral research. Ascribing personalities to animals is demonstrably more effective than raw statistics in getting the public's attention. And an analysis of the content—the work of cognitive ethology—will be relevant to the animal rights and animal law movements.

Historically, anthropomorphisms have been used to attempt to uncloak, demystify, or get traction in domains unknown (and perhaps unknowable) to humans, such as the subjective experience of an animal. In the domain of human-animal interactions, anthropomorphism might be best thought of as attributions of human qualities to nonhumans not proven to bear these qualities. The science of anthrozoology may provide such proofs. Anthropomorphism will likely continue regardless.

See also

Culture, Religion, and Belief Systems—*Religion's Origins and Animals*

Further Resources

Crist, E. (1999). *Images of animals: Anthropomorphism and animal mind.* Philadelphia: Temple University Press.

Darwin, C. (1981). *The descent of man; and selection in relation to sex.* Princeton: Princeton University Press. (Original work published 1871.)

Datson, L., & Mitman, G. (2005). *Thinking with animals: New perspectives on anthropomorphism.* New York: Oxford University Press.

Guthrie, S. E. (1997). Anthropomorphism: A definition and a theory. In R. W. Mitchell, N. S. Thompson, & H. L. Miles (Eds.), *Anthropomorphism, anecdotes, and animals* (pp. 50–58). Albany, NY: SUNY.

Hebb, D. O. (1946). Emotion in man and animal: An analysis of the intuitive process of recognition. *Psychological Review, 53,* 88–106.

Heberlein, A. S., & Adolphs, R. (2004). Impaired spontaneous anthropomorphizing despite intact perception and social knowledge. *Proceedings of the National Academy of Sciences, 19*(101), 7487–91.

Heider, F., & Simmel, M. (1944). An experimental study of apparent behavior. *The American Journal of Psychology, 57,* 243–59.

Horowitz, A. C., & Bekoff, M. (2007). Naturalizing anthropomorphism: Behavioral prompts to our humanizing of animals. *Anthrozoös, 20,* 23–35.

Kennedy, J. S. (1992). *The new anthropomorphism.* New York: Cambridge University Press.

Mitchell, R. W., Thompson, N. S., & Miles, H. L., (Eds.). (1997). *Anthropomorphism, anecdotes, and animals.* Albany, NY: State University of New York Press.

Popper, K. (1972). *Objective knowledge: An evolutionary approach.* Oxford: Oxford University Press.

Alexandra C. Horowitz

Anthropomorphism
Human Observations of Animals, Subjective vs. Objective

When we try to describe the attitude of animals toward us, there are concerns about inaccuracy resulting from the "subjectivity" in applying terms that we would normally apply to humans—terms such as "curious" or "angry." However, we cannot avoid using subjective terms, because (1) truly objective terms do not exist, and (2) *any* observation statement is affected by one's own subjective experience.

> Relationships have objective aspects that are apparent to an outside observer and subjective aspects that are specific to each participant, known in their entirety only to him or her, and shared only partially. (Hinde, 1992)

Imagine that you stand close to an artificial nest box or birdhouse in early spring. Lots of birds fly around or jump from one twig to the next, looking for food or giving some calls. However, one of them gets closer to you than the others and behaves in a different way—it literally *stares* at you. This is a clear, observable behavior, provided that you are properly trained to "see" it. The bird moves its head and tries to get a fix on you from several angles, as if it is trying to figure out what you are doing there. As I have shown in a study of titmice—birds similar to American chickadees—this behavior is a clear-cut, objective criterion allowing us to distinguish the nest owner from other birds. It is also one of the truly magic moments in which you sense you are having a one-on-one relationship with a wild animal.

This short account would be dismissed by many animal observers, because I used subjective phrases including "the bird is staring at you"; a common view is that we should describe our relationships with animals by using "objective descriptions" and "neutral language." But how can we know for certain the difference between subjective and objective descriptions?

Let's get back to the event we want to describe. We can give an accurate description of the movements of the bird's head and, if possible, of its eyes. How accurate ought this description be? Where can we stop in the certitude that we have reached the "objective" level? The request to shift from a subjective to an objective description comes from the belief that one term is better than another—or more "real" than another. But this would only hold if we were 100 percent sure that one term better corresponds to the "real" behavior. To do this, we would need someone who could stand between us (with our supposed "subjective" description of what the animal is doing) and what the animal itself *is* doing—that is, the "real" behavior. This person should "know" the real behavior without any description—for only in such a case would he or she be able to tell us whether *our* description is worse than another one. How can one know the "real" behavior, if not through another description besides one's own? Any action, either animal or human, is always described as a behavior, for example "threat." The description of a threat is based mainly upon our previous experiences and the way we categorize animals. If we categorize animals as we do machines, then the use of subjective words is inadequate and anthropomorphic. Therefore, we would better use terms that we normally employ to describe machines. However, if we categorize animals within the same domain as humans by simply admitting that humans are animals, too, then the terms used to describe machines would be inadequate when using them to describe animals. When we observe animals, we need to know what is similar *as well as* what is different between animals and humans. In doing so, we necessarily refer to our human experience. *We* experience mental phenomena such as "being curious," and we ask ourselves whether those birds up there on the tree branch experience the same phenomena as we do. Unfortunately, we do not have any prejudice-free experience that can be used to describe what we see and answer our questions; neutral language does not exist.

The art of observation is to see those things that one normally would not see. It is a highly subjective event, but what is really seen (rather than imagined) is, as such, an objective fact, however subjective the perceptual experience is (and with proper instruction, anybody can learn to see it). For many years, ethologist Adriaan Kortlandt (1995) has studied the personality of cormorants, and he has learned that each bird has a unique combination of "masculinity" and "femininity"—although such subtle characteristics are difficult to arrange into a table. He has, nonetheless, provided an excellent example of how we can grasp qualities that are typically "human" in animals and still be rigorous in the study of their behavior.

Great scientists have often perfectly understood to what extent one can use subjective terms to explain the world around us; one of them was Julian Huxley, a founder of modern ethology. Huxley studied the behavior of birds. His observations led him to conclude that birds experience emotions. In comparing birds with mammals, he wrote:

> The variety of their emotions is greater, their intensity more striking, than in four-footed beasts, while their power of modifying behaviour by experience is less, the subjection to instinct more complete. (1923, p. 108)

To fully understand these words, go to a zoo, accompanied by an expert observer, if possible, and look closely at birds such as cormorants or pelicans. Focus on just one individual, and follow its movements. You will soon realize how varied its expressions are: attention, anxiety, aggressiveness—and how quickly those expressions will change in response to the most trivial-seeming event (such as another bird landing next to the first). It takes some time and effort, but you will be rewarded.

See also

Anthropomorphism
Culture, Religion, and Belief Systems—*Religion's Origins and Animals*
Human Perceptions of Animals

Further Resources

Grieco, F. (2000). Finding out who is nesting where: a method for locating nest sites of hole-nesting species prior to egg-laying. *Avocetta, 24,* 113–119. [Also available at http://www.home.hetnet.nl/~griecof/publications.htm]

Griffin, D. R. (1992). *Animal minds.* Chicago: University of Chicago Press.

Hinde, R. A. (1992). *Developmental Psychology, 28,* 1018–1029.

Huxley, J. S. (1923). *Essays of a biologist.* London: Chatto & Windus.

Kortlandt, A. (1995). Patterns of pair-formation and nest-building in the European Cormorant *Phalacrocorax carbo sinensis. Ardea, 83,* 1–25.

Mitchell, R. W., & Hamm, M. (1997). The interpretation of animal behavior: Anthropomorphism or behavior reading? *Behaviour, 134,* 173–204.

Mitchell, R. W., Thompson, N. S., & Miles, H. L. (Eds.). (1997). *Anthropomorphism, anecdotes, and animals.* Albany: SUNY Press.

Tinbergen, N. (1951). *The study of instinct.* Oxford: Oxford University Press.

Fabrizio Grieco

Anthropomorphism
The Myth of "Sexual Cannibalism" by the Praying Mantis

We use animals for a variety of reasons, many quite mundane. However, our relationships with—and uses of—them can, at times, become subtly obscure. One such use is manipulating the opinions and behaviors of a social group toward an animal in order to increase one's intellectual stature or power within that group—or, simply, to alter the intellectual or behavioral relationships that others have with the animal. This may be done innocently, unintentionally, or for the perceived benefit of others. An example of this occurrence (on an individual level, although the idea is easily applicable to groups) is when a parent overstates the danger of dogs to keep her toddler from approaching unknown animals. At times, however, this type of manipulation can reach such extremes that it becomes psychologically and intellectually irresistible to wonder how the manipulation could have grown to such proportions. When such manipulations occur in *science,* they become even more intriguing.

In 1989, Adrian Wenner published a now-classic article in *American Zoologist,* in which he briefly presented several instances in which scientists had gone to sometimes-ridiculous extremes in promoting certain odd beliefs about animals; for instance, that deer botflies could travel at a speed of 880 miles per hour! One of the other examples that Wenner offered is the belief that female praying mantids always cannibalize the males during mating. Belief in the regularity or necessity of so-called "sexual cannibalism" among mantids is now so widespread that it has become a staple story in a variety of scientific textbooks—and an intransigently persistent urban myth, in spite of the fact that there is no evidence that mantids eating one another has anything to do with their mating behavior. So intriguing is this myth that I set out to find its roots. The journey took me back several centuries and revealed an interesting and longstanding "love/hate relationship" with mantids—much like those that people often have with large mammalian predators, such as wolves.

The mantids are a group of beautiful, sometimes very large, insects; this group contains about two thousand species. All of those species are predators that capture their prey with a rapid grasp of their large, raptorial forelegs. The earliest descriptions of mantid predatory behavior come from the Far East. In China and Japan, the mantis was used as a symbol of strength, courage, and boldness, traits that were put to use in the Chinese sport of insect fighting. Descriptions of this practice were disseminated throughout Europe by a popular genre of travel logs published during the nineteenth century. In his widely read travel documentary, *Travels in China,* John Barrow recounts, in detail, the "cruel and unmanly amusement" of insect fighting. And in his popular book, *A Sketch of a Tour on the Continent,* James Smith describes the mantis as a fierce animal that "savors little of divinity." Having himself put a male and a female together, he explains that, after mating, "[the female] . . . devoured the head and upper part of the body of her companion. But . . . a *subsequent* union took place; the life and vigor of the male being unimpaired by the loss of his head. . . ."

Such stories about mantid aggressiveness rapidly found their way into popular natural history and the general scientific literature. So, as early as 1806, one could read in Shaw's *General Zoology* that the mantis is "far from sanctity," and, in 1815, in Kirby and Spence's extremely popular *Introduction to Entomology,* the authors explained that the "cowardly and cruel" mantis engages in the practice of cannibalism out of sheer wantonness, even ". . . when in no need of other food." So, from the very beginning, descriptions of mantid behavior were shaped by a fundamental misperception. That misperception was fueled by the fact that Europeans originally thought of—and wrote about—mantids as though they were gentle, pious, and helpful, because of their perpetual prayer-like posture. They were appalled to read that the insect was actually pugnacious, voracious—and a cannibal, no less. For instance, note this turn-of-the-century description by the famous French entomologist Jean Henri Fabre:

> Ferocious creatures! It is said that even wolves do not eat one another. The mantis is not so scrupulous; she will eat her fellows when her favorite quarry, the cricket, *is attainable and abundant* . . . [But] these observations reach yet a more revolting extreme . . . [for which there is not] *the excuse of hunger* . . . to devour [one's mate] during the act surpasses anything that the most morbid mind could imagine. I have seen the thing with my own eyes, and I have not yet recovered from my surprise. (Bernard Miall, trans. *Social Life in the Insect World* [London, Leipsic: T. Fisher Unwin, 1912], 84)

Cannibalism represented the dark side of the mantis—its innate wantonness, greed, and voracity; it was not "normal" eating behavior. As explained in *The Natural History of Insects*, when prey is of *appropriate* size, such as a fly, ". . . it is curious to remark how cunningly [the mantis] endeavors to entrap [it]." However, even when well fed, "They never [cease] to attack, kill, and eat each other . . ." And in *The Penny Magazine*, ". . . several experiments have proved that they will devour each other less from hunger than from *savage wantonness.*"

So, mantid predatory behavior was originally seen as normal only when it was directed at small insects such as flies and crickets. When a mantis attacked a large creature such as another mantis, its behavior was perceived as being "abnormal" and having

A mantid (Sphodromantis lineola) eating a small lizard that it has caught. Mantids will capture large prey, if they can, as a normal part of their hunting behavior. Photo used with the permission of the photographer, Marl Gonka.

nothing to do with hunger, per se. This dichotomous misperception of mantid predatory behavior actually continues to exist—even in the scientific literature—despite almost two centuries of anecdotal and experimental reports that mantids do, in fact, normally capture a variety of very large prey (including lizards, small birds, mice, and anything else that they can hold on to). The truth is that cannibalism is actually *just one* instance of a mantis capturing a creature that is well within the normal size range for its prey; its cannibalism has to do with hunger only—and it needs no special explanation.

On the other hand, if you believe that a mantis eating another mantis is abnormal, then it would seem as though it *does* need to be explained somehow. As noted, the original explanation was made anthropomorphically; that is, by ascribing to the mantis traits such as cruelty or savage wantonness. However, as time passed and the use of anthropomorphisms faded away, another alibi for mantis cannibalism immediately took their place. This new alibi stemmed from the fact that any insect can perform many seemingly complex behaviors even if its head is removed. This is possible because, unlike vertebrates, the insect's central nervous system (roughly equivalent to your brain) is distributed along the length of its body. In the case of mantids, a decapitated male can still function well enough to mate (although, overall, its behavior is clearly impaired).

This ability of the decapitated male mantis was recognized as early as the eighteenth century. However, when it came to the attention of one L. O. Howard, he published a brief but seminal account of it 1886, in the prestigious journal *Science*. Although Howard was as struck by the female's voracity as were earlier writers, there is an interesting and important difference in *his* descriptive tone. In place of the usual disgust or disbelief expressed by others, he wrote with a voice that is both reasoned and measured. And, more importantly, he described the partially cannibalized, mating male as both well-evolved and as making the ultimate sacrifice for the greater good of the species—rather than just being a hapless victim. In fact, in a subsequent article on the same topic, published in 1892, he quips: "The nonchalance with which the male devoted himself to the sacrifice . . . indicated that [he] has no serious objection to this method of suicide." Howard's explanation helped establish the seemingly indestructible link between mantis cannibalism and sex. By the mid-1900s, various biology textbooks and PhD dissertations were rife with descriptions and explanations of how evolutionarily wonderful it was that male and female mantids have evolved this grisly partnership. Supposedly, the female turns on or "disinhibits" the male's copulatory behavior by removing his head, and, in turn, he willingly supplies her with a postcopulatory meal to ensure that his offspring's mother is well fed. But the problems with this story are many: First, there is the fact that males are not usually decapitated before copulation and then devoured thereafter; if captured, they are usually completely eaten before they have a chance to mate. Second, cannibalism is known to exist in only a handful of the approximately two thousand known species of mantid. Third, the idea that cannibalism is somehow related to sexual behavior ignores that fact that most instances occur among sexually immature mantids and between adult mantids (of both sexes) that will never mate. So, cannibalism *cannot* have anything to do with sexual behavior, per se.

The interesting question, then, is why did Howard's modern alibi for mantis cannibalism so easily replace the older anthropomorphic alibis, especially within the scientific community? That is, why did the scientific community keep seeing mantis cannibalism as some odd or special behavior that needed a special explanation, rather than just saying, for instance, "Well, mantids eat a lot of large prey, and they don't have the visual acuity to distinguish between a big grasshopper, another mantis, or a small lizard, so they simply capture whatever crosses their path. It has nothing to do with sex"? The answer to this question is that the half-eaten male mantis can still *mate*. If he simply ran around like a decapitated chicken, the phenomenon would be of little interest. (To my knowledge, no one has sought to determine the biological advantage to the headless chicken's undignified display.) But the partially cannibalized male mantis manages that most important evolutionary task. Surely, the reasoning goes, this could not be the case simply by chance.

The last step in permanently entrenching the myth of sexual cannibalism (especially within the scientific community) was the addition of a neurological rationale to the story, and this was added quite nicely between 1935 and 1963 by the influential scientist Ken Roeder. As it happened, in 1916, the French entomologist Etienne Raubad hypothesized that decapitation of the male mantis by the female releases the copulatory reflex from the inhibitory influences of the male's cerebral ganglia (the portion of its central nervous system that is in its head). Initially, Raubad's hypothesis did not reach beyond the small audience of the journal in which it was published; however, two decades later, Ken Roeder, the so-called father of American neuroethology, offered putative experimental evidence to support Raubad's idea. However, Roeder presented the idea as his own and never gave Raubad credit for it. (We know that Roeder was aware of Raubad's paper because he included it in the bibliography of his now-classic 1963 book, *Nerve Cells and Insect Behavior*.) Hence, the idea that male decapitation "disinhibits" copulatory behavior gained acceptance solely because of Roeder's status within the scientific community.

Now the myth was both "modern" and complete. It began with the idea that mantids normally eat only small prey (such as crickets), and that cannibalism is an abnormal behavior having nothing to do with hunger, per se. At first, this abnormality was seen as a product of the mantis's unsavory disposition—its savagery or the like. As anthropomorphisms fell out of favor, rather than seeing cannibalism as just one instance of an opportunistic predator getting a meal, it came to be associated with sexual behavior. This occurred despite the fact that in the vast majority of cases, cannibalism occurs in nonsexual encounters. Perhaps the reason that the cannibalism/sex association remains so strongly held is because, as Sharon May Brown (1986) put it, the idea of a decapitated male mating with the female who will ultimately devour him is "ghoulish, kinky, and bizarre." Indeed, it is.

Further Reading

Brown, S. M. (1986). Of mantises and myths. *Bio-Science, 36*, 421–23.

Prete, F. R. (1995). Designing behavior: A case study. *Perspectives in Ethology, 11*, 255–77.

Prete, F. R., Wells, H., Wells, P., & Hurd, L. E. (Eds.). (1999). *The praying mantids*. Baltimore: Johns Hopkins University Press.

Prete, F. R., & Wolfe, M. M. (1992). Religious supplicant, seductive cannibal, or reflex machine? In search of the praying mantis. *Journal of History of Biology, 25,* 91–136.
Wenner, A. M. (1989). Concept centered versus organism centered biology. *American Zoologist, 29,* 1177–1197.

Frederick R. Prete

Applied Anthrozoology and Veterinary Practice
Applied Animal Behaviorists

It seems there have always been people who have been interested in studying animals and their behavior. In the early part of the 1900s, people who studied animals in their natural environment were often called "naturalists." Charles Darwin is an example of a naturalist. Zoologists, biologists, and ecologists who studied the natural world were also interested in studying animal behavior. Konrad Lorenz was a well-known zoologist. Another group of scientists who want to understand why animals do what they do are comparative psychologists, sometimes referred to as animal psychologists. B. F. Skinner was an animal psychologist whose name has become familiar.

Scientists in these various fields did not always find themselves comfortable being forced into the ecologist, psychologist, or zoologist mold. Some began to identify themselves as "animal behaviorists," a label that previously was rarely if ever used to describe a primary area of interest. Thus, the study of animal behavior has a long history of interdisciplinary effort, bringing together people who study animals from different perspectives. The primary organization bringing together these scientists who study animal behavior is the Animal Behavior Society, which was formed in 1964.

"Animal behaviorist" has become the professional term used to refer to research scientists who have advanced degrees (a master's or doctoral degree) in the behavioral sciences and who study any species of animal, whether in their natural environments or under controlled, laboratory settings. Most animal behaviorists teach at colleges and universities or are associated with other institutions, such as primate centers, zoos, natural history museums, or government research laboratories.

Animal behaviorists study animals not only to learn more about the species they are interested in but also to learn more about concepts and theories and to apply the knowledge gained in basic research to understanding and solving behavior and social problems in both humans and other animals.

This latter endeavor is referred to as applied animal behavior. A growing number of applied animal behaviorists work with pet animals such as dogs, cats, horses, and birds to better understand, as well as modify, their behavior. A scientific paper that many believed signaled the advent of this new field was written in 1974 by Tuber, Hothersall, and Voith and was called "A Modest Proposal." It discussed how it was possible to apply the principles of animal learning and an understanding of species' typical behaviors to change pet behaviors that presented problems for their owners.

Recognizing that its members were working in applied settings, in 1990 the Animal Behavior Society (ABS, found at www.AnimalBehavior.org) began a certification program for applied animal behaviorists. The certification program is the means by which individuals demonstrate to the public that they meet the minimum standards of education, experience, and ethics required of a professional applied animal behaviorist as set forth by the Society. As of this writing, there are 48 certified applied animal behaviorists.

The field of pet dog training evolved mostly independently of the science of animal behavior. Most early dog trainers often trained dogs for the military or for hunting. Historically, those who made a career of dog training, did not, for the most part, have the benefit of academic training in the behavioral sciences. It wasn't until the 1940s that training for the family dog became popular, in the form of "obedience classes."

Since the advent of the ABS certification program in 1990 and the creation of a board certification in behavior for veterinarians by the American College of Veterinary Behaviorists (ACVB, www.DACVB.org) in 1993 (of which there are now 37 diplomates), the term "behaviorist" and related terms (dog/cat/pet behaviorist, dog/cat/pet behavior consultant, dog/cat/pet behavior therapist) have started to be used by dog trainers and others who work with pet behavior but who are not research scientists or trained in the science of animal behavior. The Animal Behavior Society and ACVB board certification programs notwithstanding, behaviorists are not licensed, and the term "behaviorist" is not a protected term. Aside from veterinarians not being able to call themselves "behaviorists" unless they are board certified, anyone can use the term in any way they choose.

This has resulted in widespread confusion among the general public and other pet professionals who have reason to seek assistance understanding and modifying the behavior of pet animals.

How is someone who is trained in the science of animal behavior different from someone who is not? Animal behaviorists are trained to objectively observe behavior and through their observations collect data or information about what they see. Animal behaviorists are trained to use scientific methods to ask questions about what they see.

The scientific method involves forming a theory or hypothesis about the "whys" of animal behavior and collecting and analyzing information about the behavior to see whether the proposed explanation can be supported by the data. This objective means of trying to understand animal behavior allows animal behavior scientists to critically evaluate claims made about the causes for behavior rather than just relying on personal experience or opinion.

This type of critical thinking and scientific evaluation benefits pet owners when they enlist the services of a certified applied animal behaviorist as compared to someone without scientific training or an advanced degree. The consumer receives the benefits of the latest and best scientific knowledge about the behavior of their pets.

People who aren't trained in the scientific method often don't think critically about why animals do what they do and may arrive at conclusions based primarily on personal opinion or experience instead of science. And although there is nothing wrong with opinion and experience, they are no substitute for carefully collected scientific knowledge.

When it comes to helping people resolve pet behavior problems, these critical thinking skills are invaluable in identifying the cause of the problem and creating a behavior modification plan that is highly likely to resolve it.

For example, it is common for people to attribute many dog behavior problems to the owner not being "dominant" over the dog. That is, the dog is seen as exerting or attempting to exert control over family members. Dogs are described as "being dominant" when they want to walk in front of the owner on the leash, don't do as they are told, lie in doorways, lean against people, lick people, and engage in many other behaviors that in reality have nothing to do with what is known scientifically about "dominance."

Problems as diverse as aggression toward strangers and destructiveness when family members are gone from the home have been attributed to dominance.

These claims reflect distortions of the scientific concepts of social dominance and role relationships and ignore the scientific literature on how dogs relate to each other and to people. During their domestication history, dogs as a species have undergone selection for tendencies to behave submissively toward and acquiesce to people.

A review of the scientific research about social dominance in wolves and a variety of other species reveals how inaccurate and incongruent popular claims about this idea really are. This is important because of not only assigning the wrong explanation to a behavior but also misunderstanding the basic behavioral concept, which can, and has, led to inappropriate and ineffective techniques to change behavior of dogs as well as inhumane and abusive treatment of dogs.

For example, in the case of so-called "dominance" problems, pet owners have been advised to roll and pin their dogs to the floor, grab them by the scruff of the neck and lift them off the floor, bite them (literally!), socially isolate them, deprive them of enjoyable activities such as cuddling on the bed, and ignore any and all solicitations for play and attention. None of these procedures affect the social hierarchy between dogs and people in the way their proponents claim. Instead, they can result in defensive aggressiveness and fear and have a negative impact on the dog's behavioral well-being.

Because of the prevalence of misunderstandings about behavior by people not scientifically trained, pet owners need to be intelligent consumers of pet services. They should investigate the credentials, academic education, and certification status of people whose services they are considering using, no matter what title such individuals may use.

Further Resources

Animal Behavior Society. http://www.animalbehavior.org/Applied/abspamplet.html [Provides information about the ABS Program for Certification of Applied Animal Behaviorists.]

Tuber, D. S., Hothersall, D., & Voith, V. L. (1974). Animal clinical psychology: A modest proposal. *Amer. Psych.* 29, 762–66.

Suzanne Hetts

Applied Anthrozoology and Veterinary Practice

Bustad, Leo Kenneth (1920–1998): Pioneer in Human-Animal Interactions

Leo Kenneth Bustad was a veterinarian and senior member of the Institute of Medicine of the National Academy of Sciences, whose pioneering work helped define, research, and organize the fields of human-animal bond, animal-assisted therapy, and animal-assisted activities.

He was born January 10, 1920, in Stanwood, Washington, and received a DVM in 1949 from Washington State University (WSU) and a PhD (Physiology) in 1960 from the University of Washington. After a successful career studying the physiologic effects of radiation exposure, he was appointed Dean of the College of Veterinary Medicine at WSU in 1973 (in post until 1983). An outstanding educator, scientist, and humanitarian, Bustad was instrumental in the creation of human-animal interaction programs at the national and international levels. He was involved in the organization of some of the first symposia and conferences in the field. Bustad published extensively on the human-animal bond and helped establish the first scientific human-animal interaction journal, *Anthrozoos* (1987). He was also instrumental in the creation of the International Association of Human-Animal Interaction Organizations (1990).

In collaboration with Linda Hines, who would later become the second president of the Delta Society, Bustad created the People-Pet Partnership (PPP), a public service program at the College of Veterinary Medicine at Washington State University. PPP exists to research and educate the public about the human-animal bond and its applications and to give veterinary students a chance to learn about and to experience the HAB firsthand. Founded in the mid-1970s, PPP was the first university-based program of its kind.

Bustad was also a founding member of the Delta Society. He joined forces with veterinarians R.K. Anderson, Stanley Diesch, and William McCulloch and psychiatrist Michael McCulloch to create the Delta Foundation in 1977. It later became the Delta Society (1981). Bustad served as Delta's first president (1981–88). The Delta Society is a national organization that focuses on improving human health through service and therapy animals, especially through its volunteer Pet Partners Program.

Influenced by his experience as a prisoner of war and also by Albert Schweitzer's philosophy, Bustad taught a Reverence for Life course for more than 25 years. In addition to educating veterinary students and veterinarians on the importance of personal and professional ethics, Bustad also strongly believed in the importance of educating youth at a very early age. To this end, he co-authored in 1986 a humane animal care curriculum entitled Learning and Living Together: Building the Human-Animal Bond. This curriculum emphasizes the interconnectedness of people, animals, and the environment. He died in Pullman, Washington, September 19, 1998.

Leo Kenneth Bustad. Courtesy of Francois Martin.

Because of Bustad's enthusiasm for human-animal interactions and their implications for society, he emerged as one of the primary voices for the field. Bustad was skilled in gaining not only the interest of the public and leading scientists, but he also attained crucial media recognition for this subject.

Bustad's work significantly contributed to the understanding of the changing role of companion animals in Western societies and its impact on veterinary education, veterinary medicine, and society in general. The American Veterinary Medical Association recognizes

Bustad's legacy by annually presenting the Bustad Award to a veterinarian who has made special achievement in the area of human-animal interactions. The Japanese Animal Hospital Association bestows a similar honor, also called the Bustad Award.

Francois Martin

Applied Anthrozoology and Veterinary Practice
Veterinarian Training and the Human-Animal Connection

Veterinarians play an important role in the healthcare of animals by preventing, diagnosing, and treating illness. More than half of all American veterinarians exclusively treat companion animals, such as dogs, cats, birds, reptiles, rabbits, and other animals that can be kept as pets. About one-fourth of all veterinarians work in mixed animal practices, where they might see pigs, goats, horses, or sheep in addition to companion animals. A few veterinarians work exclusively either with laboratory animals or with large animals such as horses, various kinds of food animals, and animals in zoos and aquariums. Veterinarians in clinical practice diagnose animal health problems, vaccinate against diseases, medicate animals suffering from infections or illnesses, dress wounds and set fractures, perform surgery, and advise owners about animal feeding, behavior, and breeding.

While the vast majority of veterinarians' clinical work involves caring for the health of animals, some veterinarians directly contribute to maintaining human health. These veterinarians use their skills to protect humans against diseases carried by animals and conduct clinical research on human and animal diseases and their cures. For example, veterinarians contributed greatly to conquering malaria and yellow fever, produced an anticoagulant used to treat heart disease, and defined and developed surgical techniques for humans, such as hip and knee joint replacements and limb and organ transplants. A few veterinarians employed by the U.S. government also help to ensure human health by working as livestock and poultry inspectors who examine slaughtering and processing plants, check live animals and carcasses for disease, and enforce government regulations regarding food purity and sanitation.

Admission into veterinary medical school is competitive because the number of applicants has risen significantly in the last several decades and there are only 28 colleges in 26 states that offer degrees necessary to practice veterinary medicine. Prerequisites for admission include undergraduate studies in organic and inorganic chemistry, physics, biochemistry, general biology, animal biology, genetics, vertebrate embryology, cellular biology, microbiology, zoology, and physiology. Some programs require a combination of calculus, statistics, and algebra and may also require core courses including literature, social sciences, and the humanities. In addition to course requirements, applicants must submit test scores from the Graduate Record Examination (GRE), the Veterinary College Admission Test (VCAT), or the Medical College Admission Test (MCAT), depending on the preference of the college to which they are applying. Most veterinary medical colleges place heavy consideration on a candidate's veterinary and animal experience along with an eagerness to work with animals. Formal experience working in veterinary clinics, animal shelters, farms, ranches, or stables is particularly advantageous.

The first few years of veterinary training consist of several introductory courses providing a basis in veterinary anatomy, biochemistry, pharmacology, pathology, parasitology, animal breeding, botany, animal feeding and nutrition, radiology, virology, microbiology, zoology, animal physiology, physics, chemistry, and other scientific subject areas. The final years of veterinary education consist of a greater proportion of practical clinical work (e.g., internal medicine, dentistry, surgery, obstetrics), in which students apply the knowledge they have learned in a supervised environment. During the clinical years, veterinary students learn about the common health problems of animals and how to diagnose patients who can't explain their own symptoms. Training to become a veterinarian involves more than learning the technical skills necessary for healing animals. Courses in practice management and career development are becoming a standard part of the curriculum because the majority of veterinarians in private practices are solely responsible for the management of their practice and employees, a task which includes promoting, marketing, and selling their services. These universities recognize that a foundation in general business knowledge is important for new graduates who go into private practice because for every animal they treat, a human will be paying for their services. Thus, aside from an affinity for animals, prospective veterinarians should develop the communication skills necessary to get along with animal owners, especially pet owners, who tend to form a strong bond with their pet.

Surveys of pet owners report that they often consider companion animals to be a member of their family and they are willing to devote a significant amount of family resources to the care and well-being of their animal. Increasingly, animals receive advanced medical, dental, and surgical care, including insulin injections, root canals, hip replacements, cataract extractions, and pacemakers. Recognizing the prevalence of increasingly strong bonds between owners and their companion animals, some veterinarians responsible for training future veterinarians are calling for courses devoted to the Human-Animal Bond (HAB) to become a standard part of the veterinary curriculum. Common interpretations of the "human-companion animal bond" refer to a relationship between a human and an animal that continues over a significant period of time and brings a meaningful benefit to the lives of both the human and the animal. This bond also suggests that both the human and the animal treats the other as not just entitled to respect but also as an object of admiration, trust, devotion, or love. Supporters of HAB studies in veterinary education suggest that veterinarians in training should develop an understanding of the importance of the human-animal connection or bond in order to become more effective at their work. Although some veterinarians believe that promoting the HAB should be a major focus in veterinary education, others believe that the HAB is a topic that cannot be taught but rather is an innate trait in people and is outside the domain of knowledge, experience, and responsibility of veterinarians. More research is necessary to better understand the nature of the human-animal bond and its implications for veterinary education and private practitioners.

An organization devoted to the study of the human-animal bond, the American Association of Human Animal Bond Veterinarians (AAHAB), has a mission to provide education, resources, and support that enhance the ability of veterinarians to create a positive and ethical relationship between people, animals, and their environment. The veterinarian's role in the human-animal bond, according to AAHAB, is to maximize the potentials of this mutually beneficial relationship between people and other animals. Deciding how to define a mutually beneficial relationship, however, can be a difficult task for young veterinary students when many veterinary experts disagree about exactly how to define animal welfare or the best interest of animals. Despite considerable disagreement about what is considered proper ethical treatment of animals among veterinary experts, young veterinarians must create for themselves an ethical standard that will serve them daily in their work. In some ways the ethical decision making part of a veterinarian's job is more difficult and complex than that of human physicians. According to their professional oath, physicians' primary concern must always be to serve the welfare and interests of the patient. Because a veterinarian's patients are animals, the informed opinion of the patient in question is never available. Veterinarians serve both the animal (the patient) and the client, who pays the fee and who, in the eyes of the law at least, may determine much of the course of treatment. Sometimes the interests of these parties conflict, and the veterinarian is caught in the middle, wanting to help both. In other words, veterinarians are called upon to steer a course between their professional opinions, their own financial interests, the interests and wishes of the owner, the interest of the animal, and, in some cases, the interest of the public.

Veterinarians have a great deal of individual responsibility because they are professionals who must make decisions that influence the fate of animals, the lives of owners, and sometimes public health. Although not every decision in veterinary medicine requires unique ethical or moral consideration, no matter what field a young veterinarian chooses, she or he will likely face dilemmas that have no obvious solution. The kinds of situations that provoke ethical or moral stress in veterinarians will be different for each individual depending on their personal values and attitudes. However, despite their unique attitudes, in the course of their training and later in their careers, veterinarians are likely to encounter situations they find ethically challenging. In fact, it is impossible to quickly capture the entire range and complexity of all of the ethical challenges different veterinarians might encounter when asked to serve both the wishes of the owner or client and what they see as the best interest of the animal or patient. Companion animal veterinarians, for example, are sometimes asked by animal owner to perform surgical procedures that are painful and distressful to the animal, such as the cosmetic cropping of a dog's ear or the declawing of cats for the convenience of the owner, and without benefit to the animal. More research is necessary to better determine the attitudes of American veterinarians about such procedures as tail-docking, ear-trimming, debarking, and declawing of animals.

Few issues in veterinary medicine present more ethical uncertainty or moral stress for novice veterinarians than the practice of euthanasia. Veterinarians are frequently called upon to euthanize or "put to sleep" animals for a variety of reasons, but the precise circumstances and justification for the euthanasia procedure can influence the level of stress experienced by the veterinarian. Even when an animal is suffering considerably and euthanasia seems clearly in its best interest, killing the animal can place tremendous burden and stress on its owner as well as on the veterinarian. The veterinarian's experience can be complicated substantially when he or she disagrees with a client's decision or reasons for choosing euthanasia. Given the wide range of attitudes toward animals in our society, it sometimes will be difficult to reach agreement about whether a client's decision to euthanize his or her animal is a "reasonable" one. Decisions about euthanasia often have to be made in light of an owner's ability to pay for a lengthy course of life-extending treatment. When an alternative to euthanasia will cost clients more than they are willing or able to pay, veterinarians must decide for themselves if they are comfortable performing the euthanasia. Sometimes completely healthy animals are brought to the veterinarian for euthanasia for reasons of convenience, such as the owners are moving and their new residence does not allow pets, family members develop an allergy, or the animal has behavior problems such as barking too loudly or refusing to use the litter box. Thus, on one extreme, a veterinarian may be asked to euthanize a healthy, well-behaved animal, but the opposite extreme may occur as well. A veterinarian may encounter a pet owner who wants to carry out potentially painful surgery on a dying animal. In this opposite extreme a veterinarian would recommend euthanasia to end the animal's suffering, whereas the owner insists that the veterinarian continue treatment; thus, an owner refusing to choose euthanasia can produce an ethical dilemma for the veterinarian as well. In fact, some novice veterinarians report that the most difficult part of their job is to have to tell someone it's time to let go of a sick pet.

Young veterinarians quickly recognize that veterinary medicine sometimes involves offering comfort and counsel to clients whose animals they euthanize. This human "counseling" aspect of veterinary medicine has until very recently been neglected in the training process as some veterinarians consider this role to be outside their domain of knowledge, experience, and responsibility as veterinarians. Those who support the incorporation of grief management into the veterinary curriculum argue that dealing with grieving clients is unavoidable and might as well be done skillfully and compassionately so that the clients will remember this thoughtfulness when it comes time to seek veterinary care for another animal in the future. Although veterinary educators are becoming increasingly sensitive to the need for formal training focused around human-animal bond issues as well as ethical decision-making, only a few schools have included such training into the formal curriculum. Regardless of the level of formal instruction they receive, novice veterinarians must learn to use their special medical expertise in light of the many ethical questions that can arise when clients contemplate various treatment alternatives for their animals.

Further Resources

American Association of Human Animal Bond Veterinarians. http://www.aahabv.org/

American Veterinary Medical Association. http://www.avma.org/

Arluke, A. (2004). The use of dogs in medical and veterinary training: Understanding and approaching student uneasiness. *Journal of Applied Animal Welfare Science, 7*(3), 197–204.

Association of American Veterinary Medical Colleges. http://www.aavmc.org/

Englart, M. R., & Stengel, M. (2003). *How do I become a veterinarian?* San Diego: Blackbirch Press.

Gorman, C. (2000). *Clients, pets, and vets: Communication and management*. Newbury: Threshold Press.

Herzog, H., Vore, T., & New, J. (1989). Conversations with veterinary students. *Anthrozoos, 2*(3), 181–88.

Kay, W. (1988). *Euthanasia of the companion animal*. Philadelphia, PA: Charles Press.

Main, D. C. J. (2006). Offering the best to patients: Ethical issues associated with the provision of veterinary services. *The Veterinary Record, 158*(2), 62–66.

Martin, F., & Taunton, A. (2006). Perceived importance and integration of the human-animal bond in private veterinary practice. *Journal of the American Veterinary Medical Association, 228*(4), 522–27.

Reeve, C., Rogelberg, S., Spitzmuller, C., & DiGiacomo, N. (2005). The caring-killing paradox: Euthanasia-related strain among animal shelter workers. *Journal of Applied Social Psychology, 35*(1), 119–43.

Rollin, B. (1999). *An introduction to veterinary medical ethics: Theory and cases*. Ames: Iowa State University Press.

Sanders, C. (1995). Killing with kindness: Veterinary euthanasia and the social construction of personhood. *Sociological Forum, 10*(2), 195–214.

Self, D., Pierce, A., & Shadduck, J. (1994). A survey of the teaching of ethics in veterinary education. *Journal of the Veterinary Medical Association, 204*(6), 944–45.

Swabe, J. (1999). *Animals, disease and human society: Human-animal relations and the rise of veterinary medicine*. London and New York: Routledge.

Tannenbaum, J. (1995). *Veterinary ethics: Animal welfare, client relations, competition and collegiality*. 2nd ed. St. Louis, MO: Mosby Publishing.

Patricia Morris

Stresses at the Vet's Office

Paul McGreevy

Evaluating the positives and negatives of existence is one way to assess the quality of life. For veterinarians, this means considering the duration and intensity of their patients' pain or pleasure. Manifestations and consequences of the diseased state are worth exploring in veterinary contexts because patients can be distressed by both treatments and prophylaxis. Even a simple trip to a veterinary clinic for a vaccination can be distressing for an animal. Injected fluids in general may cause a stinging sensation, especially if the pH they need for long-term storage is not neutral. Animals can become stressed simply from visiting veterinarian offices and clinics.

Although fears of the unknown, including death, may be of less concern to diseased animals than to humans in a similar state, the way in which disease affects their well-being merits consideration. The scent of other members of the same species that may have been fearful in the clinic can alarm even the calmest pet. By way of other examples, floors may be worryingly slippery, cats may be exposed to unknown dogs, and owners may be unusually agitated. The attending physician may palpate painful body parts in the process of performing a physical examination. If the animal has to be left at the clinic, for example, for tests, it has no way of knowing when or if it will be reunited with its owner.

As a brief case study, consider a middle-aged dog diagnosed with osteosarcoma (a type of bone cancer) of the femur. The animal's quality of life could be affected by the condition, its treatment, and the dog-owner's reaction to both. Here we should also consider sedation; the lameness itself; the way the limb is manipulated, flexed, and extended to investigate the lameness; and time spent at a veterinary clinic, for example, prior to radiography. The treatment may well involve the pain and discomfort of amputation and chemotherapy. There may also be postoperative pain and the struggle of learning to balance and walk with three legs.

There are numerous ways in which disease can have both physical and psychological consequences. It is important to consider what methods of habituation and counter-conditioning might help to reduce the anticipated distress. For example, it may help to palpate pets gently in the manner of a routine veterinary examination prior to all their meals. Similarly, habituation could include "puppy parties" and "kitty kindies" at clinics and taking dogs (of any age) to veterinary centers during quiet periods for nonclinical purposes, for example, to visit the clinic for nothing more than a meal in the waiting room. Although repeated trips to the clinic might allow pets to predict that a given visit may be only transient, fundamentally, dogs and cats cannot know that they are there for only a limited period or even for their own good. So it is important not to overlook the benefits of home visits for veterinary patients and to develop the study of psycho-pharmacological support for diseased companion animals, especially for elective procedures.

Archaeology
Archaeology and Animals

Archaeology is the scientific study of physical or cultural remains left behind by peoples or animals who lived in the past (remote or recent). By analyzing such remains, archaeologists can investigate how various human-animal relationships originated and developed over time. On the one hand, archaeology can reveal the many practical uses to which humans have put other animals—for food, clothing, tools, transport, and more. But archaeology can also give us insights into less tangible interactions—into the crucial roles that animals have played in human art, religion, and social organization for thousands of years.

Three main types of archaeological evidence can be used to reconstruct the history of human-animal interactions over time, including physical remains such as bones, shells, or horns and cultural remains such as cave paintings, intentional burials, or tools. But the most important type of evidence is the context within which physical and cultural remains were found. By carefully excavating layers of soil in a standardized fashion, archaeologists can learn the relative ages of various items found at a site (in undisturbed deposits, items found at lower depths than others are usually older) as well as how those items might have been used by the peoples who left them behind. For instance, a few pieces of burned wood (charcoal), a bison skeleton with cut marks from butchering on the bones, and some stone tools might all be interesting if found separately. But if the three are found directly associated with each other in a controlled excavation, archaeologists can use them much more effectively to help reconstruct the story they can tell us about, perhaps, a temporary camp at which several hunters butchered a bison they had just killed. Thanks to a technique called radiocarbon dating, the carbon in the charcoal can also be used to determine the time period during which the camp was occupied (radiocarbon dating is effective back to about 50,000 or 60,000 years ago; for older deposits, other dating techniques are required).

Faunal analysis, also known as zooarchaeology, can help scholars interpret bones, shells, horns, and other physical remains of animals. By examining bones and teeth, zooarchaeologists can often determine approximately how old an animal was when it died and whether it was a male or a female. This age and sex information can be crucial to understanding human-animal interactions. For instance, human hunters, like other predators, generally kill young animals and older adults. But if a number of bison bones were found together with stone tools at an outdoor site, and the bison were of all ages (not just young and old), this might indicate that the animals were driven over a cliff or into a trap and killed together there. And if the bones of domestic cattle were found at a site, and the majority were young males and older females, this would suggest the animals were being raised for meat and milk—because females can give milk throughout their adult lives, they would likely be killed at a later age than males. A predominance of bones from older male cattle, however, might indicate that they were used primarily as draft animals (traction), pulling plows or other loads, as zooarchaeologist Krish Seetah has pointed out.

The shells of mollusks (snail-like animals) such as clams or limpets can also provide a wealth of information about human hunting and occupation patterns. These shells grow continuously throughout an animal's lifetime, so an animal that lives a long time will have a large shell. Zooarchaeologists Richard Klein and Kathryn Cruz-Uribe have suggested that if the size of limpet shells found at a coastal settlement became progressively smaller and smaller over the years that the site was occupied, this would indicate that a population increase or lack of other resources might have been causing the residents to harvest smaller and smaller (younger and younger) animals to get enough food. Furthermore, in temperate waters, the shells of marine mollusks and the scales, vertebrae, and otoliths (ear stones) of fish grow most in the warm summer months, when food is relatively plentiful. Therefore, if the outermost band on a fish vertebra or otolith is wide, for instance, the fish was likely harvested in the summer, and if it is narrow, the fish was likely harvested in the winter. If all the fish vertebrae and otoliths in a kitchen midden (garbage dump) at a site have wide outermost bands, then we know that the site was most likely occupied only during the summer.

Domestication also leaves its marks on animal bones. Domesticated animals are usually smaller than their wild counterparts, and their bones are shaped differently. For instance, as Simon J. M. Davis has written in his book *The Archaeology of Animals,* the domestic pig's forehead is concave, whereas that of its wild boar ancestors is flat; modern domestic goats have spiraling horns, but those of wild goats are smoothly curved; and many dog breeds have excessively elongated or shortened skulls relative to their wolf forebears. Animals that humans used to pull their plows or for other burdens also develop occupational pathologies that may be visible in their physical remains, such as arthritic growths (exostoses) on lower limb bones and vertebrae and rope marks deeply impressed into their horns. By looking for marks of domestication like these in bones from archaeological sites, then, zooarchaeologists can determine how early various species were domesticated and how they were used. For example, as Juliet Clutton-Brock writes, "archaeological evidence indicates that the dog was the first species of animal to be domesticated and that this occurred towards the end of the last Ice Age," around 14,000 years ago. "Man's best friend" has been our best friend for a very long time indeed.

The simple presence or absence of the bones of various animals, looked at in temporal context, can also be very telling. In his book *Human Impact on Ancient Environments,* Charles Redman discusses research done by Patrick Kirch, David Steadman, and others on animal remains from islands in the Pacific. These scholars have found that since the first Polynesian settlers arrived on the Hawaiian islands around 1,600 years ago, no less than half of the native bird species were driven extinct as a result of overhunting, land clearing for agriculture that destroyed their forest habitat, and introduction of new predators such as dogs and rats. Steadman found a similar pattern on the Galápagos Islands off the coast of Ecuador, which were uninhabited until the arrival of European explorers in 1535. Although only zero to three species went extinct on the islands in the 4,000 to 8,000 years before humans arrived, twenty-one to twenty-four were lost forever in just the few hundred years after Europeans began visiting the islands.

Cultural evidence can also help us tell the story of past human-animal interactions. Another reason we know that humans have long had a close relationship with dogs is that humans have intentionally buried dogs for more than 10,000 years, either alone or with their human companions. According to Clutton-Brock, one of the earliest such burials occurred at the Ein Mallaha site in the Upper Jordan Valley of Israel, where an elderly human was buried with its hand on the chest of a four- to five-month-old puppy 12,000 years ago. And at the Koster site in Illinois, three dogs were deliberately buried in separate graves during the Early Archaic period. As Darcy Morey and Michael Wiant write in their report about these burials, "the evidence from the Koster site hints that an affectionate relationship between humans and dogs may have existed over 8,000 years ago in the North American Midwest."

Artworks can also give us a wealth of information about past human-animal relationships. Indeed, some of the first and finest artworks made by humans were the multicolored Paleolithic (Old Stone Age) paintings of animals found in caves in western Europe. Made from 32,000 to 6,000 years ago, these often-exquisite naturalistic renderings suggest that Paleolithic artists were close and sensitive observers of other animals. André Leroi-Gourhan studied hundreds of these cave paintings and found that about a third represented horses, and another third represented bison or aurochs (wild oxen, the ancestors of domestic cattle); the next most frequently painted animals were stags (5.1%), hinds (6.2%), mammoth (9.3%), ibex (8.4%), and reindeer (3.8%). Why exactly these particular animals were painted in caves, sometimes in dark and inaccessible chambers far from the entrance, remains a mystery. Some scholars have argued that the paintings were a form of hunting magic, intended to produce more game animals (fertility magic) or to control prey at a distance, making them easier to catch (sympathetic magic). But as Paul Bahn and others have pointed out, the "depicted species rarely correspond to eaten species in terms of quantity"—reindeer, for example, were one of the most important game animals at the time, yet they are relatively rare subjects for cave paintings. Perhaps, then, the paintings were used as part of tribal initiation rites, or in religious ceremonies, or as visual aids to help youngsters learn about the local landscape, game trails, and hunting territories.

A skeleton of a dog is excavated in the Ashkelon Dog Cemetery in Israel. ©Richard T. Nowitz/CORBIS.

Paleolithic peoples may have been the first to depict animals in durable media, but they were certainly not the last. The ancient Egyptians, for instance, represented a wealth of animals in wall paintings, on statues and figurines, and even in their hieroglyphs (picture writing). By looking at the animals they chose to portray and the contexts in which they were represented, we can develop a remarkably clear picture of how the ancient Egyptians viewed animals. In his book *The Animal World of the Pharaohs,* Patrick Houlihan writes that "the multitude of charmingly expressed and sympathetic renditions of animals found decorating the walls of temple and tomb-chapels over many centuries clearly reflect a caring and respectful relationship with them." Indeed, for the ancient Egyptians, many of the gods themselves manifested in animal form, and hence, those animals were not just respected, but considered to

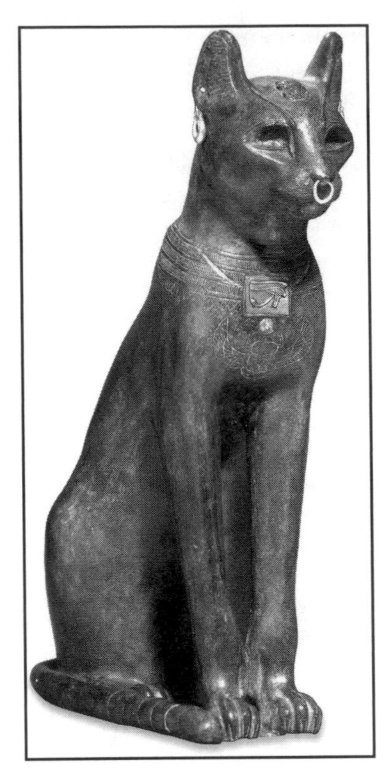

A bronze image of the Egyptian cat goddess Bastet. ©British Museum/Art Resource, NY.

be sacred. Jackals were associated with the god Anubis, for instance, falcons with Horus, and rams with Amun. In many artworks, these deities are depicted in the form of a human body topped by their sacred animal's head. But by the Late Dynastic and Greco-Roman periods, this idea of sacred animals was taken to extremes. On the one hand, the killing of house cats (which Houlihan notes were probably first domesticated in ancient Egypt) was seen as a crime punishable by death because these cats were sacred to the goddess Bastet. But on the other hand, cats and other sacred animals were believed to have the power to come back to life after death and carry prayers and offerings to the deities with which they were associated. According to Houlihan, this led to a form of early factory farming, in which animals were raised by the thousands in temple precincts to be sacrificed, mummified, and sold to religious pilgrims for use as offerings or prayer-carriers. Millions of animals lost their lives in this way, in spite of the remarkably high regard in which most ancient Egyptians held them.

From cave paintings to limpet shells and bison bones to dog burials, physical and cultural remains from archaeological sites, and the contexts within which they were found, can help us learn not just about ancient human-animal interactions, but also about how we can and should relate with animals today. Armed with insights from the past, we can work to create a brighter future for humans and other animals alike.

Further Resources

Bahn, P. G., & Vertut, J. (1997). *Journey through the Ice Age*. Berkeley and Los Angeles: University of California Press.

Brodrick, A. H. (Ed.). (1972). *Animals in archaeology*. New York: Praeger.

Casteel, R. W. (1976). *Fish remains in archaeology and paleo-environmental studies*. London and New York: Academic Press.

Claassen, C. (1998). *Shells*. Cambridge: Cambridge University Press.

Clutton-Brock, J. (1995). Origins of the dog: Domestication and early history. In J. Serpell (Ed.), *The domestic dog: Its evolution, behaviour, and interactions with people* (pp. 7–20). Cambridge: Cambridge University Press.

Davis, S. J. M. (1987). *The archaeology of animals*. New Haven and London: Yale University Press.

Houlihan, P. F. (1996). *The animal world of the pharaohs*. London and New York: Thames & Hudson.

Klein, R. G., & Cruz-Uribe, K. (1984). *The analysis of animal bones from archaeological sites*. Chicago and London: University of Chicago Press.

Leroi-Gourhan, A. (1982). *The dawn of European art: An introduction to Palaeolithic cave painting*. Cambridge: Cambridge University Press.

Morey, D. F., & Wiant, M. (1992). Early Holocene domestic dog burials from the North American Midwest. *Current Anthropology, 33*(2), 224–29.

Redman, C. L. (1999). *Human impact on ancient environments*. Tucson: University of Arizona Press.

Seetah, K. (2005). Butchery as a tool for understanding the changing views of animals: Cattle in Roman Britain. In A. Pluskowski (Ed.), *Just skin and bones? New perspectives on human-animal relations in the historical past*. Oxford: Archaeopress.

Steadman, D. W. (1995). Prehistoric extinctions of Pacific Island birds: Biodiversity meets zooarchaeology. *Science, 267*, 1123–1131.

Dave Aftandilian

Art

Animals in Art

The relationship we humans have with other animals has always been conveyed in our art. Painting, drawing, engraving, sculpture, and photography are all reflections of the society that produced them and represent what was going on in a culture at a particular place in a particular time, such as the form of the human-animal relationship (at least for the artist or for the person paying the artist to produce the work). Animals in art is a vast area of interdisciplinary study, and this essay can only land lightly on the surface of that exploding area of scholarly interest. Further, because of space limitations, this essay focuses primarily on painting and drawing in the Western world, with a steady progression through history, from the Paleolithic era to modern times.

We begin with the very first human art form: cave paintings. The compelling importance of animals to our human ancestors is clearly evident in hundreds of painted caves. Some of the most spectacular images were drawn around 30,000 BCE in a cave in France, the Chauvet. The cave walls are filled with complex scenes—confronting rhinoceri, snarling lions, herds of animals drawn as if rapidly moving through the cave—420 animal images in all (and only six human images). The Chauvet cave artists used sophisticated artistic techniques to render the carnivores in their environment spectacularly lifelike, animals that had motion, speed, strength, and power. Why did our Paleolithic ancestors go to such great lengths to paint, draw, and carve animal images inside caves? We simply do not know. Some scholars believe that the animal paintings (and numerous animal carvings, called portable art) represented a kind of hunter art in which visual representations of hunted prey ensured a successful hunt. Others argue that perhaps animals were in short supply, and the purpose of the cave art was to *make* animals rather than to kill them, and there may have been a belief that the paintings encouraged their reproduction and ensured a source of food. But regardless of the theory deployed to explain the cave art, experts agree that Paleolithic humans were expressing admiration for animals, and that admiration continued for many thousands of years.

One of the most admired animals in those early years was the bull. As a model for male power and fertility, the large, strong, brave, and libidinous bull was the most powerful image in the art of the third millennium BCE throughout most of the world. There is archaeological evidence of cattle worship in 6000 BCE in Çatal Hüyük, one of the first true cities that prospered in the area of present-day Turkey. Further, writing evolved out of the need to record cattle wealth. According to Calvin Schwabe, the first symbol used to mark things to be counted as wealth was the horned bovine head (*capital*), the first letter in the alphabet and drawn in cuneiform as an upside-down triangle, on its side with curved horns as in the Greek α and on its back as in the English *A*. Eventually, as more and more humans lived in cities, wealth, trading, and fighting increased, and humans began to use wild, ferocious animals as symbols of struggle, violence, and kingdoms at war. The earliest visual records we have of life in ancient Mesopotamia were carved into stone cylinder seals around 3500 BCE. These small cylinders of stone (which, when rolled on clay, were used to secure and record valuable materials inside a container) show that the art of the time was dominated by a motif of animals and humans in violent struggle and strife, a reflection of the ongoing cycles of battle between the uncivilized and civilized.

One thousand years later, animals were drawn in elaborate lifelike and naturalistic compositions, with beautiful stones and jewels used to illustrate and highlight the animals' physical characteristics. For example, still wildly popular, bulls were used to decorate musical instruments, with the head and horns constructed of gold sheet, eyeballs made of shell insets, and pupils, eyelids, and hair made of lapis lazuli, a beautiful deep-blue mineral. Animals were also depicted in scenes of everyday life in limestone carvings from ancient Egypt. Again, bovines are a dominant motif, with cattle shown grazing under trees or walking across the threshing floor and oxen pulling scratch plows (with the plow attached to the animal's horns with a rope, a most uncomfortable hauling position for the animal). Some of the most spectacular lifelike depictions of animals in ancient art were the lively, colorful paintings of sea creatures that decorated the mansions of Minoan Crete.

The tradition that began in the Paleolithic era of depicting animals with emotion and with agile body movements was resurrected in ancient Egypt and Greece. Egyptian palaces and tombs were decorated with illustrations of royal hunting scenes, such as King Ashurbanipal's palace at Nineveh, where the limestone walls come alive with ancient hunting scenes in enclosed parks (an ancient canned hunt in which it was impossible for

Dying lioness, Ashurbinipal's palace at Nineveh, 650 BCE. ©Werner Forman/Art Resource, NY.

an animal to escape being killed). The scenes document that lions were released from wooden enclosures and driven toward their slaughter by attendants banging cymbals; meanwhile, the king in the illustrations "hunts" from the safety of his horse-drawn chariot. The frightened, cornered animal is riddled with arrows, and her wounded body and the anguished look on her face make evident her pain and suffering.

Although hunting scenes were also part of the art of the ancient Greeks, particularly in their vase paintings, a fondness for domestic animals often found its way into ancient sculpture and grave steles. Most of the three-dimensional pieces from this ancient time are of dogs gazing lovingly up at their human companions or of birds cradled in the hands of children, but because of the Greeks' penchant for realistic depictions, they also showed animal emotions resulting from sickness or deformity, such as an ill greyhound, which was carved in marble around 200 BCE.

Visualizing the pain and suffering of animals reached its height during the Roman era and their infamous amphitheater games. Beginning in 186 BCE and ending in 281 CE, the public slaughter of animals (and of devalued humans) was a popular form of entertainment for the Romans. Exotic animals were killed by the thousands in reconstructions of actual hunting scenes; hunters with weapons chased animals about the arena, sometimes in elaborate recreations of the animals' natural environment. Romans drew visual records of their hostile relationship with animals on everyday objects such as lamps, ceramics, and gems and in works of art for the wealthy, particularly mosaic decorations for the walls and floors of villas. In her research on the visual representation of arena scenes in Roman mosaic art, Shelby Brown found that vivid depictions of blood, suffering, and death decorated the reception and dining rooms in wealthy homes. The mosaics typically show animals slaughtered or in a soon-to-be-slaughtered situation, with fear and pain emphasized in their facial expressions. A first-century CE mosaic from Zliten in Libya illustrates the visual nature of the Roman shows, which often showcased mini-performances simultaneously, such as a hunting scene, a fight between a tethered bull and bear, humans attacked by a lion and leopards, and an orchestra playing in the background.

Though horrifying and gruesome, the human-animal relationship as depicted in Roman mosaics was at least realistic—and one of the few authentic depictions of animals in visual art for the next 1,100 years. That is not to say that animal imagery was rare during that time, but rather that the images were allegorical and primarily symbolic during the Middle Ages. During the rise of Christianity, animal imagery was often used to teach morality and religious principles, and visual images were particularly useful, writes Janetta Rebold Benton, as a way of getting information to people who could neither read nor write. For example, bestiaries (or books of beasts) were popular tools for teaching Christian dogma. Bestiaries used symbolism and some real and many imaginary animals to illustrate lessons in morality and religion. For example, one common medieval animal image is the pelican, who is depicted piercing her own breast to feed her young with her blood, a symbol of self-sacrifice modeled on Christ. Most of the animals in medieval art are framed by fantasy, religion, and symbolism, but in the mid-1200s historian and artist Matthew Paris sketched a picture of an elephant from his first-hand observations of the animal while she lived in the Tower of London's menagerie. This was the first naturalistic depiction of an animal in art since the last of the Roman mosaics of the early Common Era.

Hunting continued to be a central motif in art depicting the relationships between humans and other animals, and the activity also continued to be a staged, spectacular event for the amusement of humans. The medieval hunt was highly ritualized and reserved for the nobility. Tapestry weavings, illustrated manuscripts, and calendar pictures show the ceremony and preparation associated with hunting, the stages of the hunt (the chase, the kill, and the breaking or dismemberment and gutting) of the prey (usually the noblest of all prey was the stag), and the special role of "hounds" in the event.

As the Middle Ages drew to a close, animal art became more and more authentic. Some of the most spectacular images from the time were scenes of everyday life drawn in the margins of devotional books. Illustrated in the mid-1300s, the Luttrell Psalter is one of the best known illuminated books from the Middle Ages, and it shows a variety of animals in the daily activities of rural life in England: oxen pull plows, horses haul carts of freshly cut barley from the fields, dogs harass peddlers, women feed chickens and tend sheep, and foxes prey on ducks.

The Renaissance ushered in a passion for naturalistic representations, and many artists were also considered scientists devoted to the exact imitation of nature. This naturalistic-objective view of animals and the natural world would dominate Western art for the next 700 years. For the first time, animals (and animal body parts) were drawn for their own sake, without busy backgrounds filled with humans, other animals, or the countryside. Albrecht Dürer drew stunningly realistic drawings of rabbits, bird wings, and stag heads. Leonardo da Vinci was a master of many talents, and one of them was his ability to sketch animals as individuals—active, alert, with emotion and feeling—particularly horses. Leonardo was an avid sketcher of horses: horses drawing chariots, pulling carts, carrying riders, and rearing up on hind legs.

The living standards for humans increased dramatically at this time in Europe. There were surpluses of food from enhanced agricultural knowledge, from intensive land use, and from raising animals for their flesh (also for their wool and skin). In some areas, the amount of meat consumed more than doubled between the fourteenth and fifteenth centuries. The increasingly commodified relationship between humans and other animals can be seen in the aesthetic illustration of food animals that emerged in the sixteenth century. Paintings depicted food abundance in the form of huge displays of fruit, vegetables, and butchered animals. Typical examples include the kitchen and market scenes painted by Pieter Aertsen (who was still working in a cultural context that required that art have some connection to religion) and his nephew, Joachim Beuckelaer (for whom a direct link to religion in his art was no longer imperative—he could paint everyday scenes and title them as such).

The paintings of food abundance of the sixteenth century were soon followed by the gamepiece, or paintings of dead animals in hunting scenes. In this genre of animal painting, dead game art of the seventeenth century conveyed messages of aristocratic privilege, particularly in Holland, where those who aspired to the fashion and leisure pursuits of the aristocracy purchased gamepieces for their homes as a general representation of rising social status. The gamepiece was a display of animals that were hunted only by the nobility—stag, roe deer, boar, pheasants, and swan—all carefully and elaborately positioned on the canvas, sometimes in human poses such as lying on their backs with legs crossed and heads canted to the side or lolling over the side of a table.

Focused primarily on food, markets, and kitchens until the mid-1600s, gamepiece art gradually took on a trophy-like motif that highlighted not only dead animals but also the weapons of the hunt, sporting dogs, and outdoor scenes. The work of Jean-Baptiste Oudry in the eighteenth century illustrates how the gamepiece assumed an outdoor decorative theme. Many of his paintings are of a single dead animal, often a wolf or a deer, lying on the ground of an elaborate outdoor pavilion, a hunting rifle propped against a stone wall and hunting dogs standing nearby. Hunting dogs were important in Oudry's dead animal art; he painted numerous pieces in which packs of snarling dogs attack animals—wolves, foxes, deer, and even a wild pig and her piglets and a swan sitting on her nest.

Live animals were also popular artistic subjects in the seventeenth and eighteenth centuries. Domestic animals such as cows, horses, dogs, and birds were painted in the fields, meadows, and, in the case of dogs, in the homes where they lived. Prize cattle portraiture was popular, and wealthy breeders commissioned portraits of their pedigreed animals, with artists often overemphasizing the size of the animal's body to establish his value (which in turn reflected the status of the animal's owner). Dogs were also selectively bred, particularly hunting hounds, and their portraits were also painted for wealthy owners. The fondness for companion animals, ever more popular since antiquity and an established part of even middle-class households in sixteenth-century England, was expressed in live animal portraiture in the eighteenth century, particularly dogs who were posed with loving family members.

Animals were also deployed in art targeted for the "common people." In response to the growing opposition to animal cruelty, considered by many eighteenth-century intellectuals to be a problem primarily of the working class, the abuse of animals was depicted visually in illustrated magazines and printed materials. In 1751 William Hogarth published *The Four Stages of Cruelty*, a set of four prints that showed how the torture of animals inevitably progressed to cruelty to other humans. In the first two prints, the impoverished and orphaned antihero, Tom Nero, is depicted in the act of abusing animals, moving from the torture of dogs as a child to the beating of a lame horse as a young man.

In the third and fourth frames, Tom has murdered a young woman, and he is hanged for the crime. Typical of the treatment of those who had the misfortune of dying poor in the eighteenth century—and a fate greatly feared by the common people—Tom's body is dissected in a public forum. Hogarth wrote that his art was intended to reach the largest possible audience at the cheapest possible price in the hope that the depiction of animal cruelty would prevent some of the horror being inflicted on animals in London at the time.

"Second Stage of Cruelty," from The Four Stages of Cruelty *by William Hogarth, 1751. ©INTERFOTO Pressebildagentur/Alamy.*

A growing interest in natural history, exploration of foreign lands, and the invention of the camera in 1839 ushered in an enthusiasm for an art form that was capable of recording unusual, exotic, and spectacular animals—photography. Photographs of animals in their natural habitats were essential to the construction of realistic dioramas in the increasingly popular natural history museums. For example, Donna Haraway recounts how the dioramas in the American Museum of Natural History's African Hall were built from photographs taken of the best animals to be found in the wild. Lions, elephants, giraffes, and—most desirable of all—gorillas were prize trophy specimens who were photographed and then immediately killed so that they could be stuffed and posed according to the "action" documented by the camera. Amy Fitzgerald and I have found that trophy art continues to be a popular pastime for hunters and taxidermists, all devoted to posing dead trophy animals as if they were alive—eyes open looking alertly toward the camera, legs tucked neatly under the body, some even staged as if performing live behaviors such as eating straw. Photographs of disparaged animals (foxes, bobcats, coyotes) are exceptions to this general rule of trophy art as a form devoted to the representation of dead animals as if they were alive; the death of these predators is celebrated in the art, with bodies flung across human shoulders like bags of dirty laundry or held triumphantly upside down for the camera.

Photography is also used to depict animals in a postmodern era of blurring boundaries between human and animal, subject and object, self and other. Britta Jaschinski, a celebrated animal photographer, has captured the essence of animal captivity in her photographs of zoo animals, photographs that emphasize how animals are disembodied and artificial under forced imprisonment. Alternatively, William Wegman has been photographing his Weimaraner companion dogs dressed as other species, such as human models wearing the latest fashion accessories, in an attempt to blur the distinction between human and animal bodies. This is the world of postmodern animal art—as Steve Baker argues, an attempt at producing an unsettled view of animals—often producing images and exhibits that are shocking and strange, such as pigs and sharks preserved in formaldehyde and displayed as a museum exhibit. In recent years, a vast electronic world of digital images and virtual reality has joined painting, sculpture, photography, and museum exhibits in representing animals in art. In this postmodern arena of electronic technology, there are few "real" animals, writes Randy Malamud, but rather animals that are not only inauthentic, but often not even really there.

Further Resources

Baker, S. (2002). *The postmodern animal*. London: Reaktion.

Benton, J. R. (1992). *The medieval menagerie: Animals in the art of the Middle Ages*. New York: Abbeville Press. [In addition, an excellent visual resource of animals in the Middle Ages can be found at http://expositions.bnf.fr/bestiaire/index.htm, Bestiary in medieval illumination.]

Brown, S. (1992). Death as decoration: Scenes from the arena on Roman domestic mosaics. In A. Richlin (Ed.), *Pornography and representation in Greece and Rome* (pp. 180–211). New York: Oxford University Press.

Clark, K. (1977). *Animals and men: Their relationship as reflected in Western art from prehistory to the present day*. New York: Morrow.

Haraway, D. (1989). *Primate visions: Gender, race, and nature in the world of modern science*. New York: Routledge.

Jaschinski, B. (1996). *Zoo*. London: Phaidon.

Kalof, L. (2006). *Looking at animals in human history*. London: Continuum.

Kalof, L., & Fitzgerald, A. (2003). Reading the trophy. *Visual Studies, 18*, 112–22.

Klingender, F. (1971). *Animals in art and thought to the end of the Middle Ages*. (E. Antal & J. Harthan, Eds.). Cambridge: MIT Press.

Malamud, R. (1998). *Reading zoos: Representations of animals and captivity*. New York: New York University Press.

Schwabe, C. W. (1994). Animals in the ancient world. In A. Manning & J. Serpell (Eds.), *Animals and human society: Changing perspectives* (pp. 36–58). London: Routledge.

Sullivan, S. A. (1984). *The Dutch gamepiece*. Totowa, NJ: Rowman & Allanheld.

White, R. (2003). *Prehistoric art: The symbolic journey of humankind*. New York: Abrams.

Linda Kalof

Art

Animals in Celtic Art

The human-animal relationship and animism are strongly visible in all Celtic artwork and decoration. Although the true origin of the Celtic peoples lies deep in the mists of prehistory, their art and mythology have been most preserved in Ireland. Celtic art is based on a deep connection with the natural world, a related intellectual curiosity, and a sense of comradeship and interdependence with all creatures, viewed as fellow travelers on life's journey.

In all Celtic art, one can identify their kinship with and strong affinity for animals. Their belief in the interconnectedness of all life—humans, animals, and nature—is characterized in their art by intertwined cords, intricate knot work, lattice, unbroken lines and spirals, and their distinctive curvilinear artistic style, all of which symbolize the thread of life.

The Celts were surrounded on all sides by the natural world and were therefore continually reminded of its presence. They recognized the essential aliveness and importance of animals and nature, not just in a biological sense, but as a community of mobile sentient entities, a community parallel to that of the human world and of which the human world was an integral part. Hence, the behavior and characteristics of a flock of birds, a solitary stag, a river, or a thundercloud would all be incorporated and explained in their art, which was communicative as well as ornamental.

Animal imagery is apparent in every aspect of Celtic art and decoration. Animals were considered sacred, and in Celtic artwork, animals are symbolic of possessing special attributes of divinities. Celtic saints included animals. Often their saints would adopt the form of an animal. In Celtic mythology, animals were also believed to pass to and from the realm of magic.

The human-animal relationship held such importance that the Celts recognized horses, blackbirds, foxes, badges, and in fact all animals as their close friends and disciples. For the Celts, there was inscrutable mystery relating to the boundaries that existed between humans and the natural world—the animal world in particular. Celtic Druids were believed to be able to talk with animals, and as such, the most common way of gaining knowledge from animals was to talk with them and to interpret their actions. The Druids believed that animals could speak the language of humans and were able to pass on their wisdom to humans through speech. Birds were largely associated with speech and prophetic knowledge and were thus used extensively in their art and decoration. A special understanding of the speech of animals was believed to yield a great advantage. Some Celtic heroes were believed to have gained supernatural knowledge from the speech of birds, enabling them to be warned of impending danger, and they were said to have also been told secrets by the birds. They believed that the interpretation of birdcalls could determine the approach of visitors.

Celtic belief was conveyed and communicated through art. Celtic art is strongly rooted in the earth, and deeply reflected therein is their belief in the "tree of life," the seven created beings of the Celtic world: animal (zoomorphic), bird, reptile, insect, fish, plant, and human (anthropomorphic). Because it was forbidden in the Celtic belief to duplicate creation in a perfect state through art, it is not unusual to see art in which animals appear with human hands and feet, or even to see a calf's head on an eagle's body or a human being with animal parts. Often such art looks like a beautiful mosaic of design and color, until, upon closer examination, one identifies a head, or a tail, or a plant and sees that several such images are interwoven with each other, becoming part of an interlacing pattern. It is interesting that all animal designs are logically completed and conform to the laws of nature, no matter how the artist may have interlaced and contorted his motif.

The importance of the human-animal relationship in Celtic belief is depicted in a manner of great artistry. Often their animals are semi-realistic and mythical, and their forms and movements are depicted with great ornamental rendering. The Celts had a great interest in the beauty and movements of animals, particularly the horse, which was the animal most associated with the Celts and which was represented on artifacts and coins, indicative of the high regard the Celts had for their equine companions. Deities were frequently painted on horseback. A common Celtic design is that of a man with a bird in each hand. The deer was sacred and also considered to be one of the oldest creatures, and in the symbolic message of art, deer would appear as otherworldly messengers. The wisest and most knowledgeable creature in the Celtic belief was the salmon, also linked to sacred mysteries. Crows and ravens were associated with battle. The serpent represented the cyclic nature of life because of the shedding of its skin. Dogs were symbols of loyalty, faithfulness, unconditional love, courage, and strength and were associated with heroic figures or deities. Dogs were also associated with healing; their saliva was thought to have curative properties. The dragon was another mighty magical animal. Eagles were usually linked to death gods. Boars symbolized courage and strong warriors.

Animals were of all importance to the Celts: each well, river, or grove was said to have its own guardian spirit, in the form of an animal. The Celts were also skilled craftspeople in jewelry, metalwork, engraving, and sculpture. These art forms and their decorative styles were used on everything from battle chariots to household utensils.

The animal symbols that the Celts used in these diverse art forms are our kin that are still all around us today: animals, birds, insects, fish, trees, hills, lakes, groves, and all the other manifestations of the life force on earth—all interconnected and all in relationship with one another.

Further Resources

Celtic Mythology. http://www.celtic-myth.com/

MacCulloch, J. A. (1911). *The religion of ancient Celts.* Edinburgh: T. & T. Clarke.

Urban, J. Celtic History and Information. http://www.jurbanrings.com/celticinfo.html

Wilde, L. W. (2004). *Celtic inspirations.* London: Duncan Baird Publishers.

Zaczek, I. (1997). *Chronicles of the Celts.* New York: Sterling Publishing.

Gwyneth St. Francis

Art
Insects in Art

Humans have evolved relatively recently, in a world long replete with insects. Sharing a world with nearly a million described species of insects (and millions potentially remaining to be described), humans have formed relationships with insects in many ways. Sometimes humans and insects influence each other minimally; other times we profoundly impact each other. Look for evidence of our relationship with insects, and it will appear all around you, particularly in the works of artists. Human culture is largely marked by the works of artists, so art can offer clues to the history of our relationship with insects.

It is easy to see why artists might be inspired to depict an insect in their work. We marvel at the diversity of colors, forms, and behaviors of insects. As a species, we domesticate insects and exploit insect products (e.g., silk, beeswax, honey, cochineal, lacquer); include insects in our diet, language, and recreational activities; base forensic investigations on carcass-feeding insects; use insects as indicators of habitat quality; use insects as model organisms for scientific research; and pollute or radically alter habitats in an attempt to attract, repel, or extinguish species of insects. Some insects, in turn, pollinate our crops, use our homes for shelter, parasitize our bodies, spread pathogens and allergens, and feed on our resources and excrement and remains. Human-insect interactions are so pervasive that insects often appear as central players, even as symbols, within politics, science and technology, religion, mythology and folklore, literature, poetry, music, the performing and visual arts, and recreation. These cultural products represent some categories of "cultural entomology," as Charles Hogue defined it, and our relationship with insects has been recorded through the visual arts in ways and for a duration that is unmatched by any other source of information.

History of Insects in Visual Art

Artists are influenced by their environment, and there are few places on the planet that are not inhabited or affected in obvious ways by insects. Response to the presence and impact of insects has a long history within the visual arts. Long before written languages existed, humans scratched, sculpted, drew and painted materials that serve as evidence of humans' associations with insects. Our earliest evidence of human-insect relations was inscribed on a bison bone by a Cro-Magnon person—a 20,000-year-old rendering of a cave cricket. A chalk drawing dated 8000 BP depicts a woman gathering honey (Cave of the Spider, Spain). Similar images of bee-handling were painted on rocks in Africa, and advanced beekeeping was depicted in an ancient Egyptian tomb (2626 BP). Prehistoric petroglyphs and pictograms from Europe, South Africa, and North America often feature insects, as do millennia-old Native American pottery, Greek ceramics, African amulets, and Asian paintings. Kano Kagenobu produced a scroll of insect sketches in the styles of Asian artists spanning nearly 1,000 years, leading up to his contemporaries of the early nineteenth century.

Artists have included insects in their works for a variety of reasons. Some sought and seek to adorn shelters or garments or to record observations. Others have imbued insects with spiritual or supernatural qualities. Supernatural associations with insects are often found in the form of totems and symbols. Totemism, the association of a nonhuman organism with a blood relative of a human or clan of humans, is evidenced by totem poles, pictograms, and aboriginal Australian paintings, each of which include insect examples. Totems of ancient Mexico include bees and butterflies—Mayan icons that feature prominently throughout Mexico's history of art. Investigating these examples of material culture can offer a glimpse into the mindset of ancient societies and their connection with insects.

Insects as Symbols

Insect symbols can hold great importance within societies, offering clear and powerful visual representations of ideas. Insects have symbolized many qualities, including change, industry, royalty, social harmony, might, pestilence, disease, death, and so on.

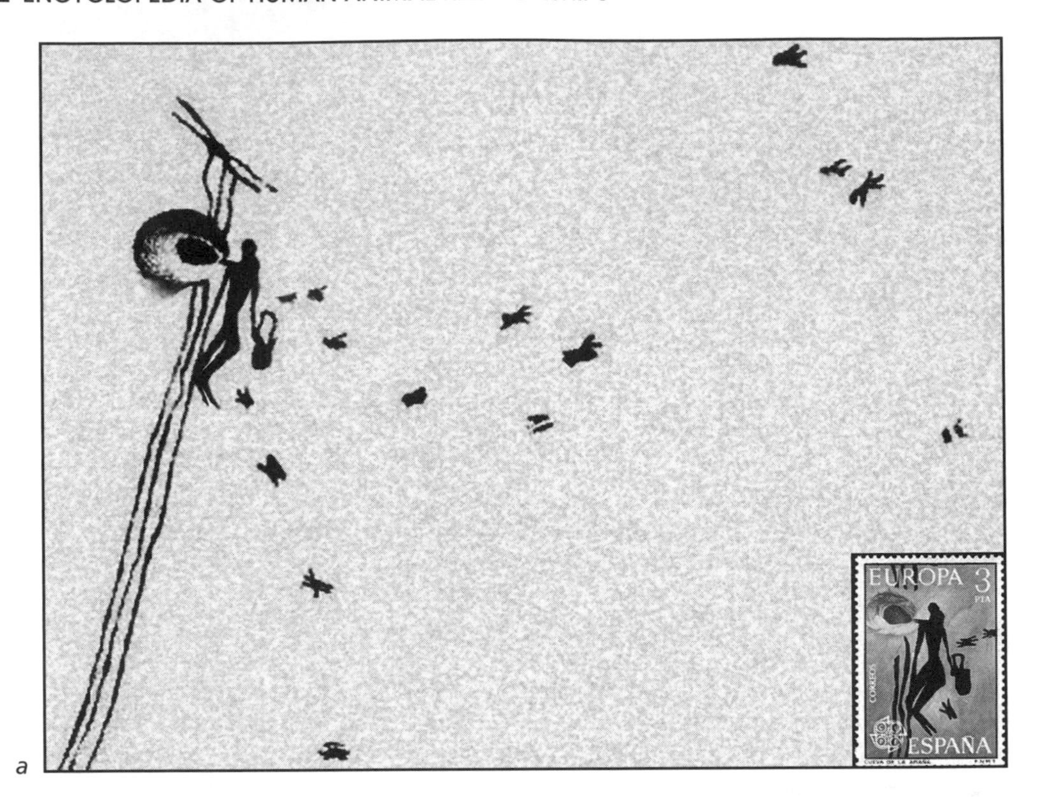

a

b

Humans use insects and their products for survival and pleasure: (a) a honey-gatherer was drawn on a cave wall in Spain, documenting an association of humans with honeybees over 8,000 years ago (detail of cave image, adapted from illustration, http://www.mdbee.com/articles/cavepainting.html) and note recent Spanish stamp adapted from it; (b) a life-size aquatic insect sculpture, "tied" by Bill Logan today, is an extreme example of realistic fly ties commonly used for fishing.

Although certain insects inspire specific and somewhat universal symbolic associations (e.g., flies with pestilence, disease, or death), symbols are often contextually and culturally dependent.

One of the earliest examples of an insect symbol is that of the "sacred" scarab beetle, rendered on papyrus scrolls and tomb walls of ancient Egypt. Egyptians worshipped these dung beetles for their symbolic rebirth (metamorphosis, ending with eclosion into adulthood) and transport of the sun across the sky (rolling of dung ball to provide a source of food for young). Insects appeared as Egyptian hieroglyphs by at least 5100 BP and are visual relics of many cultures' creation myths and folk tales. Insects can be found in pictograms in the North American Southwest, in Japanese crests, on European armor, and in medieval manuscripts. During the Northern Song dynasty (960–1127), representations of nature often carried meaning, so if a cicada was pictured on a scroll, it may have signified immortality because of a cicada's extended subterranean existence as an immature. Insect symbols commonly appear in Mexican art, from Mayan icons to José Guadalupe Posada's satirical symbols of evil and perversity. In modern cultures today, we can find insects symbolically on stamps and currency, in advertisements, and even as visual representations of sports teams or military units.

Insects and Art Themes

Insects make good symbols because they often induce emotional reactions in humans. Exploiting these reactions, artists have often included insects in their art to more clearly or powerfully express an idea. Politics, war, and environmental devastation are examples of themes that have inspired artists to generate insect art.

Recent examples of politically motivated insect art include anti-Marxist ants, cockroach executions, a militaristic beetle, and ant flags. The anti-Marxist ants are the product of Alberto Faietti, who has produced books of insects glued to pages. In one work, "the third letter of an ant community to karl marx," Faietti commented on human socialism by forming elaborate text and equations, ending with the word "NO," formed by the bodies of ants. The cockroach executions belong to the photographic and video series by Catherine Chalmers. Cockroaches appear to be electrocuted, hanged, or burned in staged executions that provoke questions about human execution practices. The militaristic beetle is the product of "Bansky," an elusive artist based in the United Kingdom. Bansky secretly hung a piece of antiwar artwork in the American Museum of Natural History—a harlequin beetle equipped with sidewinder missiles and a satellite dish.

Yukinori Yanagi's art, unlike the previous examples of politically motivated insect art, displays live insects. In most works, entire ant colonies are mounted in transparent displays of colorful sand arrangements, which form either political flags or paper currency. The ants displace the sand grains, breaking down the political boundaries initially set up by Yanagi.

Environmentally motivated insect art is a category that has a more recent history than politically motivated insect art, because of awareness of large-scale environmental disasters recently affecting our planet. Loss of biodiversity, natural resource depletion, and habitat destruction have all inspired artists to generate insect art. Andy Warhol, although he denied that his pop art had any depth or substance, created a series of prints of animals at risk of going extinct, including a butterfly, that have been displayed by conservationists and others concerned about species losses. Cornelia Hesse-Honegger paints insects with mutations that she has found near Chernobyl, Three Mile Island, and other nuclear installations. Artists create such examples of environmentally minded art in large part to generate attention needed to conserve nature and protect the diversity of life.

Politics, war, and environmental damage are only a few of the themes that have inspired artists to incorporate insects in their works. Sometimes, themes of natural beauty inspire the artist-artisan to create works for display or utilitarian value. Cultures within Asia that treasure singing insects, for example, produce art featuring sound-producing insects or elaborately decorated cages within which they are kept.

Insects serve as tools, including symbols, to evoke emotions. When emotional subjects such as politics, war, and environmental degradation are used as art themes, insects can help to elicit emotions that lend power to these themes.

History of Western Insect Art

Although a great cultural heritage of insect art is found throughout the world, the most widely documented and accessible materials are associated with Western art. Prior to the seventeenth century, insects appeared in Western art as religiously charged symbols of the soul, ephemeral life, or Jesus Christ—appearing in religious texts and paintings of biblical scenes. Insects also appeared in portraits. Sometimes the realistic depiction of a fly on a portrait indicated that the subject of the portrait had died.

With the Reformation of the Catholic Church in the sixteenth century, iconoclasts temporarily instated a ban on religious imagery, resulting in a separation of Western art and religion. As a result, insects were used by artists for different purposes following the Reformation. Seventeenth-century Belgium and the Netherlands spawned genres of art that were less religious, resulting in waves of landscapes and still lifes. Marcel Dicke (2000) spent three years documenting 1,942 examples of Western art featuring insects, after which he concluded that the seventeenth century constituted the richest period in Western history, both for total number of pieces featuring insects and for diversity of insect orders represented. Most still lifes featured insects, and some, as Dicke calculated, included over 100 insects per painting. Examples of artists include Jan van Kessel the Elder (1626–79), Rachel Ruysch (1664–1750), Jan van Huysum (1682–1749), and later, Paulus Theodorus van Brussel (1754–95). Several still-life artists of the same period, most notably Maria Sibylla Merian, extended their talents to develop the science of entomology. Merian and scientific illustrators since her time have combined esthetics and realism to promote insect science through centuries of insect paintings and drawings.

Most eighteenth-century Western insect art continued in the form of still lifes. Nineteenth-century impressionism rarely focused on small arthropods, although James McNeill Whistler used a butterfly as his signature, and Vincent van Gogh prominently painted insects as subjects in the last years of his life. The turn of the nineteenth century through the twentieth century, on the other hand, exploded with respect to the total number of insect art pieces. Art nouveau, surrealism, and modern and contemporary art all prominently feature insects.

Art nouveau embraced natural motifs, and artists included insects in painting, sculpture, and utilitarian design such as glassware, furniture, and ceramics. Emile Gallé produced insectal furniture, and some of René Lalique's most famous jewelry consisted of sculpted insects and animal chimeras that display fabricated elements of insects. "[Lalique] was well-acquainted with the teeming multitude of insects of the meadow that leap from stalk to stalk—butterflies and wasps, grasshoppers, bumblebees, and beetles" (Brunhammer 1998). The decorative insect arts achieved a zenith with Lalique and another French designer, E. A. Seguy. Seguy produced realistic and decorative assemblages of butterflies and other insects to promote the application of nature to the decorative arts.

Surrealism promoted liberation of the mind by attaining a state deemed truer than reality. Insects evidently feature prominently in this surreal state because insects abound in the works of Salvador Dalí, James Ensor, René Magritte, and other surrealists.

Modern and Contemporary Insect Art

Modern and contemporary artists from around the world continue to include insects in their work. The diversity of insect inclusions in modern-day art is overwhelming, but one way to examine modern or contemporary insect art is to organize the art by degree of realism. A few examples follow.

Realistic Insects

Realistic renderings of insects in art share a history hearkening back at least to Albrecht Dürer's *Stag Beetle* (1505). Realistic insects in contemporary art include paintings by Mark Fairnington, life-size sculptures by Bill Logan and Tom Friedman, and magnified insect sculptures by Patrick Bremer and Lorenzo Possenti.

Realistic Insects in Unnatural Settings

Surrealists' works often fall in this category, and M. C. Escher's illusory visions included ants and sequences of metamorphoses. David Prochaska depicted insects on a series of cans of insecticide, April Vollmer creates woodcuts of insect arrangements, and Karen Anne Klein produces still-life thematic compositions that almost invariably feature realistic insects.

Loose Renderings of Insects

As with chalk drawings of early honey-gatherers and Van Gogh's impressionist pieces with insects, Joseph Beuys, Graham Sutherland, and Jean Émile Laboureur may have interpreted the lives of insects accurately, but they depicted the actual insects with less scientific rigor.

Fantastical or Abstract Insects

Finally, the abstract representation of insects dates to stone drawings, insect totems, and symbols and is often extracted more from the mind of the artist than from the source of inspiration. Some modern examples of insect abstractions include works by André Masson, Joan Miro, Alexander Calder, Sue Johnson, and Francisco Toledo. Charles Burchfield suggested the presence of singing insects by painting waves representing sound generated by the insects.

Insect Art Media

Insect art can be made with any material, including the insects themselves. Beginning with bison bone and chalk on stone, insects have been depicted with materials ranging from metal cast from the lost beeswax process to tattooed human flesh. Painted insects adorn Mimbres pottery from the North American Southwest, Greek ceramics, and Egyptian frescoes, along with most insect art throughout history. Sculpted insects can be carved, as with Edo Period insect wood carvings in Japan or Antonio Canova's marble sculpture

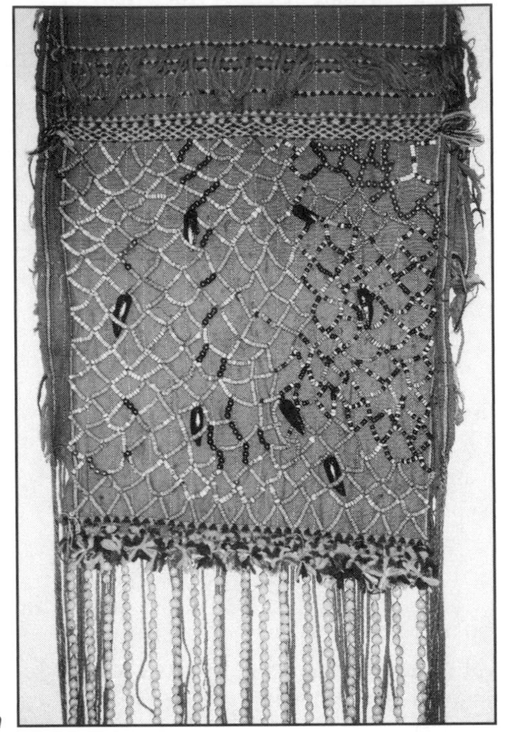

a

b

c

Art and design often incorporate nature: (a) a "Singing Shawl" incorporates the green metallic forewings of beetles (Pwo Karen people, Northern Thailand/Northeast Myanmar, collection of Victoria Z. Rivers); (b) E. A. Seguy promoted nature's beauty in design; and (c) Karen Anne Klein augments still lifes with insects. Courtesy of Barrett Anthony Klein.

Cupid and Psyche. Sculpted insects can also be cast in a variety of materials, woven, or constructed from such random ingredients as Play-Doh, "fuzz," and plastic hair (Tom Friedman's sculptures). Many insect artists use photography (e.g., works by Jacques Kerchache, Gregory Crewdson, and Catherine Chalmers) or film.

Insect bodies can themselves be used as art media. Henry Dalton, a Victorian microscopist, arranged individual scales from butterfly wings into microscopic still lifes, and Jean Dubuffet assembled collages of entire butterfly wings. Some traditional people in the Amazon rainforest enhance their beauty by wearing jewelry composed of damselfly or beetle wings. Others display intact insect corpses in their works: as mentioned earlier, Alberto Faietti adhered insects to pages of books. Kazuo Kadonaga reared 110,000 silkworm moths until the immatures attached their cocoons throughout wood crates, where they were ultimately killed and displayed. Jennifer Angus decorates entire rooms with geometric insect arrays, and Jan Fabre coats structures with beetles or beetle wings, including a ceiling and chandelier in the Royal Palace of Belgium. Other artists display live insects. Yukinori Yanagi produces live ant displays, as noted already, and Hubert Duprat manipulates immature caddis flies so that the cases they build are constructed of jewels supplied by the artist.

Insects have always been a part of the cultural history and creativity of our species. The spectrum of insect art recorded on the walls of ancient caves and on artifacts of past civilizations and the exhibits and festivals dedicated to insect-related art found in the streets and galleries of modern societies are a testament to this long and interesting relationship between insects and human artists. Insects provide inspiration, subject matter, and sometimes even the raw materials for the artisan. The ubiquitousness of insects and the boundless imagination and creativity of humans should ensure the continued relevance of insects in the artistic ventures of humans well into the future. Ultimately, all art can reveal aspects of our history, and insect art documents the history of our relationship with insects.

Further Resources

Berenbaum, M. R. (1995). *Bugs in the system: Insects and their impact on human affairs*. Don Mills: Addison-Wesley.

Brunhammer, Y. (1998). *The jewels of Lalique*. New York: Flammarion.

Dicke, M. (2000). Insects in Western art. *American Entomologist, 46*(4), 228–36.

Heyden, D., & Czitrom, C. B. (1997). Los insectos en el arte prehispánico. In *Insectos y artropodos en el arte Mexicano* (2nd ed.). Mexico City: Cámara Nacional de la Industria Editorial Mexicana.

Hogue, C. L. (1987). Cultural entomology. *Annual Review of Entomology, 32*, 181–99.

Hogue, J. N. (2003). Cultural entomology. In V. H. Resh & R. T. Cardé (Eds.), *Encyclopedia of insects* (pp. 273–81). San Diego: Academic Press.

Klein, B. A. (2003). Par for the palette: Insects and arachnids as art media. In É. Motte-Florac & J. M. C. Thomas (Eds.), *Insects in oral literature and traditions* (pp. 175–96). Paris: Peeters.

Kritsky, G., & Cherry, R. (2000). *Insect mythology*. San José: Writers Club Press.

Manos-Jones, M. (2000). *The spirit of butterflies: Myth, magic, and art*. New York: Abrams.

Barrett Anthony Klein

Art
Rock Art and Shamanism

Animals, and figures with animal-like features, are prominent in rock art throughout the world, providing clues to how people of varied times and places understood themselves and their relationship with animals. During more than a century of serious rock-art research, many theories have been proposed for what the art means and why it was created. One of the most popular of recent years suggests these figures were drawn by shamans to record their visions and journeys to the other world. Animal paintings or etchings, even when apparently realistic in style, may well have carried deep spiritual meanings. Sometimes this model accounts nicely for certain otherwise-peculiar features of the art. Sometimes the theory is less convincing.

What Is Shamanism?

Shamanism is a very ancient view of the cosmos that remained widespread among traditional peoples on several continents until very recently. Some consider it the first religion. Others believe it is not actually a religion at all, but a method or technique. This is the view of Mircea Eliade, whose 1951 book *Shamanism: Archaic Techniques of Ecstasy* is still the scholarly classic on the subject. Eliade notes that these techniques were associated with widely different understandings of the cosmos and varied specific practices. Others believe he has included too many varied customs and belief systems under one label. Michael Harner, the anthropologist whose 1980 book *The Way of the Shaman* is considered a key work in the modern Western revival of shamanism, agrees with Eliade that shamanism is basically a set of techniques.

The label "shaman," which comes originally from the Tungus people of Siberia, has often been used interchangeably, though confusingly, with the terms medicine man, sorcerer, magician, or witch doctor. Shamanism centers on communication with spirits, including those animating nature. The shaman is often a specialist practitioner who at a young age discovered an unusual openness to the spirit world as conceived by his or her people. This natural ability is typically supplemented with rigorous, even dangerous, training.

Jean Clottes and David Lewis-Williams point out in *Shamans of Prehistory* that the "induction, control and exploitation of altered states of consciousness are at the heart of shamanism the world over" (1998). Techniques for reaching altered states vary from ingesting psychoactive substances to rhythmic, even frenzied dancing and drumming. Sensory deprivation (as in the lonely dark and quiet of a cave) or intense pain might also be used. Trance states can be caused by pathological conditions as well, such as schizophrenia and certain kinds of epilepsy, suggesting that some shamans have been mentally ill and turned what might have been a handicap to great advantage.

Regardless of how a trance state is induced, David Lewis-Williams and T. A. Dowson have proposed that a person will typically experience the same three stages in reaching it. In the first stage, the shaman experiences curves and spirals and various geometrical forms. These are also widely represented in rock art and are one source of evidence often put forward in support of a shamanic interpretation. During the second stage, people try to make sense of the geometric visions. What they "see" will depend somewhat on cultural background and state of mind. The shaman will then, in stage three, feel as though irresistibly drawn into a vortex or a tunnel and while "in" the tunnel will see the first recognizable things such as humans or animals.

People in stage three feel they can fly, or dive to the depths of the sea, or travel underground or into the realm of spirits. They can change into animals or birds. These are technically hallucinations—seeing things that are not there. Everyone knows the shaman, physically, is still with them, though in a trance. But as shamans explain upon returning (through words or art), their spirits were elsewhere, perhaps in the form of an antelope or a hawk or halfway between a human and an animal form, learning something needed for curing an illness or retrieving a patient's wandering (or stolen) soul.

Shamanism and Rock Art

Researchers have offered many explanations for the meaning of rock art. Any number of them might be partly correct. Because art is a local matter (as anthropologist Clifford Geertz has argued), with meanings unique to one time and place, we can never hope to fully understand what an artist on the far side of the world, and perhaps from millennia ago, was trying to convey. It is also very unlikely that there is one universal explanation behind rock art in all its varied manifestations around the world. Indeed, there are probably different purposes even for different drawings in the same cave. Some animals may well have been drawn on rock surfaces as part of a hunting magic ceremony, for aesthetic reasons, or just for fun.

That said, the shamanic model for interpreting rock art has become popular among scholars for good reason. The idea first gained respect when David Lewis-Williams established a solid connection between drawings and ideas expressed by rock artists of South Africa. There is equally good evidence that some rock art of California also functioned primarily for shamanic purposes. David Whitley points out that in relatively isolated areas of the southern Sierra Nevada, the ethnographic record related to production of rock art "probably represents the single best corpus of ethnographic data on rock art in North America." Much of the art was produced by shamans

> at the conclusion of their vision quests to illustrate the spirits they had seen and the supernatural events, such as curing, rainmaking, and sorcery, they had participated in during their altered states of consciousness. The shaman's rock art site was a sacred place that served as his portal into the supernatural: during his altered states of consciousness the cracks in the walls of the site were believed to open, allowing him to enter the sacred realm. (Whitley 2000)

Most of the rest of the rock art in the area was created as a result of puberty initiation rites that involved vision quests. Thus, virtually all the animals depicted in the art of this region represent spirits and spirit helpers, and not animals as game, pets, decorations, or subjects of doodling.

Because shamanism is known to be very old and widespread and because it is demonstrably the basis for at least these two major rock art complexes, it has become very popular as an explanation for rock and cave art in all parts of the world, including the famous Paleolithic cave art of Europe. But we must be careful not to assume all rock art is based on shamanism. British archaeologist Paul Bahn, one of the world's leading rock art specialists, is particularly bothered by what he considers a dramatic overuse of this concept, the phenomenon he likes to call "shamania." Indeed, nearly every element of this model—from the ethnographic connection to the three stages of trance and their relation to spirals and other markings—have been questioned both in general and for each specific example of rock art. Lewis-Williams and others have responded to the objections (2004, for example), and the debate continues.

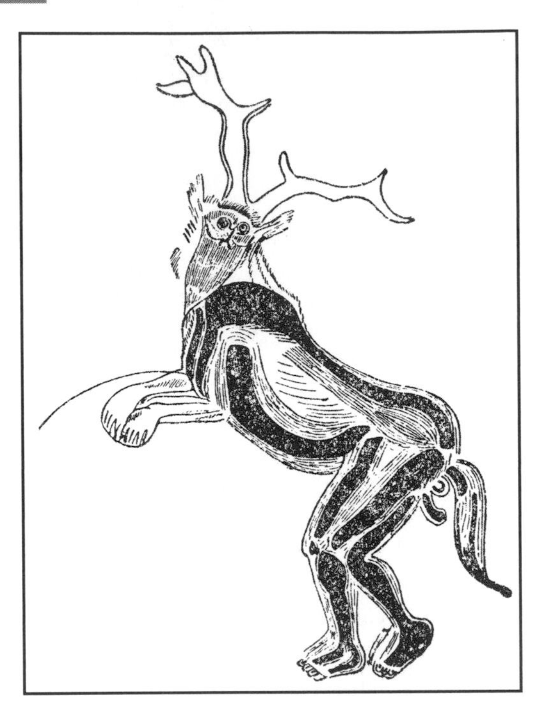

Detail of reproduction from Le Trois-Frères *cave near Les Eyzies, France. ©The Print Collector/Alamy.*

What Does This Teach Us about Animal-Human Relations?

Shamanism was varied in its practices. Also, it may be that shamanism does not account for as much rock art as advocates would have us believe. But we can still make valuable generalizations. First, among traditional peoples who engaged in shamanic practices, animals were not just physical beings, but beings with spirits or souls. This would also be true for plants and nonliving things such as rocks, but animals nearly always stand out because they exhibit agency, intention, and even personalities. Second, the boundaries between humans and other animals were fluid. The shaman could transform into an animal. A spirit might make use of a flesh-and-blood animal who would then come to someone in a dream or vision. The categories of person, animal, and spirit might be clear enough under ordinary circumstances, but they are not unbridgeable.

Shamans from around the world have described the experience of being transformed into animal form and traveling throughout the spirit world. One of the appealing aspects of the shamanic theory of rock art is that it so naturally explains the otherwise enigmatic fact that strange beings, such as those that appear to be half human and half some other animal, are so common.

The reverence for animals engendered by a shamanic cosmos must have affected how people related to the physical animals around them on a daily basis. But it did not typically lead to any ban on hunting except when additional ideas like totemism were also involved. Indeed, as Julian Baldick points out, in regions of Asia, it was not uncommon for shamans to engage in animal sacrifice. Economic and ritual "uses" of an animal may be combined in other ways as well—for example, in the use of eland blood in the paint South African shamans used to record their visions.

See also

Culture, Religion, and Belief Systems—*Animism*
Culture, Religion, and Belief Systems—*Shamanism*
Culture, Religion, and Belief Systems—*Totemism*
Culture, Religion, and Belief Systems—*Totems and Spirit Guides*

Further Resources

Bahn, P. G. (1998). *The Cambridge illustrated history of prehistoric art.* Cambridge: Cambridge University Press.

Bahn, P. G., & Vertut, J. (1988). *Images of the Ice Age.* Leicester: Windward.

Baldick, J. (2000). *Animal and shaman: The ancient religions of central Asia.* New York: New York University Press.

Chippendale, C., & Taçon, P. (Eds.). (1998). *The archaeology of rock art*. Cambridge: Cambridge University Press.

Clottes, J. (Ed.). (2003). *Return to Chauvet Cave: Excavating the birthplace of art, the first full report*. (P. G. Bahn, Trans.). London: Thames & Hudson.

Clottes, J., & Lewis-Williams, D. (1998). *The shamans of prehistory: Trance and magic in the painted caves*. (S. Hawkes, Trans.). New York: Abrams.

Eliade, M. (1964). *Shamanism: Archaic techniques of ecstasy*. (W. R. Trask, Trans.). Bollingen Series LXXXVI. New York: Pantheon Books.

Geertz, C. (2000). Art as a cultural system. In *Local knowledge: Further essays in interpretive anthropology* (2nd ed.). New York: Basic Books.

Grim, J. A. (1983). *The shaman: Patterns of religious healing among the Ojibway Indians*. Norman: University of Oklahoma Press.

Harner, M. (1980/1990). *The way of the shaman*. New York and San Francisco: HarperCollins and HarperSanFransicso.

Helvenston, P. A., & Bahn, P. (2002). *Desperately seeking trance plants: Testing the "three stages of trance" model*. New York: RJ Communications.

Helvenston, P. A., & Bahn, P. (2003). Testing the "three stages of trance" model. *Cambridge Archaeological Journal, 13*(2), 213–24.

Kehoe, A. (2000). *Shamans and religion: An anthropological exploration of critical thinking*. Long Grove, IL: Waveland Press.

Lewis, I. M. (2003). *Ecstatic religion: A study of shamanism and spirit possession* (3rd ed.). London: Routledge.

Lewis-Williams, D. (2002). *The mind in the cave: Consciousness and the origins of art*. London: Thames & Hudson.

Lewis-Williams, D., & Dowson, T. A. (1988). Signs of all times. *Current Anthropology, 29*(2), 201–45.

Lewis-Williams, J. D. (1990). *Discovering Southern African rock art*. Cape Town: David Philip.

———. (2004). Neuropsychology and upper palaeolithic art: Observations on the progress of altered states of consciousness. *Cambridge Archaeological Journal, 14*(1), 107–11.

Pearson, J. L. (2002). *Shamanism and the ancient mind: A cognitive approach to prehistory*. Walnut Creek, CA: AltaMira Press.

Schlesier, K. H. (1987). *The wolves of heaven: Cheyenne shamanism, ceremonies, and prehistoric origins*. Norman: University of Oklahoma Press.

Whitley, D. S. (2000). *The art of the shaman: Rock art of California*. Salt Lake City: University of Utah Press.

Paul K. Wason

Bestiality
Bestiality

Until approximately the sixteenth century, the term "bestiality" referred either to a broad notion of earthy and often distasteful otherness or to sexual relations between humans and nonhuman animals.

The earliest and most influential condemnations of bestiality are the Mosaic commandments contained in Deuteronomy, Exodus, and Leviticus. Deuteronomy, for example, declared, "Cursed be he that lieth with any manner of beast" (27:21), and Exodus ruled that "whosoever lieth with a beast shall surely be put to death" (22:19). Besides mandating death for humans, Leviticus dictated that the offending animal must also be put to death. It is hard to know the precise intentions of those who originally condemned bestiality, but in Judeo-Christianity, there have been three principal beliefs about the origins of its wrongfulness: (1) it is a rupture of the natural, God-given order of the universe; (2) it violates the procreative intent required of all sexual relations between Christians; and (3) it produces monstrous offspring that are the work of the Devil.

In some societies, such as in colonial New England from the Puritan 1600s until the mid-nineteenth century, bestiality was regarded with such alarm that even the very mention of it was condemned. It was therefore also referred to as "that unmentionable vice" or "a sin too fearful to be named" or "among Christians a crime not to be named." Nowadays, bestiality is variously described as "zoophilia," "zoöerasty," "sodomy," and "buggery," and its meaning is almost always confined to human-animal sexual relations.

Since the end of World War II, especially, bestiality has been one among several categories of nonreproductive sexual practices toward which society in general has tended to exercise a growing tolerance. Indeed, in the last fifty or so years, those offenders whose sexual activities with animals have been reported to legal or medical authorities have faced considerably lesser charges—such as breach of the peace or offending against public order. Instead of criminal prosecution, offenders have typically been sent either for counseling or for psychiatric treatment, or, with probably the greatest deterrent effect, they have been subject to public ridicule in their local communities.

Information about the incidence and prevalence of bestiality is quite unreliable, especially given its private nature and the social stigma attached to it. Bestiality can occur in a wide variety of social contexts. These include adolescent sexual experimentation, typically by young males in rural areas; eroticism (sometimes termed "zoophilia," practiced by "zoos"), a rare event where animals are the preferred sexual partner of humans; aggravated cruelty, especially by young males or in cases of partner abuse; and commercial exploitation, as in pornographic films or in live shows of women copulating with animals in bars or sex clubs.

The prevalence of bestiality probably depends on such factors as the level of official and popular tolerance, opportunity, proximity to animals, and the availability of alternative sexual outlets. Some sexologists have claimed, with the use of interviews and questionnaires, that 8 percent of the male population has some sexual experience with animals, but that a minimum of 40 to 50 percent of all young rural males experience some form of sexual contact with animals, as do 5.1 percent of American females. But because of the poor sampling techniques of such studies, these figures should be treated with great caution.

Further Resources

Dekkers, M. (1994). *Dearest pet.* (P. Vincent, Trans.). London: Verso.

Kinsey, A. C., Pomeroy, W. B., & Martin, C. E. (1948). *Sexual behavior in the human male.* Philadelphia: W. B. Saunders.

Kinsey, A. C., Pomeroy, W. B., Martin, C. E., & Gebhard, P. H. (1953). *Sexual behavior in the human female.* Philadelphia: W. B. Saunders.

Liliequist, J. (1991). Peasants against nature: Crossing the boundaries between man and animal in seventeenth- and eighteenth-century Sweden. *Journal of the History of Sexuality, 1*(3), 393–423.

Miletski, H. (2002). *Understanding bestiality and zoophilia.* Germantown, MD: Imatek.

Piers Beirne

Sexual Assault of Animals

Piers Beirne

Historically, sexual relations involving humans and animals have tended to be condemned and investigated—or, in the interests of "tolerance," ignored—exclusively from an anthropocentric perspective. Yet sexual relations with humans often cause animals to suffer great pain and even death, especially in the case of smaller creatures such as rabbits and hens.

Today, both the feminist movement and the animal rights movement have started to rethink the moral and ethical status of bestiality. Sexual relations between humans and nonhuman animals are beginning to be seen as wrong for the same reasons we see sexual assault by one human against another human as wrong—because it involves coercion, because it produces pain and suffering, and because it violates the rights of another being.

It is impossible to know whether animals can ever consent to sexual relations with humans, so it is best to treat all such cases as forced sex. Sexual relations involving humans and animals are therefore more appropriately termed "animal sexual assault."

Further Resources

Adams, C. J. (1995a). Bestiality: The unmentioned abuse. *The Animals' Agenda, 15*(6), 29–31.

———. (1995b). Woman-battering and harm to animals. In J. Donovan & C. J. Adams (Eds.), *Animals and women: Feminist theoretical explorations* (pp.55–84). Durham, NC: Duke University Press.

Beirne, P. (2002). On the sexual assault of animals: A sociological view. In A. N. H. Creager & W. C. Jordan (Eds.), *The animal/human boundary: Historical perspectives* (pp. 193–227). Rochester, NY: Rochester University Press and Davis Center, Princeton University.

———. (2002). Peter Singer's "Heavy Petting" and the politics of animal sexual assault. *Journal of Critical Criminology, 10*(1), 43–55.

Singer, Peter P. (2001, March/April). Heavy petting. *Nerve.* Available at http://www.nerve.com/Opinions/Singer/heavyPetting

Bestiality
Zoophilia and Bestiality

Bestiality

Zoophilia, having an attraction to and affection for animals, which incorporates many shades of meanings, and bestiality, human sexual relations with animals, are controversial forms of social relations that are only beginning to be better understood at this time. The practice of human sexual relations with nonhuman animals, now commonly referred to as "bestiality," appears to have existed within most every culture throughout history, and it is a theme represented in a wide variety of people's literature, art, and myths the world over. Sex between humans and nonhuman animals can be dated back tens of thousands of years and is depicted on a number of prehistoric artifacts and cave paintings. According to Midas Dekkers and Hani Miletski, who each offer copious histories of the subject, many cultures from ancient times up until the present have themselves shown forms of tolerance for bestiality for religious, ritualistic, and even recreational reasons. Still, practices of bestiality have also been, and remain, extremely controversial on the whole for societies rooted in Abrahamic religious traditions, which maintain important prohibitions against the practice. This is especially true of the Judeo-Christian West, where some people have always tolerated bestiality in certain places and at certain times, but overall, bestial relations are associated with social stigma, and religious forces have successfully advanced the idea in the popular mind that bestiality is a great moral transgression never to be condoned.

It is interesting to note, then, that as secular culture became instantiated throughout much of Europe and North America during the nineteenth and twentieth centuries, legislative concern with and the penalties for bestiality largely lessened. Whereas for the ancient Egyptians, Hebrews, and Romans, as well as throughout pre-Enlightenment Christian Europe, the human and nonhuman parties charged with bestiality could expect a punishment of torture and death, modern European nations such as the Netherlands, Sweden, and Belgium maintain that human sexual relations with nonhuman animals are ostensibly legal. Still, bestiality does remain officially criminalized in many countries throughout the world. In 1962 Illinois became the first U.S. state to decriminalize sexual relations with animals, but the practice has been outlawed at the federal level in America and is once again considered criminal throughout all fifty states, either in the form of explicit statutes against it or implicitly through a combination of animal cruelty and welfare laws. On the other hand, whether the commitment exists on the part of U.S. authorities to actively enforce prohibitions against bestiality and then vigorously prosecute and penalize offenders once they have been caught is questionable. In the absence of such a commitment, the pressure of socio-religious mores against bestiality remains the primary force preventing its wider adoption as an acceptable American practice.

As noted, these mores are hardly insignificant in many Western nations, where they have also given rise to stereotypes that often suggest bestiality is behavior more typical of "uncivilized," rural, or agrarian-based peoples. Possibly lending credence to these stereotypes, early sex studies such as those famously conducted by Dr. Alfred Kinsey concluded that bestiality was primarily a rural phenomenon, and it has been found that bestial relations are considered a much more normal part of sexual maturation in nomadic herding societies, such as throughout Africa and the Middle East, than in primarily urban cultures. Yet other research over the last fifty years has demonstrated that—if it is not always possible to know who engages in such relations, or how often they do so, because of bestialists' usual desire for anonymity—a broad spectrum of people in modern society have a statistically significant interest in bestiality themselves. Perhaps this has never been better confirmed recently than when, in 2005, the *Seattle Times* published an article on a local man who had died from a perforated colon after having sex with a horse, and a columnist for the paper, Danny Westneat, later found that because of unprecedented downloads of the article via the Internet, it was perhaps the most read piece of journalism ever published during the paper's 109-year printing history.

Indeed, current understandings of bestiality owe much to how the Internet has been utilized pivotally to exchange information, to educate others about lifestyle orientations associated with human sexual relations with nonhuman animals, and to trade pornography depicting the same. Additionally, some argue that the Internet has begun to establish and grow communities dedicated to varieties of bestiality, though the degree to which these communities occur offline and are constituted by face-to-face interactions remains unclear. The first digital community of major importance was the Usenet newsgroup alt.sex.bestiality, which arose in the early 1990s. Over the last decade, there has been a tremendous proliferation of bestiality Web sites, chat rooms, electronic mailing lists, and peer-to-peer file-sharing networks, and thus, it has never been easier to view explicit pictures and videos of men and women of a wide range of races and ages engaging in sexual intercourse with dogs, horses, cattle, snakes, birds, fish, rodents, reptiles, cats, sheep, goats, and other species.

Zoophilia

The Internet has also been vital in developing a subculture of self-identified "zoos," which is slang for "zoophiles," a word combining the Greek words for animal and friend or lover, respectively. Although zoophiles may engage in acts of bestiality, many highlight that their zoophilia implies an emotional affinity for their relationships with nonhuman animals that goes beyond the merely sexual or erotic and does not require the presence of these elements. Thus, one can be a zoophile without engaging, or even wishing to engage, in sexual relations with nonhuman animals, just as one can engage in such relations without thereby being a zoophile. This being said, many zoophiles will have sexual relations with nonhuman animals and yet differentiate themselves from other bestialists through the addition of an emotional component for their nonhuman counterparts.

Unfortunately, the term "zoophilia" has additional ambiguities beyond often being conflated with bestiality proper. Initially, it was scientifically defined in the nineteenth century by the psychiatrist Richard von Krafft-Ebing, who used it to denote that a person had an erotic attraction to animals' fur or excitement upon viewing the copulation of animals. He did not believe zoophilia involved the desire for intercourse with nonhuman animals, and this was not included in his definition. In the twentieth century, conversely, zoophilia was identified by the American Psychological Association (APA) as a form of paraphilia, or clinical form of perverse sexual and psychological dysfunction in individuals because of their attraction to and desire for intercourse with nonhuman animals. The most recent version of the APA's professional handbook (*DSM-IV*) continues to think of zoophilia as a human sexual and psychological affinity for nonhuman animals, but no longer considers it as a form of disorder unless an individual's attraction to nonhuman animals causes personal distress. Finally, in the recent work of Hani Miletski, Andrea Beetz, and Colin Williams and Martin Weinberg, zoophilia has been studied and defined as a form of "zoosexuality" that exists as a full-fledged sexual orientation for some people in a manner akin to other sexual orientations. In defining this as a sexual orientation, these researchers have called for more complex and nuanced understandings of zoophilia. Accordingly, they have noted how zoophiles can be distinguished from "zoosadists," or those who derive pleasure from inflicting pain in sexual and nonsexual ways upon nonhuman animals.

In many ways, the distinction between zoophilia and zoosadism is at the center of much of the recent debate about human sexual relations with nonhuman animals. Piers Bierne has arguably been the leading advocate in challenging this differentiation. Bierne believes that zoophiles incorrectly assume that their nonhuman animal partners can signal forms of consent, which would thereby transform acts of bestiality into zoophilic relationships based on reciprocity and mutual affection. Instead, according to Bierne, nonhuman animals can never consent in this way, and additionally, human sexual relations with them almost always involve some degree of coercion. Further, he believes that zoophilic and bestial relations with nonhuman animals often result in the latter suffering injury and even death. Therefore, he concludes that through a strong commitment to animal welfare, human sexual relations of any kind with nonhuman animals should be considered "interspecies sexual assault" and that this should form the rightful basis for social intolerance for those people.

Relatedly, a growing number of human-animal studies are establishing links between nonhuman animal abuse and interpersonal violence. Although causal relationships have yet to be proven, evidence now clearly exists that those who have been abused physically or sexually in adolescence are more likely to commit, or have committed, abuses upon either human or nonhuman animals, or both. Taken altogether, this research serves to further dispute positive notions of zoophilia in favor of a critical focus on zoosadism that is often guided by a concern for the humane treatment of nonhuman animals.

Not all who are presently interested in establishing stronger welfare for nonhuman animals are against zoophilic ideas, however. For instance, the philosopher Peter Singer—himself a leading animal rights advocate—has been described as a recent champion of bestiality. Though hardly condoning zoosadist practices, his recent work does challenge long-held cultural stigmas against zoophilic bestiality. Further, he notes that nonhuman animals often copulate as humans do and that some, such as domesticated canines, appear to commonly make humans the objects of their own nonhuman sexual advances and desires. Therefore, Singer illustrates that zoophilia should not necessarily be understood as a solely human persuasion. Yet Singer's main purpose in considering bestiality is to highlight that the boundaries between human and nonhuman animals are changing and should no longer be so sharply defined in a rational society. Additionally, he believes that as increased attention to human creatureliness is achieved, it is best balanced by an evolved ethical awareness of the sentience that many nonhuman animals share with humanity. When this occurs, in his opinion, sexual relations across species no longer deserve to be marked by traditional religious or other moral taboos that guard against affronts to human dignity. It is telling, however, that Singer's refusal to condemn bestiality generated widespread controversy from all sectors of society, and he was forced to issue a form of mea culpa from Princeton University, further explaining his position, even as other philosophers such as Neil Levy concluded that Singer was hardly insane and, at least partially, correct.

In conclusion, the steady growth of zoo subculture, the rise of a number of transdisciplinary scholarly studies on bestiality, and a changing legal status for animals in many nations may very well point to changing conceptions of human identity that support Singer's view of existence as a continuity between human and nonhuman animals. Zoophilia, if not bestiality, then, represents a marginal but potentially illuminative practice of how new conceptions of equality, reciprocity, and love could be made manifest between human and nonhuman species. However, it is certain that Western society's long-standing taboo against bestiality remains powerfully proscriptive for a great many people, and so, as conceptions about bestiality and zoophilia continue to emerge and confront society, they are sure to generate and be met with heated disapproval. Therefore, bestiality and zoophilia should be considered highly controversial and multifaceted forms of social relations that are only beginning to be better understood. As such, they are deserving of further exploration and study.

Further Resources

Beetz, A. (2004). Bestiality/zoophilia: A scarcely investigated phenomenon between crime, paraphilia, and love. *Journal of Forensic Psychology Practice, 4*(2), 1–36.

Beetz, A., & Podberscek, A. L. (2005). *Bestiality and zoophilia: Sexual relations with animals*. West Lafayette, IN: Purdue University Press.

Bierne, P. (2000). Rethinking bestiality: Towards a concept of interspecies sexual assault. In A.L. Podberscek, E. S. Paul, & J. A. Serpell (Eds.), *Companion animals and us: Exploring the relationships between people and pets* (pp. 313–31). Cambridge: Cambridge University Press.

Dekkers, M. (1994). *Dearest pet: On bestiality*. (P. Vincent, Trans.). New York: Verso.

Kinsey, A. C., Pomeroy, W. B., & Martin, C. E. (1948). *Sexual behavior in the human male*. Philadelphia: W. B. Saunders.

Levy. N. (2003). What (if anything) is wrong with bestiality? *Journal of Social Philosophy, 34*(3), 444–56.

Miletski, H. (2002). *Understanding bestiality and zoophilia*. Bethesda, MD: East West.

Singer, P. (2001, March/April). Heavy Petting. *Nerve*. Available at http://www.nerve.com/Opinions/Singer/heavypetting

Westneat, D. (2005, December 30). Horse sex story was online hit. *Seattle Times*. Retrieved on February 17, 2007, from http://seattletimes.nwsource.com/html/localnews/2002711400_danny30.html

Williams, C., & Weinberg, M. (2003). Zoophilia in men: A study of sexual interest in animals. *Archives of Sexual Behavior, 32*(6), 523–35.

Richard Kahn

Biomimicry
Biomimicry: Innovation Inspired by Nature

What Is Biomimicry?

Biomimicry is a design discipline that emulates natural forms, processes, and ecosystem strategies in order to solve human design challenges. An example is a solar cell that mimics photosynthesis. The core idea is that life has been on earth for 3.8 billion years, and in that time, organisms have grappled with and solved many of the problems humans now face. The thirty-plus million organisms on earth have evolved physical attributes and behavioral strategies that are well adapted to the limits and opportunities of this biosphere. Mimicking these well-adapted designs may help humans leapfrog to more material-saving, energy-efficient, and chemically benign technologies.

Nature-inspired innovation—also known as biomimetics, bionics, and bio-inspired design—is practiced by academic researchers, industrial innovators, and governmental agencies throughout the world, with leading centers in Germany, the United Kingdom, the United States, China, and Japan. The areas of the most intense activity include materials, nanotechnology, sensors, computing, robotics, medicine, chemistry, building, alternative energy, and transportation. Governmental funding has traditionally come from the departments of defense, space, health, and energy.

Historical examples of biomimicry date back to Leonardo da Vinci's bird-inspired flying machines, the femur-inspired Eiffel Tower, the tongue- and ear-inspired telephone, Buckminster Fuller's tensegrity architecture, and more recently, the invention of Velcro, modeled on the physical-attachment mechanisms of seed burs.

Today's biomimics are learning to grow crops like a prairie (natural systems agriculture), harness energy like a leaf (artificial photosynthesis), manufacture tough materials in benign conditions like an abalone (nanolayered self-assembly), find drug plants like a chimp (zoopharmacognosy), compute like a cell (biomolecular electronics), optimize via natural selection (evolutionary computing), and close the material loops in business like a mature forest (industrial ecology). Some of the best-studied organismal models in biomimetics include geckos (adhesion without adhesives), ants (foraging strategies to optimize telecommunications), lotus leaves (self-cleaning surfaces), flies (micro air vehicles), brittlestars (distortion-free lenses for optical computing), and sea mice (photonic crystals for fiber optics).

The possibilities have just begun to be explored, however. Julian Vincent, head of the Center for Biomimetic and Natural Technologies at the University of Bath has recently completed a survey of biological "inventions." He compared a survey of 3 million human patents to nature's approaches and found that there is only a 10 percent overlap between how we solve problems and how the rest of the natural world solves the same problems. This means that 90 percent of the time we seek nature's advice, we will be surprised.

How Does Biomimicry Differ from Other Bio-Approaches?

Biomimicry introduces an era based not on what we can extract from organisms and their ecosystems, but on what we can learn from them. This approach differs greatly from bioutilization, which entails harvesting a product or producer from the wild (e.g., cutting wood for floors, wild crafting medicinal plants). It is also distinctly different from bio-assisted technologies, which involve domesticating an organism to accomplish a function (e.g., bacterial purification of water, cows bred to produce milk). Instead of harvesting or domesticating, biomimics consult organisms; they are inspired by an *idea*, be it a physical blueprint, a process step in a chemical reaction, or an ecosystem principle such as nutrient cycling. Borrowing an idea is like copying a picture—the original image can remain to inspire others.

For those of us in Western industrial culture, looking to nature for advice marks a new way of viewing and valuing other organisms. When we begin to see nature as a teacher rather than a warehouse, our respect for life and its adaptive ability grows. As more people practice biomimicry and realize what we might learn from living systems, the argument for conserving biodiversity becomes self-evident.

Finally, learning *from* instead of just *about* nature calls for a fundamentally different scientific approach, involving the study of an organism, a subsequent attempt to emulate, and often a return to the organism with a new set of questions. This has been called "a deepening conversation with the organism" by plant geneticist Wes Jackson, who studies prairie patterns to come up with a more robust agriculture. This shift in stance, from conqueror to student, marks a new relationship between humans and the rest of the natural world.

The Practice of Biomimicry

The practice of biomimetic invention can proceed from biology to design or from design to biology.

In the biology-to-design approach, a biological phenomenon suggests a new way to solve a human design challenge. Wilhelm Barthlott (1997) of the Nees-Institute, University of Bonn, studied how leaves such as the lotus manage to remain free of contaminants without the use of detergents. His papers described how a landscape of small bumps and waxy crystals causes water to ball up. Dirt particles teeter on the nano-mountains and are easily picked up by the water, like a snowball lifting leaves from a lawn. Barthlott and colleagues worked out how to replicate the geometric profiles of the lotus into commercial products, such as a building façade paint that exhibits a nano-rough surface when it dries. Rainwater cleans the building. Today, dozens of self-cleaning products such as glass, roofing tiles, and textiles bear the lotus-effect symbol.

In the design-to-biology approach, the innovator starts with a human design challenge, identifies the core function, and then reviews how various organisms or ecosystems are achieving that function. An example is the quest for a new way to reduce microbial growth without causing antibiotic resistance. Peter Steinberg of the University of New South Wales used a characteristic biomimicry approach. He identified an environment that was teeming with microbes and then searched for organisms within that environment that had no biofilm on their surfaces. He found his "champion adapter" in the murky waters of Botany Bay Australia. *Pseudomonas aeruginosa,* a red kelp called "sea purse," remains free of microbes by releasing furanones, molecules that interfere with the bacteria's communication signaling mechanisms. When bacteria are "jammed" by furanone, they are unable to receive a quorum of signals from other bacteria, and without positive "quorum sensing," they don't begin biofilm formation. Steinberg's company, Biosignal Ltd. of Eveleigh, Australia, has mimicked these repellant compounds and licenses them to companies producing nontoxic antifouling paints, contact lenses, and surface treatments for hospitals.

Three Levels of Mimicry

There are three levels of mimicry: (1) mimicking natural form, (2) mimicking natural process, and (3) mimicking ecosystem strategies.

Examples of mimicking natural form include the following:

- Lung-inspired fuel cell optimization: Morgan Fuel Cell's (UK) patented 'Biomimetic' bipolar plate technology (electrodes) drew its inspiration from the branching structures in animal lungs and plant tissues. The bipolar plates of the fuel cell contain two large conduits that feed into a system of capillaries. As with the lung, this maximizing of surface area for gas exchange allows gases to flow through the plate in a far more efficient way than has ever been achieved before. The biomimetic bipolar plates are cheaper to produce, and they boost peak power by 16 percent, while improving water management, enhancing reliability, and reducing back pressure.

- Butterfly-inspired pigment-free color: The feathers, scales, and exoskeletons of iridescent birds, butterflies, and beetles have structural features that cause light to diffract and interfere in ways that amplify certain wavelengths. This creates brilliant colors to the viewer through the use of structure rather than the addition of a chemical pigment. Imagine, instead of painting a product, simply adding surface layers that play with light. Thin-film interference of this sort can create color that is four times brighter than pigment, that never needs repainting, and that avoids the toxic effects associated with pigment mining and synthesis. The first products from this research include Morphotex, a pigment-free fiber produced by Teijin (Japan), and a low-energy, sunlight-readable PDA screen from Qualcomm (USA).
- Beetle-inspired water harvester: A fog-catching device patterned on the Namibian beetle's prodigious water-harvesting abilities captures ten times more water than existing fog-catching nets. The beetle's ability to pull water from fog is a result of bumps on its wing scales that have water-loving tips and water-shedding sides. QinetiQ (UK) has developed plastic water-harvesting sheets that mimic the beetle's bumps, useful for capturing water in cooling towers and industrial condensers, arid agricultural systems, and buildings in fog-rich areas.
- Mollusk-inspired fan: A three-dimensional logarithmic spiral is found in the shells of mollusks, in the spiraling of tidal-washed kelp fronds, and in the shape of our own skin pores, through which water vapor escapes. Liquids and gases flow centripetally through these geometrically consistent flow forms with far less friction and more efficiency. PAX Scientific (USA) has designed fans, propellers, impellers, and aerators based on this shape. Computational fluid dynamics and particle image velocimetry tests showed the technology's streamlining effect can reduce energy requirements in fans and other rotors by between 10 and 85 percent, depending on the application; the fan blade design also reduces noise by up to 75 percent. The first air-handling products scheduled for release are fans in computers, auto air conditioners, and kitchen range hoods. The Pax streamlining principle could also lead to improvements in industrial mixers, water pumps, marine propellers, and devices for circulating blood in the body.
- Whale-inspired aircraft wings: Unlike commercial aircraft wings with a straight leading edge, the leading edge of humpback whale flippers are scalloped with prominent knobs called tubercles. In wind-tunnel experiments conducted by Phillip Watts of Applied Fluid Engineering, Inc. and Dr. Frank Fish of West Chester University, the scalloped flipper proved a more efficient wing design than the smooth edges used on airplanes. In tests of a scalloped versus a sleek flipper, the scalloped flippers had 32 percent lower drag and 8 percent better lift properties, and they withstood stall at a 40 percent steeper wind angle. This discovery has the potential to optimize not only airplane wings but also the tips of helicopter rotors, propellers, and ship rudders. The improved stall angle would add a margin of safety while making planes more maneuverable, and the drag reduction would improve fuel efficiency.

- Termite-inspired air conditioning: Architect Mick Pearce collaborated with engineers at Arup Associates to build a mid-rise building in Harare, Zimbabwe, that has no air conditioning, yet stays cool thanks to a termite-inspired ventilation system. The Eastgate building is modeled on the self-cooling mounds of *Macrotermes michaelseni,* termites that maintain the temperature inside their nest to within one degree of 31°C, day and night—while the external temperature varies between 3°C and 42°C. Eastgate uses only 10 percent of the energy of a conventional building its size, it saved $3.5 million in air-conditioning costs in the first five years, and it has rents that are 20 percent lower than a newer building next door. The TERMES project, organized by Rupert Soar of Loughborough University, is digitally scanning termite mounds to map the three-dimensional architecture with a level of detail never before achieved. This computer model will help scientists understand exactly how the tunnels and air conduits manage to exchange gases, maintain temperature, and regulate humidities. The designs may provide a blueprint for self-regulating human buildings.

Examples of mimicking natural processes include the following:

- Leaf-inspired solar cells: Plant biologists and engineers at many labs are looking to leaves to help them make smaller and more efficient solar cells. A leaf has tens of thousands of tiny photosynthetic reaction centers that operate at 93 percent quantum efficiency, producing energy silently with water, sunlight, and no toxic chemicals. Mimics of these molecular-scale solar batteries could one day be used to split water into clean-burning hydrogen and oxygen, or as computer switching devices that shuttle light instead of electrons. Konarka (USA), Dyesol (Australia), NexTech Materials, Ltd (USA), and many other companies have commercialized dye-sensitized solar cells that mimic photosynthesis to maximize light harvesting and increase the efficiency of conversion of sunlight to electricity.
- Abalone-inspired ceramics: On the underside of the red abalone (*Haliotis rufescens*) shell is a remarkable iridescent ceramic that is twice as tough as our high-tech ceramics. Mother-of-pearl, also called nacre, is composed of alternating layers of calcium carbonate (in a special crystal form called aragonite) and Lustrin-A protein. The combination of hard and elastic layers gives nacre remarkable toughness and strength, allowing the material to slide under compressive force. The "bricks" of calcium carbonate are offset, and this brick-wall architecture stops cracks from propagating. Several groups have mimicked nacre's structure, using materials such as aluminum and titanium alloy to create a metal laminate tough enough for armor. Dr. Jeffrey Brinker's group at Sandia National Laboratories used a self-assembly process to create mineral/polymer-layered structures that are optically clear but much tougher than glass. Unlike traditional "heat, beat, and treat" technologies, Brinker's evaporation-induced, low-temperature process allows liquid building blocks to self-assemble and harden into coatings that can toughen windshields, bodies of solar cars, airplanes, or anything that needs to be lightweight but fracture-resistant. The complex nano-laminate structure of these bio-composite materials is characterized and related to their mechanical properties.
- Diatom- and sponge-inspired silicon manufacture: Silicon chips are now processed in energy-intensive, toxic ways. Marine sponges, on the other hand,

form silica dioxide structures at ambient conditions with the help of a protein called silicatein. Researchers at the University of California, Santa Barbara, have created a mimic of this protein called a "cysteine-lysine block copolypeptide." Lab results confirm that these molecules are able to direct formation of ordered silica structures, just as silicatein does. This demonstrates the possibility of developing a non-toxic, low-temperature approach to computer chip manufacture.

- Microbe-inspired replacement for platinum catalysts in fuel cells: One reason fuel cells are so expensive is the use of platinum in the membrane that conducts the hydrogen chemistry. Cyanobacteria catalyze this same reaction using an enzyme created from common and biocompatible metals. Cedric Tard and Christopher Pickett of the John Innes Centre in the United Kingdom have successfully mimicked the active site of the hydrogenase protein. The resulting iron-sulfur framework functions as an electrocatalyst for proton reduction, a potentially important step toward inexpensive materials to replace platinum in the anodes of fuel cells.
- Microbe-inspired mining: Dr. Irving DeVoe spent years studying how microbes capture essential elements such as iron, magnesium, chromium, selenium, copper, and even gold from water. He then realized that this same process could help humans mine in nondestructive ways. He learned how to make analogues of the molecules by mimicking the active sites that have the high affinity for various metals. Now, instead of digging into the earth's crust and heap-leaching metals with harsh chemicals, his company, MR3 Systems (USA), mines wastewater streams, gathering and purifying the metals that are traditionally seen as pollution.
- Anhydrobiosis-inspired vaccine storage: Current vaccines spoil easily without refrigeration, and 50 percent fail to reach patients because of a break in the "cold chain." The quest for thermally stable storage led Bruce Rosner of Cambridge Biostability Ltd (UK) to study anhydrobiosis, the process by which organisms such as tardigrades and resurrection ferns are able to remain in a long-term desiccated state. These organisms replace the water in their cells with a protective sugar called trehelose. By coating vaccines with trehelose and suspending them in vials of inert liquid, Biostability was able to create multivalent vaccines that remain stable for years, despite freezing or high temperatures. Because the liquid formulations are anhydrous, they are inherently bacteriostatic, eliminating the need for antiseptics. Cambridge Biostability has already stabilized vaccines for conjugate meningitis A, hepatitis B, and tetanus toxoid. They are now developing programs for measles, pentavalent childhood vaccines, heptavalent botulinum, and anthrax vaccines.

Examples of mimicking natural ecosystems include the following:

- Forest-inspired industrial economies: On the broader, macroeconomic scale, some leading-edge planners, industrialists, and entrepreneurs are studying the material cycling that occurs in mature ecosystems such as prairies, forests, and coral reefs. These industrial ecologists are trying to envision how we could shift our economy from a linear, throughput kind of economy to a closed-loop, diverse, highly interconnected system in which only solar ambient energy is coming in, all the "nutrients" are juggled forever in cascading loops, and very little waste results.

- Soil community-inspired residential wastewater treatment: The Biolytix Filter is a compact septic system that mimics the structure and function of decomposer organisms along a river's edge. In the Biolytix system, worms, beetles, and microscopic organisms convert solid sewage and food waste into structured humus, which then acts as the filter that polishes the remaining water to irrigation grade. The treated water is then distributed through shallow tubes to irrigate lawn and landscape. The system uses one-tenth the energy of conventional sewage-treatment systems, needs no chemicals, and produces irrigation water that is safe for the environment.
- Prairie-inspired farming: Prairies hold the soil, resist pests and weeds, and sponsor their own fertility, all without our help. Prairie-like polycultures using edible perennial crops and biofuel candidates such as switchgrass would over winter, making plowing or planting every year obsolete. Mixtures of plants would give farms resilience, reducing the need for oil-based pesticides. Instead of an extractive agriculture that mimics industry, prairie-inspired farming is a self-renewing agriculture that mimics nature while sequestering significant amounts of carbon.

Formalizing the Discipline

The process of borrowing nature's blueprints is not new, but has enjoyed resurgence in part because of society's search for more sustainable methods of agriculture, manufacturing, chemistry, health care, and business. Also contributing to renewed interest are lab techniques and imaging technologies that allow us to more fully characterize how nature's materials, processes, and ecosystems work. At the same time, our ability to mimic life's devices, especially at the nano and micro levels, makes emulation more possible.

According to a study by Richard Bonser at the Centre of Biomimetics at the University of Reading in the United Kingdom, the number of global patents containing the term "biomimetic" in their title has increased by a factor of 93 since 1981, compared to a factor 2.7 increase for non-biomimetic patents. Industry is accelerating this trend by seeking the consulting services of biologists. In the United States, a research and consulting firm called Biomimicry Guild offers the services of "biologists at the design table" to innovators who want to practice biomimicry, but do not have biological expertise. Industrial networks such as BIONIS (UK) and BIOKON Bionics Competence Network (Germany) are catalyzing commercial interest in Europe. New resource tools are also being created for bio-inspired innovators. A "Google-like search engine for nature's solutions called BioMuse allows an engineer to ask questions such as "How would nature desalinate?" and learn about the strategies of mangroves, sea glands of seabirds, kidneys, and more.

As the approach gains recognition, new interdisciplinary centers are forming, including the Biodesign Institute at Arizona State University, the Center for Biologically Inspired Design at Georgia Tech, the Center for Biologically Inspired Materials and Material Systems at Duke University, the Biomimetics Research Center at Doshisha University, and the Swedish Center for Biomimetic Fiber Engineering at the Royal Institute of Technology. New journals devoted to biomimetics, such as the *Journal of Bionic Engineering*, and an increasing number of symposia and conferences are further signs of the maturing of the discipline.

Perhaps most encouraging is the beginning of a movement to tie biomimetic design back into conservation. The Biomimicry Institute (USA) is promoting the program

Innovation for Conservation, which asks companies to donate a percentage of their proceeds to conserve the habitat of the organism that inspired the product or process.

In the end, biomimicry has the potential to change our worldview as well as our designs. The process of quieting human cleverness, listening, and then echoing what we hear is a process that deepens human respect for the rest of the natural world. The real legacy of biomimicry will be more than products and processes that help us fit in here. It will be gratitude and, from this, an ardent desire to protect the genius that surrounds us.

Further Resources

Anastas, P. A., & Warner, J. C. (1998). *Green chemistry: Theory and practice.* New York: Oxford University Press.

Bar-Cohen, Y. (2005). *Biomimetics: Biologically-inspired technologies.* Boca Raton, FL: CRC/Taylor & Francis.

Barthlott, W., & Neinhuis, C. (1997). Purity of the sacred lotus, or escape from contamination in biological surfaces. *Planta, 202,* 1–8.

Benyus, J. (2002). *Biomimicry: Innovation inspired by nature.* New York: Morrow (reprint from 1997).

Bonser, R. (Unpublished data). Centre of Biomimetics, University of Reading. Available at http://www.rdg.ac.uk/Biomim/

Forbes, P. (2006). *The gecko's foot: Bio-inspiration: Engineering new materials from nature.* New York: Norton.

Neinhuis, C., & Barthlott, W. (1997). Characterization and distribution of water-repellent, self-cleaning plant surfaces. *Annals of Botany, 79,* 667–77.

Soule, J., & Piper, J. (1991). *Farming in nature's image: An ecological approach to agriculture.* (Foreword by W. Jackson). Washington, DC: Island Press.

Van der Ryn, S., & Cowan, S. (1996). *Ecological design.* Washington, DC: Island Press.

Vincent, J. F. V. (1990). *Structural biomaterials* (Rev. ed.). Princeton, NJ: Princeton University Press.

Vogel, S. (2003). *Comparative biomechanics: Life's physical world.* Princeton, NJ: Princeton University Press.

Janine M. Benyus

Biomimicry
*How Gecko Toes Stick**

Geckos can run up a wall or across a ceiling with ease because of their remarkable toes. But gecko toes aren't sticky in the usual way, like duct tape or Post-it notes. Instead, gecko toes bear a hierarchy of structures that act together as a smarter adhesive.

The pad of a gecko toe is crossed by ridges covered with hairlike stalks called *setae*, which branch into hundreds of tiny endings. Gecko toes stick to nearly every material under nearly any conditions (even underwater or in a vacuum), and neither stay dirty nor stick to one another. Geckos can attach and detach their adhesive toes in milliseconds while running on smooth vertical and inverted surfaces, a feat no conventional

*Adapted from an article that was originally published in *American Scientist*, Volume 94: 124–132, "How Gecko Toes Stick."

adhesive can match. And unlike sticky pressure sensitive adhesives, gecko toes don't degrade, foul or attach accidentally to the wrong spot. My colleagues and I have been studying these remarkable animals for over a decade.

Sticky Fingers

The ability of geckos to stick to surfaces has attracted scientific scrutiny since the time of Aristotle, but the microscopic setae on gecko toe pads were only documented in the 1870s. The underside of a gecko toe typically bears a series of ridges, or *scansors*, which are covered with uniform ranks of setae. By the early 1900s, scientists using light microscopes observed that the setae themselves had branches. It took the development of electron microscopy in the 1950s to reveal hundreds of split ends and flat tips called *spatulae* on each seta.

A single seta of the tokay gecko (*Gekko gecko*) is roughly 110 micrometers long and 4.2 micrometers wide. Each of a seta's branches ends in a thin, triangular spatula connected at its apex. The end is about 0.2 micrometer long and 0.2 micrometer wide.

Although the tokay gecko is the best studied (and one of the biggest) gecko species, more than a thousand species of geckos encompass a variety of sizes and shapes of spatulae, setae, scansors, and toes. Some geckos even have setae on their tails. Remarkably, similar structures have evolved independently in certain iguanian lizards (genus *Anolis*) and scincid lizards (genus *Prasinohaema*).

In the laboratory, a tokay's two front feet with a pad area of 227 square millimeters (smaller than a dime) were able to withstand 20.1 newtons (about 4.5 pounds) of force parallel to the surface, according to the work of Duncan J. Irschick at Tulane University and his colleagues. There are about 14,400 setae per square millimeter on the foot of a tokay gecko. However, in isolation, single setae proved to be much less—and much more—sticky than predicted, depending on test conditions. The fact that their stickiness can be so variable led me and my coworkers to conclude that control of attachment and detachment is mechanical rather than chemical.

Setae are not sticky by default—in fact, they are quite difficult to attach unless they are properly oriented and manipulated. We showed that a small touch to a surface ("preload"), followed by a 5 micrometer rearward drag was needed to attach a seta. Following preload and drag, we measured forces much greater than predicted from measurements of the whole animal: a gecko's worth of setae (6.5 million) attached at once could lift 133 kg!

The surprisingly large forces generated by single setae made us wonder how geckos manage to lift their feet so quickly—in just 15 milliseconds—with no measurable detachment forces. A few years ago, we observed that simply increasing the angle between the setal shaft and the substrate to 30 degrees causes detachment. As this angle increases, we think that increased stress at the trailing edge of the seta causes the bonds between seta and substrate to break. The seta then returns to an unloaded default state. Thus, gecko adhesive can be thought of as the first known programmable adhesive: Preload and drag steps turn on and modulate stickiness; increasing the shaft angle to 30 degrees turns off stickiness.

Although scientists have spent many years documenting the setal structures of geckos, finding out how they stick has been harder. Numerous hypotheses have been rejected, including suction, interlocking (like Velcro), static electricity, and glue. In 2002, we discovered that gecko setae do not bond chemically: there is no "gecko glue" molecule needed. Rather, gecko setae adhere by weak intermolecular forces called van der Waals forces, named after Dutchman Johannes Diderik van der Waals who won the

1910 Nobel Prize in physics. The force that carries his name occurs whenever two surfaces come close at the nanoscale. Van der Waals forces are not strongly dependent on the chemical nature of the materials. Gecko setae thus have the property of material independence: They can adhere strongly to a wide range of materials with little regard for surface chemistry. This was an exciting discovery because it implied that gecko-like adhesives could be engineered from a variety of materials.

Non-Stick Surface

Paradoxical as it may seem, there is growing evidence that gecko setae are themselves strongly anti-adhesive. Setae do not stick spontaneously to surfaces, but instead require a mechanical program for attachment. And unlike adhesive tapes, gecko setae do not self-adhere: Pushing the setal surfaces of a gecko's feet together does not cause them to stick. Furthermore, gecko setae do not seem to stay dirty. How is it that sticky gecko feet remain quite clean around everyday contaminants such as sand, dust, leaf litter, pollen, and plant waxes? Insects, which face similar challenges, must restore soiled adhesive pads to normal function by spending much of their time grooming. By contrast, geckos do not groom their feet. Although some plant and animal surfaces self-clean (with water droplets), no self-cleaning adhesive had ever been shown until we documented it in geckos in 2005. The key to this phenomenon seems to be adhesion energy, or molecular stickiness of the surfaces. We developed mathematical models of self cleaning that suggested that in order to shed debris, the adhesion energy of all spatulae adhering to a dirt particle must be equal to or less than the adhesion energy between the same particle and the surface.

The Gecko Muse

With such remarkable properties, it is unsurprising that materials scientists are trying to create artificial, gecko-like adhesives. Using a nanostructure to create an adhesive is a novel and bizarre concept. It is possible that had it not evolved, humans never would have invented it. For the booming nanotechnology industry, such products would be valuable for picking up, moving and aligning ultraminiature circuits, sensors or motors. For bigger applications, such as robots that could explore the wreckage of a fallen building or the surface of another planet, artificial gecko setae would endow the machines with unprecedented freedom of movement. Because a gecko-like nanostructure could be applied directly to the surface of a product, such adhesives could replace screws, glues, and interlocking tabs in manufactured goods. More whimsically, they might enable fumble-free football gloves or revolutionary rockclimbing aids. (This last idea is not new. Shivaji, a legendary Hindu ruler of 17th-century India, reportedly used adhesive lizards as grappling devices to scale a sheer cliff and mount a surprise attack on his enemies.)

Several groups of scientists have made good progress toward fabricating synthetic spatulae in the years since our team published the first such effort in 2002. However, by gecko standards, today's best synthetic setae are still primitive. Two materials, one by Andre K. Geim and colleagues at the University of Manchester, the other by Michael T. Northen and Kimberly L. Turner at UC Santa Barbara, have adhesion coefficients (a ratio of adhesive force to preload force) that are about half a percent and one percent, respectively, of real gecko setae. In late 2005, Ali Dhinojwala and others from the University of Akron and Rensselaer Polytechnic Institute published their description of a carbon-nanotube carpet that generated adhesive force even greater than that of gecko setae. However, the product only works at a nanometer scale, rather than the centimeter scale of real gecko toes.

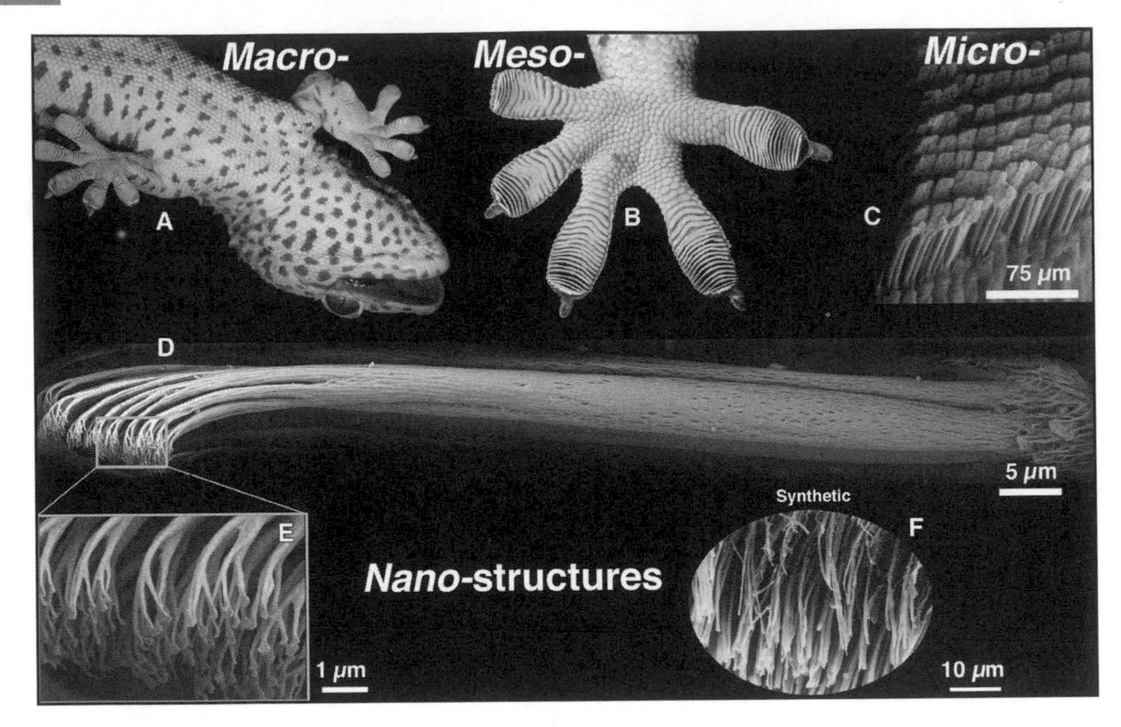

Gecko toe pads operate under perhaps the most severe conditions of any adhesive. The underside of a gecko toe is striped with ridges covered with rows of microscopic, hairlike stalks, as shown in this scanning electron micrograph. The stalks end in hundreds of tips, each just 0.2 micrometers wide, which make intimate contact with the surface. The functional properties of gecko feet are as extraordinary as their structure, enabling geckos to run up walls and across ceilings with seeming indifference to gravity. ©Andrew Syred/Science Photo Library.

Clearly, better designs will require deeper exploration of real gecko setae. And as technology and the science of gecko adhesion advance, it may even become possible to tune the design to create completely new properties.

Many questions remain for scientists who study mechanisms of gecko adhesion. What is the effect of surface roughness on friction and adhesion? How can scientists better model the hierarchical contributions of spatulae, setae, scansors, toes, and legs? How do spatulae and setae work in more than a thousand other gecko species (assuming they don't go extinct before scientists can study them)? What is the molecular structure of setae? Answers to these basic biological questions are key to the development of bio-inspired adhesives that may someday rival their natural counterparts. Then maybe we will be able to scamper across the ceiling too.

Further Resources

American Scientist Online. http://www.americanscientist.org/IssueTOC/issue/821

Autumn, K. 2006. Properties, principles, and parameters of the gecko adhesive system. In Biological Adhesives, eds. A. Smith and J. Callow, pp. 225–255. Berlin Heidelberg: Springer Verlag.

Autumn, K. and Hansen, W. 2006. Ultrahydrophobicity indicates a nonadhesive default state in gecko setae. Journal of Comparative Physiology A-Sensory Neural & Behavioral Physiology.

Autumn, K., Y. A. Liang, S. T. Hsieh, W. Zesch, W.-P. Chan, W. T. Kenny, R. Fearing and R. J. Full. 2000. Adhesive force of a single gecko foot-hair. *Nature* 405:681–685.

Autumn, K., M. Sitti, A. Peattie, W. Hansen, S. Sponberg, Y. A. Liang, T. Kenny, R. Fearing, J. Israelachvili and R. J. Full. 2002. Evidence for van der Waals adhesion in gecko setae. *Proceedings of the National Academy of Sciences of the U.S.A.* 99:12252–12256.

Autumn, K., S. T. Hsieh, D. M. Dudek, J. Chen, C. Chitaphan and R. J. Full. 2006. Dynamics of geckos running vertically. *Journal of Experimental Biology* 209: 260–272.

Geim, A. K., S. V. Dubonos, I. V. Grigorieva, K. S. Novoselov, and A. A. Zhukov. 2003. Microfabricated adhesive mimicking gecko foot-hair. *Nature Materials* 2:461–463.

Irschick, D. J., C. C. Austin, K. Petren, R. Fisher, J. B. Losos, and O. Ellers. 1996. A comparative analysis of clinging ability among padbearing lizards. *Biological Journal of the Linnean Society* 59:21–35.

Northen, M. T., and K. L. Turner. 2005. A batch of fabricated dry adhesive. *Nanotechnology* 16:1159–1166.

Kellar Autumn

Biosynergy
Biophilia

Coined by distinguished Harvard University professor and Pulitzer Prize–winning author Edward O. Wilson, *biophilia* refers to the hypothesis that humans have an "innate [i.e., hereditary] tendency to focus upon life and life-like forms and in some instances to affiliate with them emotionally" (Wilson, 1984, p. 1). The biophilia hypothesis argues that humans evolved to respond more strongly to natural stimuli—especially animals—than to artificial ones such as cars and computers. Biophilia implies that we find pleasure in other animals; researchers sometimes use the term biophobia to talk about people reacting more strongly in negative ways to animals than to artificial things. For example, humans (and most other apes) have a proclivity to fear snakes but not frayed electrical wires.

Biophilia is not thought to arise from a single gene or even to represent a single behavior or tendency. Instead, biophilia is thought to represent a collection of inborn, mental, and behavioral tendencies. The components that make up biophilia do share one common feature, however: they are all hypothesized to be products of our evolution in intimate relationship with other life, of the countless millennia in which other animals and nature governed our survival, for good or ill. As psychologist Martin Seligman explained, it is as if evolution has prepared us (and other animals) to learn some things very quickly. For example, among the most common phobias are fears of snakes and spiders, two animals whose smallish size belies their danger. In contrast, evolution has not prepared humans to learn as quickly and enduringly about artificial things such as guns and wall outlets. Not only do we learn about living things more easily, but we may also pay more attention to them, as if our minds are designed to respond to life. Why? Probably because ancient people who were naturally more wary of these two animals must have fared better—and thus lived to have more children—than people who were not. At the other end of the spectrum, humans are innately inclined to nurture baby-like things, including baby animals. In between these extremes, there are myriad ways that we respond to animals and nature, presumably because our survival once depended on them.

Human cultures, of course, are shaped by the people in them. Cultures also change in response to the environment. Wilson and professor Charles Lumsden contend that cultures and the people in them influence each other's evolution. Therefore, we should not be surprised to find evidence of biophilia in cultural elements. Returning to the example of snakes, art historian and biologist Balaji Mundkhur points to evidence of a lasting fascination with them that may have led to their prevalence in myths, religions, and idiomatic expressions. To name a few examples, snakes play profound roles in the first extant written work, *Gilgamesh,* in many divine pantheons, and in the Semitic creation story. Snakes' emotional power is strong; colonists rallied themselves under the symbol of a rattlesnake during the American Revolution, notably represented by the Gadsden flag with its "Don't tread on me" motto. From mascots to gods, animals embody many of the ways we envision ourselves.

If humans are designed to respond emotionally, cognitively, and even culturally to animals and nature, then it may be that we are healthier and more psychologically complete when we are around animals and nature—that when we are closer to nature, we are closer to ourselves. Indeed, there is considerable scientific evidence that people from very different cultures all prefer natural scenes (grassy, partially wooded scenes and water, in particular) over urban or distressed scenes such as clear-cut forests. Professor Roger Ulrich suggests that simply spending time in natural settings can help people recover from stress.

In addition, there is little doubt that animals and nature can improve our mental and physical health. From lowering stress and blood pressure to surviving heart attacks, contact with affectionate animals can help us heal. Hospital patients with views of trees recover more quickly than patients with views of brick walls. Many police cruiser trauma kits include teddy bears. This may represent further support for the biophilia hypothesis, but more research is needed to say so with any confidence. Animals do enrich our lives, but only more research can help us decide why.

Although there is reason to support at least some aspects of biophilia, it remains a hypothesis. Biophilia has received strong popular support, but there is not yet sufficient proof that it is correct or incorrect. Most of the research used to support or refute it is borrowed from other areas. In addition, given that there may not be one, simple source for biophilia, research may well support some parts better than others.

Whatever the underlying reasons for its existence, our emotional connection with animals and nature reveals yet another way in which we impoverish our very selves with environmental destruction and wholesale extinctions. We are losing more than "just" robust ecosystems, undiscovered wonders, unfound medical cures, and so on; we are also losing a great source of psychological health, beauty, and inspiration. It may be that we sicken psychologically as the earth weakens environmentally. Industrialized humans do not rely as directly on animals for our survival as our ancestors did, but the innate affiliation with nature that we have inherited may yet move us to prevent the environmental damage that threatens our survival now.

Of course, there is variability in inherited traits. We have all inherited the need to socialize with other people (presumably because humans who worked in groups out-competed those who did not), but some of us are much more social than others. Likewise, although biophilia is thought to be present in all of us, it is expected to vary between people. An important question, then, is the extent to which biophilia can be nurtured. And if it can be nurtured, will it be enough to help save and protect the animals and habitats we rely on for our well-being? Maybe, as Michael Soulé points out, "the question is not whether humans love the earth and its biota, but whether we love them enough."

Further Resources

Beck, A., & Rowan, A. (1994). The health benefits of human-animal interactions. *Anthrozoös, 7,* 85–89.

Frumkin, H. (2001). Beyond toxicity: Human health and the natural environment. *American Journal of Preventive Medicine, 20*(3), 234–40.

Kellert, S. R., & Wilson, E. O. (Eds.). (1993). *The biophilia hypothesis.* Washington, DC: Island Press.

Robinson, I. (1995). *The Waltham book of human-animal interactions: Benefits and responsibilities.* Oxford: Pergamon.

Wilson, E. O. (1984). *Biophilia: The human bond with other species.* Cambridge: Harvard University Press.

William Ellery Samuels, Debbie Coultis, and Lieve Meers

Biosynergy
Biosynergy: A Personal Essay

Biosynergy is the fundamental and overarching process that sustains life on Earth. Its seed and its spark have been born and imbedded in every living cell, organism, and ecosystem since the beginning of time. Biosynergy is the inner force that compels each and every individual being to collaborate with others for the greater good.

Biosynergy is so ubiquitous as to be invisible, so vastly important as to feel unspeakable. So rife with promise and hope as to seem impossible in this era of human hegemony and global chaos. It is the impetus for persons to fall in love, for dogs and cats to nuzzle and mark their human companions, and for field researchers and explorers to rescue orphan apes in the rain forest and for the apes to let them. Biosynergy is the ultimate interface of life forms that scientists long to observe. Biosynergy is the sunrise awakening of wilderness that naturalists long to experience. Biosynergy is the call of the wild that adventurers long to follow. This essay aims to help you make biosynergy visible, effable, and possible. To do so, you will need to set aside resistance, defiance, and pessimism. Biosynergy, the synergy of life, has tremendous implications for the future of all life on earth. Exploring and understanding biosynergy will foster an inclusive optimism that many believe we direly need in difficult and troubled times.

It was in the Gunung Leuser reserve in the far north of Sumatra that I first experienced the profound process that I labeled "biosynergy." The morning air was crisp and cool at the Ketembe research site, and we had struck out early in search of orangutan. Two hours spent along mud trails laced with strands of archetypal light and inscrutable green mists that drift in layers through the ever-rising canopy conformed me, body and mind, to a vitally mystic reality. We found three of our great ape cousins that day—a mother with babe in arms and a young daughter stared down at us from the branches and offered us figs, an orangutan welcome. It was an absolutely profound interspecies event. Still, it was not the exhilarating encounter with apes that taught me the ultimate lesson of biosynergy. It was the leeches.

A day earlier, I had stopped at a bend in a narrow trail, studied the spaghetti lines that crisscrossed my soiled and crumpled map, and was about to tilt a water bottle toward my lips when I saw the wormy creatures. Twenty, at least, were wiggling out of the leaf litter and heading my way. The prior evening, I had extracted thirteen of these bloodsuckers from my toes and ankles upon return from a trek in the forest. Now a company of annelids was streaking toward the scuffle and heat of my feet, gnashing their teeth in the excitement of a probable feast. I deftly hopped over the encroaching circle and stood fast. In an instant they had whirled around and were again heading my way. I leapt back over them again. They turned again. I bent down, offered my forefinger to the leader of the pack, and watched him crawl on board and attach his jaws to my skin. In a minute or two, he had doubled in size, filled with my blood. I removed him and looked closely at the small red dot on my finger. In thanks for my liquid offering, he had injected his anticoagulant.

"Tit for tat—you feed me; I clean up after," he seemed to be saying.

Biosynergy. The realization spiraled through me. I was part of an ecosystem that was in a state of synergy, with all life forms engaged in mutual service. More than service, in mutual attraction, fascination, interdependence, harmony. I had entered to explore, and the biosynergy of the place had transformed me from observer to participant, from interloper to inhabitant, from utilizer to synergizer.

In the twenty-five years since that moment of immersion and discovery, I have named, framed, sorted, and reported the experience of biosynergy on six continents. No lesson, no essay, and no book can reveal all there is to know about the relationships among life forms. We explorers move through our worlds collecting momentary specimens of experience, assembling fleeting reactions and ideas, concocting personal interpretations and theories. At the end of the day, we sit around campfires and coffee tables to compile it all into short stories. At the end of each great adventure, we sit at desks and stand in lecture halls patching our stories together into epic tales that strive to illuminate a tiny particle of life's eternal truth. How audacious is the person who would expose the whole truth in a few thousand words! Yet we must try.

It is the duty of the scientist to document the broad context in which important phenomena are studied. It is also the scientist's obligation to conform measurement procedures to statistical fundamentals and to examine and factor out the influence of observer and experimenter bias on data collection, analysis, and interpretation. When these vital precepts are ignored, doctrinaire theories about the behavior of life on earth are "proven" by the biased collection of confirming data in small guarded environments. Contrary observations are explained away with obtuse and arbitrary arguments. Still, no matter how open-minded we scientists try to be, we must admit that we cannot achieve objective measurement of human-animal relationships. We filter events and situations through human senses, infuse them with human emotions, analyze them with human minds, judge them with human values, and envisage them with human intuitions. And within the category "human," our objectivity is further skewed by culture, education, and personal history. At best, we can amalgamate the collective subjective perceptions of individual humans who relate to nonhuman animals. Knowing that we cannot speak for the other animals, we still speak about them in our own diverse human ways.

My research into this preeminent natural phenomenon has focused on indicators of biosynergy that emerge when humans and other animals break through psychosocial barriers to create synergistic relationships. My encounters with leeches, orangutans, and the myriad interlaced life forms in Sumatran rain forests led to extensive exploration of other wild places. Condors and crickets, gorillas and guinea fowl, rhinos and rattlesnakes, and leatherback turtles and leopards all became part of my life story. Thousands of animals have helped me to experience and define the phenomenon of biosynergy.

In 1994 I began to document the histories of persons who had become devoted to wildlife and wilderness. This enabled me to determine the essential characteristics of profound interspecies events (PIEs) in which previously disconnected humans and non-humans initiate interspecies synergy. My analysis revealed a remarkable similarity between the sequence of elements that compose the most complete PIE and the reports of "near death experiences" (NDE) investigated by Elizabeth Kubler-Ross. Like the NDE, these "near life experiences" have extraordinary transforming effects on the humans who undergo them. The elements of the most complete PIE are as follows:

1. Initial insurmountable difficulty for the human to gain access to the animal.
2. Perseverance—patience and faith—by the human in pursuit of a connection.
3. Reversal of mistrust by the animal with regard to the pursuing human.
4. An arresting first contact, followed by successively closer and longer interaction.
5. Intervening forces that separate the pair, leaving one or both highly endangered.
6. Heroic acts by one or both members of the pair to reach, protect, or save the other.
7. Profound shifts in perception of self, other, or species by one or both members of the pair.

Experiencing any one of these elements can evoke empathy for other species in a person who is emotionally receptive. Experiencing all the elements together can change a person's view of life from ego- and human-centered to eco- and life-centered. Unlike the NDE, which is a seemingly long event compressed into seconds of real time, the chain of elements that compose the ultimate PIE can take hours, days, or even months to unfold. Yet when conditions are right for the humans and animals involved, the profound experience of epiphany often feels instantaneous.

The fastest route back to nature occurs when animals we consider dangerous, distant, or uninterested befriend us. Humane treatment by wild creatures that seek a friendly encounter with a human produces the most profound interspecies events—in my research I call this the "SAFE" scenario. My first interspecies epiphany occurred at the UCLA Brain Research Institute over forty years ago when I had to capture a macaque monkey that had escaped from his cage and was ransacking my lab. This had never happened to me before. Handling a scared monkey in cage or experimental chamber was one thing: catching an escapee was another. I entered the room and peered through the haze and clutter. Snicky, a large young male, starred down at me from atop a bookcase, hair on end, eyes wide, teeth bared. Half terrified and thinking him hostile, unsure what to do, I mechanically smacked my lips at him, our usual morning greeting. He shuddered through a kind of tension meltdown and suddenly jumped from the shelf, leapt into my arms, and held on like a worried child. In the distance, he had seemed so huge, imposing, wild. Now in my arms he was small, vulnerable, dependent. I sat on the linoleum floor holding this animal in my arms for the first time in our years together. He groomed my skin while I cleaned the scab that edged his cement skullcap, checked his implanted electrodes to be sure they had not loosened, and examined his dilated eyes. I remember thinking, "After all I've done to him, he wants my friendship more than his freedom." I cried. This profound experience turned me away from medical research forever. I vowed never again to experiment on friends.

Since that time I have reviewed reports and conducted interviews with hundreds of wildlife professionals and animal aficionados whose worldviews were transformed by humane emotion-laden interspecies encounters. Squid and shark, dolphin and whale, tortoise and tarantula, tiger, wolf, and bear, and all the nonhuman primates from the tiny bush baby to the great gorilla have befriended people and changed them. Most noteworthy is the alteration of these people's worldviews. We are propelled by our epiphanies into a world in which other animals are experienced as kindred spirits, a world where synergy among all living beings is the natural way of life on the planet.

I also discovered that early interspecies epiphanies divide into distinct types that correlate with the types of professional work that people eventually undertake. The predominant PIEs reported by people who are not involved in animal work are the SAFE scenarios similar to the one described previously. For laypeople, these events involve domesticated animals more often than wild animals. Persons who work as animal caregivers, zookeepers, animal trainers, and veterinarians also tend to report having profound friendly encounters with other animals when they were children and young adults. Zookeepers' early PIEs often involved wild animals. Our research in schools and communities in west and central Africa has shown that telling the general public stories about PIEs in which wild animals seek friendly encounters with humans have the most payoff in the development of conservation values. In general, people are more likely to argue for the protection of endangered animals because the animals are like people than because the animals are rare and appear to be going extinct.

Wildlife scientists' appetites for discovery appear to have been stimulated by profoundly intriguing interactions with other animals in which they felt that they had been shown a secret clue to understanding animals and nature. This is the kind of epiphany that scientific explorers seek, in which animals *exhibit natural reactions that illuminate crucial hypotheses* (the "ENRICH" scenario). Although forest people and game hunters had seen tool use among wild apes for decades before Jane Goodall went to Gombe Stream, her systematic observation of chimpanzees fishing for termites caused a revolution in thinking among students of evolution and marked a major turning point in her life. Jane's life experience has been filled with every kind of interspecies epiphany. For many years, she thrived on the discovery of animal secrets and the testing of human hypotheses about the nature of our ape cousins. By the time I met her, she had been befriended by countless wild and orphan chimpanzees and had turned from great ape research to the care and protection of all animals and all of nature.

The last time I spoke with Jane Goodall, she said that her most profound moments now came when she walked in the rain forest, engulfed by the majesty of that ever-mysterious and ever-changing world. Thus, she illuminated the third type of natural epiphany reported by laypersons and animal professionals alike—an awe-inspiring naturalistic event in which the person is *shown an extraordinary element of nature* (the "SEEN" scenario). A tremendous amount of writing has been done about this type of experience. After many long discussions about interspecies epiphanies, orangutan researcher and protector Birute Galdikas wrote, "Looking into the calm, unblinking eyes of an orangutan we see, as through a series of mirrors, not only the image of our own creation but also a reflection of our own souls and an Eden that once was ours. And on occasion, fleetingly, just for a nanosecond, but with an intensity that is shocking in its profoundness, we recognize that there is no separation between ourselves and nature." The life stories of Birute Galdikas, Jane Goodall, and gorilla expert Dian Fossey are brilliant beacons to light our path toward kinship with every living being in this world.

In his studies of human attitudes toward nature, Stephen Kellert describes a utilitarian value structure that stresses the physical benefits of nature for human sustenance, protection, and security. We have found extensive testimony to the psychological benefits. Naturalists, scientists, animal caregivers, adventurers, and conservationists—as well as people in countless fields not related to animals—have confirmed the healing effects of interspecies contact and nature immersion. Poet Wendell Berry expresses most poignantly our need for contact with nature, in a poem about nature, "The Peace of Wild Things." In Berry's poem, we glimpse the wonder of natural epiphanies, and we are reminded of their necessity. If we despair for the world and are fearful of our human future, it is very important to realize that synergy with other animals is more than a palliative. Biosynergy is a requisite for all life on earth to survive and to thrive.

Common sense throughout the ages is being validated over and over by scientific studies that demonstrate the personal and social dysfunction of individual humans and societies that are detached from and destructive of the natural life with which we are genetically, ecologically, and spiritually entwined. In the United States, fear of the wild and of strangers has escalated to the point where even children in rural communities spend nearly all their time indoors. In suburbs and cities, practically nobody goes into their backyards; most young people avoid streets and open space. Children spend nearly forty-five hours weekly engaged with electronic media and less than thirty minutes a day with their parents and guardians. The result is a deficit in nature and nurture that has adverse effects on the well-being of us all.

Evidence is mounting that the greater part of human conflict emanates from our alienation from the rest of nature. Edward O. Wilson postulated that we suffer from inability to satisfy our need for biophilia, the innately emotional affiliation of human beings to other living organisms. It has become apparent that biophilia is not enough. For our lives to be whole and healthy, affiliations with other species must be reciprocally supportive. To survive and thrive, living beings need synergistic relationships with other organisms in their ecosystem, and ultimately in the biosphere. Like the leech and me in the Sumatran rain forest, we must give to one another in return for what we take. We have come a long way from that realization to this definition:

bi·o·syn·er·gy *n.*

1. The interaction of two or more biological agents or forces so that their combined effect is greater than the sum of their individual effects. **2.** Cooperative interaction among species, especially among the individuals and groups in an ecosystem, that creates an enhanced combined effect. **3.** The theory that organisms cooperate with passage of time in the same ecosystem, mainly as a result of reciprocal altruism, so that biosocial structure and dynamics change to assure the vigor of all life forms. [Greek, from *bios*, life. From Greek *sunergia*, cooperation, from *sunergos*, working together.] (Rose 2004)

The hypothesis of "reciprocal altruism" is key to this treatise. It is supported by volumes of observation, analysis, and theory. In recent decades, scientists have reinforced the importance of social and genetic co-evolution in the development of earth life. Animal and plant societies thrive through processes of adaptive conformity and cooperation, attenuated by suppression or rejection of aberrant individuals and rebellious groups. The most prolific life forms on the planet are those that undertake intricate social collaboration that enables adjustment to environmental and ecological change. Ants and humans are among the most celebrated and studied of communal species. Both foster the harmonic growth of other organisms to create soothing landscapes, safe and healthy habitats, verdant gardens, and fertile farms. Both honor the greater good above the idiosyncratic demands of selfish individuals. As with all animal groups, ants and humans do best when practicing biosynergy with others.

A few philosophers cling to the belief that individual organisms and species are driven entirely by a need to win dominance over others in a vast battle for survival of the fittest. This competitive view of life emerges from fear-induced selective perception that causes people to fixate on aberrations and to ignore commonalities. It manifests in stolen car chases that hijack TV news, pathology and crime studies that plunder scientific research coffers, and paranoia about outsiders that drives global politics. It culminates in the unnatural selection of corporate and national leaders who trade on terror and fulfill the dog-eat-dog prophecy by transforming democracies into oligarchies in the pursuit of private wealth and power. Insecure people are entranced by the pathos and excitement of superheroes engaged in worst-case scenarios. Terrified people are unable to look beyond the highly advertised infernos to see that the vast majority of life on earth is not burning. It is this rampant paranoia, with its concomitant destructiveness, that renders the illumination of biosynergy so very important.

But just as media and its messages have plunged our worldview into a deeply pessimistic outlook on life, so can new ideas and discoveries return us to a brighter reality. Richard Rorty makes an elegant case for our human potential to change:

> For beyond the vocabularies useful for prediction and control—the vocabulary of natural science—there are the vocabularies of our moral and our political life and of the arts, of all those human activities which are . . . aimed at . . . giving us self images that are worthy of our species. Such images are not true to the nature of species or false to it, for what is really distinctive about us is that we can rise above questions of truth or falsity. We are the poetic species, the one which can change itself by changing its behavior—and especially its linguistic behavior, the words it uses.

Rorty challenges humankind to redefine itself. By shifting paradigms from biodominance to biosynergy, we are challenging humankind to redefine the fundamental and ultimate vision of life on earth.

Albert Einstein wrote that the enlightened person "looks upon individual existence as a sort of prison and wants . . . to feel the sublimity and marvelous order which reveal themselves both in nature and in the world of thought." Natural scientists, adventurers, wild animal caretakers, and conservationists have been blessed with opportunities to feel that sublimity and marvelous order. They have risked engagement with wildlife and wilderness and have initiated the reunion of humanity and nature. George Schaller speaks from deep experience when he says that "the recent decades have been a turning point, indeed a revolution in our relationship with animals. Humans have begun to overcome cross-species barriers, achieving intimacy with hump-backed whales, chimpanzees, lions, mountain sheep, wolves . . . the gorilla, of course, is more than an animal. These apes are part of human heritage. Our kin." In this essay and in this encyclopedia, we strive to bring that kinship experience to the world of thought.

These worlds where we evolve—the primordial ooze, the sea, the beachhead, the swamp and the savanna, the mountain and valley, the forest and desert—are home to all life. From birth to death, all that grows and moves on this planet reveals itself in glorious living murals, ballets, and symphonies. All visions and voices contribute. All eyes and ears are tuned to the magic interplay of nature. The biota of earth is interwoven in an ever-changing biosynergy, like threads on a multidimensional loom, the tapestry of nirvana. It is our calling as humans to illuminate the images of life, enrich the dance of reunion, and sing in the voice of all beings. And it is this voice—the voice that dares to sing all songs of all life—that guides us toward the fulfillment of our natural destiny.

Further Resources

Axelrod, R. (1984). *The evolution of cooperation*. New York: Basic Books.

Einstein, A. (1979). *The world as I see it*. New York: Citadel Press, pp. 7, 26.

Galdikas, B. M. F. (1995). *Reflections of Eden: My years with the orangutans of Borneo*. New York: Little-Brown, p. 403.

Kellert, S. R. (1996). *The value of life: Biological diversity and human society*. Washington, DC: Island Press.

Kellert, S. R., & Wilson, E. O. (1993). *The biophilia hypothesis*. Washington, DC: Island Press.

Louv, R. (2005). *Last child in the woods: Saving our children from nature deficit disorder*. Chapel Hill, NC: Algonquin Books.

Rorty, R. (1982). Mind as ineffable. In R. Q. Elvee (Ed.), *Mind in nature* (p. 60). New York: Harper & Row.

Rose, A. L. (1996). Orangutan, science, and collective reality. In R. Nadler, B. Galdikas, N. Rosen, & L. Sheeran (Eds.), *Orangutan—The neglected ape* (pp. 29–40). New York: Plenum Press.

———. (2002). Conservation must pursue human-nature biosynergy in the era of social chaos and bushmeat commerce. In A. Fuentes & L. D. Wolfe (Eds.), *Conservation implications of human and nonhuman primate interconnections* (pp. 158–184). Cambridge: Cambridge University Press.

———. (2004). *Biosynergy and the future of humankind*. San Diego: International Leadership Forum, Western Behavioral Sciences Institute.

———. (2006). On tortoises, monkeys, and men. In K. Solisti & M. Tobias (Eds.), *Kinship with the animals: Expanded edition* (pp. 14–31). San Francisco: Council Oak Books.

Rose, A. L., & Auw, A. (1974). *Growing up human*. New York: Harper & Row.

Rose, A. L., et al. (2003). *Consuming nature: A photo essay on African rainforest exploitation*. (Photography by Karl Ammann). Los Angeles: Altisima Press.

Schaller, G. B. (1995, October). Gentle gorillas, turbulent times. *National Geographic*, 65–83.

Trivers, R. L. (1971). The evolution of reciprocal altruism. *Quarterly Review of Biology*, *46*, 35–57.

Wilson, E. O. (1984). *Biophilia*. Cambridge, MA: Harvard University Press.

Wright, R. (1994). *The moral animal: Why we are the way we are: The new science of evolutionary psychology*. New York: Vintage Books.

Anthony L. Rose

Bonding
Agility Training

Power, Play, and Invention

Playing agility with my Australian Shepherd Cayenne helps me understand a controversial, modern relationship between people and dogs—training to a high standard of performance for a competitive sport. Training together is an historically located, multispecies, subject-shaping encounter in a contact zone fraught with power, knowledge and technique, moral questions, and the chance for joint, cross-species invention that is simultaneously work and play. How can dogs and people in this kind of relationship be means and ends for each other in ways that call for reshaping our ideas about and practices with companion animals?

Introducing the notion of "anthropo-zoo-genetic practice," the Belgian philosopher and psychologist Vinciane Despret (2004) studies how animals and people become available to each other, become attuned to each other, such that both parties become more interesting to each other, more open to surprises, smarter, more inventive. The question between animals and humans becomes, "Who are you?" and so, "Who are we?" So, how *do* dogs and people learn to pay attention to each other in a way that changes who and what they become together?

For the sport of agility, picture a grassy field about 100 square feet. Fill it with fifteen to twenty obstacles arranged in patterns according to a judge's plan. The sequence of obstacles and difficulty of patterns depend on the level of play, from novice to masters. Obstacles include various kinds of jumps and tunnels; weave poles, consisting of six to twelve poles in line through which the dog slaloms; pause tables; and contact obstacles called teeter totters; which include 5.5- to 6.5-foot-high A-frames and elevated dog walks. These last obstacles are called contact obstacles because the dog must put at least a toenail in a painted zone at the up and down ends of the obstacle. Leaping over the contact zone earns a "failure to perform" penalty. Dogs jump at a height determined by their height at their withers.

Human handlers, but not the dogs, walk through the course for about fifteen minutes before the dog and human run it. The human is responsible for knowing the sequence of obstacles and for figuring out a plan for human and dog to move quickly, accurately, and smoothly. The dog navigates the obstacles, but the human has to be in the right position at the right time to give good information. Advanced courses are full of trap obstacles to tempt the untimely or the misinformed; novice runs test fundamentals for getting through a course accurately and safely. Both humans and dogs on a well-trained team know their jobs, but most errors are caused by inept human handling. The errors might be bad timing, overhandling, inattention, ambiguous cues, bad positioning, failure to understand how the course looks to the dog, or failure to train basics. Qualifying runs in the higher levels require perfect scores within a demanding time limit. Teams are ranked by accuracy and speed.

Agility began in 1978 in the United Kingdom when a trainer of working trial dogs, Peter Meanwell, designed a dog-jumping event to entertain spectators waiting for the main action at the Crufts dog show. In 1979 agility returned to Crufts as a regular competitive event. Agility spread from the United Kingdom across the world. The United States Dog Agility Association was founded in 1986. In 2000 the International Federation of Cynological Sports (IFCS) was founded on the initiative of Russia and Ukraine to unite dog sport organizations and hold international competitions. The first IFCS world championship was held in 2002. The growth in the sport has been explosive, with thousands of competitors in many organizations, all with somewhat different rules and games.

Workshops, training camps, and seminars abound. On any weekend year-round, a hotspot of agility such as California will have several agility trials, each with 200 to 300 teams. Most dog-human teams train formally once a week and informally constantly. The year I kept count, I spent $4000 on everything it took to train, travel, and compete; that is considerably less than many humans spend on the sport. In the United States, white women about forty to sixty-five years old dominate the sport numerically, but people of various races, genders, economic classes, and ages play, from preteens to folks in their late seventies.

Many breeds and mixed-ancestry dogs compete, but the most competitive dogs in their height classes tend to be Border collies, Australian shepherds, Shelties, and Parson Jack Russell Terriers. High-drive, focused, athletic dogs and high-drive, calm, athletic people tend to excel. But agility is a sport of amateurs in which most teams can have a great time and earn qualifying runs and titles if they work and play together with serious intent, recognition that the dogs' needs come first, a sense of humor, and a willingness to make mistakes interesting.

Positive training methods, offspring of behaviorist operant conditioning, are the dominant approaches in agility. Having begun her training career with marine mammals in 1963 at Hawaii's Sea Life Park, Karen Pryor is highly skilled in teaching and explaining positive methods to the amateur and professional dog-training communities, as well as many other human-animal communities (Pryor 1999; for a successful competitor's approach, see Garrett 2005).

Positive methods work by marking desired actions called "behaviors" and delivering an appropriate reward to the behaving organism in time to make a difference—positive reinforcement. Reinforcement is anything occurring in conjunction with an act that tends to change that act's probability. Timing is all; even three seconds after the interesting "behavior" is way too late to get or give good information in agility training. Not just out there in the world waiting for discovery, a "behavior" is an inventive construction put together by people, organisms, and apparatus. Out of the flow of bodies moving in time, bits are carved out and solicited to become more or less frequent as part of building other patterns of motion through time. A behavior is a natural-technical entity that traveled from the psychology lab to the agility training session.

Restraint, coercion, and punishment—such as ear pinching—are actively discouraged in agility training. Strong negative words like "no!" are kept for dangerous situations and emergencies. Use of strong negative reinforcers and punishments by novice trainers like me is foolish as well as unnecessary. A dog will visibly shut down in the face of a tense or negative human and hesitate to offer anything interesting with which to build great runs. Positive reinforcement, properly done, sets off a cascade of happy anticipation and inventive spontaneous offerings for testing how interesting the world can be. Positive reinforcement improperly done reduces the stock of liver cookies, chew toys, and popular confidence in behavioral science.

Human beings must acknowledge that one's partner is a member of another species, with his or her own exacting species interests and individual quirks, and not a furry child, a character in *Call of the Wild,* or an extension of one's intentions or fantasies. People fail this recognition test depressingly often. Training together is extremely prosaic; that is why training with a member of another biological species is so interesting, difficult, full of situated difference, and moving. My notes repeatedly record agility people's remarks that they are learning about themselves and their human and dog companions in new ways. Playing a competitive cross-species sport provokes strong, unexpected emotions and preconception-breaking thinking about power, status, failure, skill, achievement, shame, risk, injury, control, companionship, body, memory, and joy. A human being must learn something about one's partner, oneself, and the world at the end of each training day that she or he did not know at the beginning. The dog, in turn, becomes shockingly good at learning to learn, thereby fulfilling the highest obligation of a good scientist.

The two-foot-long yellow contact zone painted on the up and down ends of A-frames was a site of pedagogical trouble for Cayenne and me. Our problem was simple: we were not communicating; we did not yet *have* a contact zone entangling each other. The result was that she regularly leapt over the down contact, not touching the yellow area with so much as a toepad before racing to the next obstacle, much less holding her two rear paws on the zone and two front paws on the ground until I gave the release cue ("all right"). I could not figure out what she did not understand; she could not figure out what my ambiguous cues and ever-changing criteria of performance meant.

Then I remembered lessons from other kinds of contact zones. Colonial theorist Mary Pratt adapted the term "contact zone" from linguistics, where "'contact language' refers to improvised languages that develop among speakers of different native languages who need to communicate with each other consistently . . . A 'contact' perspective emphasizes how subjects are constituted in and by their relations to each other . . . It treats the relations . . . in terms of co-presence, interaction, interlocking understandings and practices, often within radically asymmetrical relations of power" (Pratt 1992, pp. 6–7). Cayenne and I definitely have different native tongues, and much as I reject the analogy of colonization for human-dog relations, I know how much control of Cayenne's life and death I hold in my inept hands.

I also remembered those contact zones called ecotones, where biological species assemblages come together outside their comfort zones. These are the richest places to look for ecological, evolutionary, and historical diversity. Then I turned to the phenomenon studied in developmental biology called reciprocal induction, through which cells and tissues of an embryo mutually shape each other through cascades of communications. Contact zones are where the action is.

Complex relations of authority pertain in the reciprocal inductions of cross-species training. Agility is a human-designed sport, not spontaneous play. I have good evidence for judging that Cayenne loves to do agility. She is a focused working dog; her whole mind and body come alive when she gets access to her scene of work. However, I would be a liar to claim that agility is a utopia of equality and spontaneous nature. The courses and rules are designed by human beings, not dogs. The human decides for the dog what an acceptable criterion of performance will be. But there is a hitch: the human *must* respond to the authority of the dog's actual performance. The real dog—not the fantasy projection of self—is mundanely present. Fixed by the specter of yellow paint, the human must learn to ask fundamental questions: Who are you, and so who are we? Here we are, and so what are we to become?

Casualties of taking this question seriously were some of my favorite stories about freedom and nature. Fixing mistakes on the A-frame forced me to confront the pedagogies of training, including their relation to narratives and practices of freedom and authority. (See *Contact Training, Clean Run Special Issue* [Chicopee, MA, Clean Run Productions, November 2004].) Some radical animal people are critical of any human training "of" (I insist "with" is possible) another critter. They regard what I see as polite manners and beautiful skill acquired by the dogs I know best to be evidence of human control causing the degradation of animals. Wolves, say the critics of trained animals, are more noble (natural) than dogs precisely because they are more indifferent to the doings of people; to bring animals into close human interaction infringes their freedom. From this point of view, training is nonnatural domination made palatable by liver cookies.

Behaviorists are notoriously cavalier about what constitutes natural (biologically meaningful) behavior. If the probability of an action can be changed, no matter how meaningless the bit of action is to the organism, then that action is fodder for the technologies of operant conditioning. Yet the coming into being of something unexpected, something new and free, something outside the rules of function and calculation, *is* what training with each other is about. Training requires calculation, method, discipline, and science; but training is for opening up what is not known to be possible, but might be, for all the intra-acting partners.

I used to look down on behaviorism as a pale science at best, hardly biology at all, and really an ideological, determinist discourse. Then, Cayenne and I needed what skilled behaviorists could teach us. I became subject to a knowledge practice I had despised. I had to get it that behaviorism was not my caricature of a mechanistic pseudo-science fueled by niche-marketed food treats, but a flawed, historically situated, and fruitful approach to questions in the fleshy world. I needed not only behaviorism, but also ethology and the more recent cognitive sciences.

Preoccupied with the baleful effects for dogs of the *denial* of human control and power in training relationships, I have under-stressed another aspect of the human's obligation to respond to the authority of the dog's actual performance. A skilled human competitor in agility, not to mention a decent life companion, must learn to recognize when *trust* is what the human owes the dog. Dogs recognize when the human being has earned trust; human beings are not as good at reciprocal trust. I lose many qualifying scores for Cayenne and me because, in the sport's idiom, I "overhandle" her performance. Because I am not confident, I do not see that *she* has mastered difficult patterns. I do not need to be as fast as she is (good thing!); I merely need to be as honest.

Trainers cannot forbid themselves the judgment that they can communicate meaningfully with their partners. The philosophic conceit that we have only representations and therefore no access to what animals think and feel is wrong. Human beings do—or can—know more than we used to know; and the right to make that judgment is rooted in historical, flawed, generative cross-species practices, including sciences and sports. To claim to be unable to communicate with and to know each other, however imperfectly, is a denial of mortal entanglements for which we are responsible and in which we respond. Technique, calculation, method—all are indispensable and exacting. But they are not response, which is irreducible to calculation. Response is understanding that subject-making connection is real. Response is face to face in the contact zone of an entangled relationship.

Despret (2004) suggests that "the whole matter is a matter of faith, of trust, and this is the way we should construe the role of expectations, the role of authority, the role of events that authorize and make things become" (p. 121). She describes studies of skilled human riders and educated horses in which analysis of "unintentional movements" shows that homologous muscles contract in both horse and human at precisely the same time, a phenomenon called "isopraxis." "Both, human and horse, are cause and effect of each other's movements. Both induce and are induced, affect and are affected. Both embody each other's mind" (p. 115). A good run in agility has similar properties.

So I learned to be at ease with the natural cultural art of cross-species sports training. But surely, I imagined, Cayenne could be free off the course to roam the woods and visit the off-leash parks. (I had taught her a reliable recall that authorized that freedom.) I watched how my fellow agility competitor and friend Pam Richards trained with Cayenne's littermate brother, Cappuccino; and I was secretly critical of how relentlessly she worked with Capp to fix his attention on her and hers on him in the activities of daily life. I knew Capp was aglow with pleasure in his doings, but I thought Cayenne had the greater animal happiness. I knew Pam and Capp were achieving things in agility out of our reach, and I was proud of them. Then Pam took pity on us. She offered to show me in detail what we did not know. I became subject to Pam so that Cayenne could become free and lucid in ways not admitted by my existing stock of freedom stories.

Pam backed us up, forbidding me to send Cayenne to the A-frame in competition until she and I knew our jobs. She showed me that I had not "proofed" the obstacle performance in about a dozen fundamental ways. And so I set about actually teaching what the release word meant instead of fantasizing that Cayenne was a native English speaker. I started thinking practically about adding distractions to make the "two-on, two-off" performance that I had chosen for us surer, in circumstances approximating the intense world of trials. Finally, Pam said I was sufficiently coherent and Cayenne sufficiently knowledgeable that we could do the A-frame in competition—if I held the same standard of performance there that had become normal in training. Consequences, that sledge hammer of behaviorism, were the point. Rewarding a contact-zone performance that did not match our hard-won criterion, besides condemning us to a loss of mutual confidence, would prove that I had less respect for Cayenne than for my fantasies.

We had not advanced out of novice competition because of the A-frame contact zone. After we had retrained each other more honestly, I walked Cayenne off the course at a real trial when just once she leapt over the zone, and I zipped her into her crate without a glance. We got a year of perfect contacts after that. My friends cheered us over the finish line in our last novice event as if we had won the world cup. "All" we had done was achieve a little coherence. Now, Cayenne sails through this performance with glee written all over her coursing body.

But what about Cayenne's independent animal happiness off the course, in comparison with the bond of attention between Pam and Capp? Here, Pam and I have changed each other's notions of freedom and joy. I had to face the fact that many more "I pay attention to you; you pay attention to me" games had to fill our days. I had to deal with my sense of paradise lost when Cayenne became vastly more interested in me than in other dogs. This still feels like a loss as well as a spiritual and physical achievement for both Cayenne and me. Ours is not an innocent, unconditional love; the love that ties us is a natural-cultural practice that redoes us molecule by molecule.

Pam, for her part, tells me she admires the sometimes chaotic fun in Cayenne's and my doings. She knows that can exact a price on performance criteria. The gods rejoice when Pam and I accompany our dogs to a grassy field and urge them to ignore us. Pam's partner Janet will even leave a riveting women's basketball game on TV to revel in the unmatchable joy when Cayenne and Cappuccino play together. When Cayenne solicits her littermate long and hard enough, with all the metacommunicative skill at her command, they increase the stock of beauty in the world.

Agility is built on the tie of cross-species work *and* play. A puppy who does not know how to play is seriously disturbed. Most adult dogs know how to play too, and they choose partners selectively throughout their lives if they can. Agility people know that they need to learn to play with their dogs, if only to take advantage of the tremendous tool that play is in training. Play builds powerful affectional and cognitive bonds, and permission to play is a priceless reward for dogs and people. Agility people want to cavort with their dogs for sheer joy too. Nonetheless, many agility people have no idea how to play with a dog; they require remedial instruction, and some pay for it in special seminars. Discouraged dogs who have given up on their people's ability to learn to play with them politely and creatively are not rare. Without the skills of play, adults of both the canine and hominid persuasion are developmentally arrested, deprived of key practices of ontological and semiotic invention.

I suggest people must learn to meet dogs as strangers first in order to unlearn the crazy assumptions and stories we all inherit about who dogs are. Respect for dogs demands it. So how do strangers learn to play with each other? First, a story told by Barbara Smuts.

> Safi taught Wister to jaw wrestle, like a dog, and she even convinced him to carry a stick around in his mouth, although he never seemed to have a clue what to do with it. Wister enticed Safi into high-speed chases, and they'd disappear over the hills together, looking for all the world like a wolf hunting her prey. Occasionally, apparently accidentally, he knocked her with a hoof, and she would cry out in pain. Whenever this occurred, Wister would become completely immobile, allowing Safi to leap up and whack him several times on the snout with her head. . . . Then they would resume playing. (Smuts 2001, p. 13)

Safi was bioanthropologist Barbara Smuts's eighty-pound sheepdog mix, and Wister was a neighbor's donkey. Meeting in a remote part of Wyoming, dog and donkey lived near one another for five months. Wister was no fool; he knew his ancestors were lunch for Safi's ancestors. Around other dogs, Wister took precautions, braying loudly and kicking threateningly. When he first saw Safi, he charged her and kicked. But with a history of befriending critters from cats to ferrets, Safi set to work on her first large herbivore buddy, soliciting skillfully and repeatedly until Wister took the great risk of an off-category friendship.

Dogs are the kind of predators who know how to read the kind of prey that donkeys are in detail and vice versa. But the adults Safi and Wister played together by raiding their predator-prey repertoire, disaggregating and recombining it, changing the order of action patterns, adopting each other's behavioral bits, and making things happen that did not fit anybody's idea of function, practice for past or future lives, or work. Dog and donkey had to craft atypical ways to interpret each other's specific fluencies and to reinvent their own repertoires through affective, meaningful interaction.

Among beings who recognize each other, something delicious is at stake. As Smuts put it after decades of scientific field studies of baboons and chimps, cetaceans, and dogs, co-presence "is something we taste rather than something we use. In mutuality, we sense that inside this other body, there is 'someone home', someone so like ourselves that we can co-create a shared reality as equals" (Smuts 2001, p. 16). In the contact zones I inhabit in agility, I am not so sure about "equals"; I dread the consequences for significant others of pretending not to exercise power that shapes relationships despite denials. But I am sure about the taste of co-presence and the shared building of worlds.

The power of human language is its potentially infinite inventiveness (called "discrete infinity"). The inventive potency of play redoes beings in ways that should not be called language. It is not potentially *infinite* expressiveness that engages play partners, but unexpected and nonteleological inventions that can only take mortal shape within the noninfinite and dissimilar natural-cultural repertoires of companion species.

Anthropologist and biologist Gregory Bateson studied other mammals, including monkeys and dolphins, for their practices of metacommunication. Metacommunication is communication about communication, the *sine qua non* of play (Bateson 1972, p. 179). Language cannot engineer this delicate matter; rather, language relies on this other semiotic process, on this gestural, never literal, always implicit, corporeal invitation, to risk co-presence, to risk another level of communication. Bateson was not looking for denotative messages; he was looking for semiotic signs that said other signs do not mean what they otherwise mean. These kinds of signs make relationships possible, and "preverbal" mammalian communication is mostly about "the rules and contingencies of relationship" (Bateson 1972, p. 367). In studying play, Bateson was looking for things such as a bow followed by "fighting" that is not fighting and is known not to be fighting by the participants. Play can only occur among those willing to risk letting go of the literal. The world of meanings loosed from their functions is the game of co-presence in the contact zone. Dogs are extremely good at this game; people can learn.

Biologist Marc Bekoff studies the play of canids, including dogs. Granting that play sometimes serves a functional purpose, Bekoff argues that that approach neither accounts for play nor leads one even to recognize its occurrence. Instead, Bekoff and J. A. Byers offered a definition of play that encompassed "all motor activity performed postnatally that appears to be purposeless, in which motor patterns from other contexts may often be used in modified forms and altered temporal sequencing" (Bekoff & Byers 1981; 1998). Like language, play rearranges elements into new sequences to make new meanings. Play also requires something not explicit in Bekoff and Byers's definition in the 1980s: namely, *joy* in the sheer doing. I think that is what "purposeless" means. Like co-presence, joy is something we taste, not something we know denotatively. Play is not making a living; it discloses living.

Time opens up. Unexpected conjunctions and coordinations of creatively moving partners take hold of both and put them into an open that feels like a suspension of time, a high of "getting it" together in action, or what I am calling joy. No liver cookie can compete with that! Agility people joke about their "addiction" to playing agility with their dogs. How can they possibly justify the time, money, constant experiences of failure, public exposure of foolishness, and repeated injuries? And what of their *dogs'* addiction? How can their dogs possibly be so intensely ready all the time to hear the release word at the start line that frees them to fly in coordinated flow with this two-legged alien across a field of unknown obstacles? For people and for dogs, there is a lot that is not fun about the discipline of training, not to mention the rigors of travel and boredom while waiting for one's runs at an event. Yet the dogs and the people egg each other on to the next run, the next experience of what play proposes.

After a good run, Cayenne prances; she shines from inside out; by contagion, she *causes* joy all around her. So do other dogs, other teams, when they conjoin in a good run. With string cheese and affirming attention, Cayenne is pleased enough with a mediocre run. Mediocre runs or not, I have a good time too. But Cayenne and I both know the difference when we have tasted the open. We both know the tear in the fabric of our joined becoming when we rip apart into functional time and separate movement after the joy of inventive isopraxis. That is why we do it. That is the answer to my question, "Who are you, and so who are we?"

Bekoff (2004) suggests that animals' abilities to initiate, facilitate, and sustain joint "fair" play, in which partners take the risk to propose something seriously out-of-order, underlie the evolution of justice, cooperation, forgiveness, and morality. Remember Wister's letting Safi whack him when the donkey had accidentally struck the dog's head?

A golden retriever competes in an agility training competition in California, 2004. Courtesy of Shutterstock.

When I am incoherent instead of responsive in training, Cayenne gives me the gift of her readiness to engage again. I know that I am "anthropomorphizing" (as well as theriomorphizing) in writing these things, but *not* to say them is both inaccurate and impolite. Bekoff directs our attention to the world-changing evolution of trust. Consider the implications if the experience of cognitive, affective, sensual joy in the nonliteral open of play underlies the possibility of responsibility for and to each other—strangers that we always are—in all of our undertakings.

Play proposes. Agility is an ordinary game, in which the syncopated dance of rule and invention reshapes players. The open beckons; the world is not finished; the mind and body are not a giant computational exercise, but a risk in play. That is what I learned as a developmental biologist and a humanist scholar; that is what I learn again in the contact zones of agility.

See also

Communication and Language—Human-Horse Communication

Further Resources

Bateson, G. (1972). *Steps to an ecology of mind.* Chicago: University of Chicago Press.

Bekoff, M. (2004). Wild justice and fair play: Cooperation, forgiveness, and morality in animals. *Biology and Philosophy, 19,* 489–520.

———. (2007). *The emotional lives of animals.* Novato, CA: New World Library.

Bekoff, M., & Byers, J. A. (1981). A critical reanalysis of the ontogeny of mammalian social and locomotor play. In K. Immelmann, G. W. Barlow, L. Petrinovich, & M. Main, (Eds.), *Behavioural Development* (pp. 296–337). Cambridge, UK: Cambridge University Press.

Bekoff, M., & Byers, J. A. (Eds.). (1998). *Animal play: evolutionary, comparative, and ecological approaches.* New York: Cambridge University Press.

Despret, V. (2004). The body we care for: Figures of anthropo-zoo-genesis. *Body and Society, 10*(2–3), 111–34.

Fender, B. (2004). History of agility, part 1. *Clean Run, 10*(7), 32–37.

Garrett, S. (2005). *The education of an unlikely champion.* Chicopee, MA: Clean Run Productions.

Pratt, M. L. (1992). *Imperial eyes.* New York: Routledge.

Pryor, K. (1984/1999). *Don't shoot the dog: The new art of teaching and training.* New York: Bantam.

Smuts, B. (2001). Encounters with animal minds. *Journal of Consciousness Studies, 8*(5–7), 1–17.

For more information on the sport of agility, see http://www.cleanrun.com/ and the monthly magazine *Clean Run.*

Donna J. Haraway

■ Bonding
Animal Beings I Have Known: A Personal Essay

Rusty was my childhood friend. He was also my teacher. He was a mutt—perhaps a spaniel-poodle mix, with a glossy black coat that had coppery glints in the sun. He did not even belong to me, but lived in a hotel around the corner. His owners knew where he went when they let him out at 6:30 every morning—they were glad, I think, for they did not have time for a dog. He always went back to the hotel for his meals and then stayed with me until we told him to go home at about 10 PM or so. I took him with me to all places where dogs were allowed and often smuggled him into places they were not. Rusty was never more pleased than when learning some new skill—and not just the usual "sit!" "wait!" and "come!" He wore the paint off all the doors with his vigorous response to "Close the door." He would jump through a hoop and follow me up a tall ladder. He was one of those unusual dogs who loved to be dressed up, so long as no one laughed at him. Sometimes we nearly died trying to suppress our amusement. And he is the only dog, out of a succession of amazing canine companions, who would sulk if I reprimanded him for behavior that, in his book, was not wrong. He would go and sit with his nose almost touching the wall and remain thus until I went over, knelt down, and apologized. When it was very hot, he took himself off on a ten-minute walk to the sea, had a swim, and returned wet and cool.

I heard of his death when I was working in London, and it was one of the most terrible times of my life. Each week, I longed for Fridays, when I would catch the coach to go home for the weekend, knowing I would then experience the rapture of our greetings. Indeed, for weeks after he died, I could hardly bring myself to go home because everything there reminded me of Rusty. Even now, fifty years later, there are times when I feel that I shall climb the stairs and see him sitting in his favorite place looking out of the landing window, watching the people and cars go by.

It was my friendship with Rusty, and also with the cats, horses, and guinea pigs with whom I shared my childhood, that enabled me to stand up to the ethologists and other scientists who told me, when I first got to Cambridge University, that I should not talk or write about personality, mind, or emotions in other-than-human animals. But knowing what I knew about Rusty, and knowing that chimpanzees are our closest living relatives, I refused to subscribe to the somewhat reductionist views of the scientific establishment of the early sixties.

My relationships with chimpanzees in the Gombe National Park, Tanzania, have been very different from those I have enjoyed with dogs, for the chimpanzees were not dependent on me in any way. They were living their own wild lives, and our interactions were those between free beings of different worlds. They are so like humans, with personalities so different one from another, that although I have loved some individuals, I have actively disliked others! I have never counted any of them as "friends" for friendship is a two-way thing. Rusty was my friend: he depended on me for walks and love, and I depended on him for love and because he understood my moods. My relationship with the chimpanzees has been based on mutual acceptance and mutual trust, but we have never depended on each other for anything. Indeed, I have loved some of the chimpanzees—Flo, Melissa, Galahad, and, today, Gremlin and Freud—but my love was not reciprocated. That is not the relationship one wants when studying a wild chimpanzee.

Only with one individual, in all the forty-five years, did I come close to something like friendship: that was with David Greybeard, the first of the chimpanzees to lose his fear of the strange white ape who had so suddenly appeared in their world. One experience defined our relationship. It happened about a year after I had met him, when I was following him through the forest. At one point, as I crawled after him through a tangle of thorny vines that seized my hair, my clothes, and the buckles of my sandals, I was sure I had lost him. But when I got through, I found him sitting almost as though he was waiting for me. I sat near him and, spying a ripe red fruit lying on the ground, picked it up and held it toward him on outstretched palm. He turned his head away. I moved my hand closer. At this, he turned back, looked directly into my eyes, took and dropped the fruit—and then very gently squeezed my hand. This is the gesture used by chimpanzees to reassure one another. David's message was clear: he did not want the fruit but understood my good intention. For me it was an extraordinary moment that I remember now, all these years later, as though it happened yesterday. For I believe that we communicated, chimpanzee and human, in an ancient language that predates words—a language that links us to some common ancestor in our shared evolutionary past.

And so I give thanks to these two beings. To Rusty who taught me the meaning of unconditional love and helped me to understand the true nature of animals. To David Greybeard who reached out to me from his world, across the chasm once believed to separate humans from the rest of the animals. Between them, and with the help of all the other amazing individuals of many species whom I have known, they have brought me into the joyous knowledge that we are all members of the same wonderful, awe-inspiring, and utterly fascinating animal kingdom.

Further Resources

The Jane Goodall Institute for Wildlife Research and Conservation. http://www.janegoodall.org/

Jane Goodall, DBE
Founder, The Jane Goodall Institute, and UN Messenger of Peace

Bonding
Animal Personality and the Human-Animal Bond

In his 1998 book *Chimpanzee Politics,* primatologist Frans de Waal claims that he knows that chimpanzees have personalities because he often dreams about them. What de Waal is expressing is a conviction that, with close familiarity, the individuality and awareness of other animals becomes evident. When asked to describe another individual, human or nonhuman, we most often refer to what makes them similar to or different from other individuals—in other words, their personality. When describing their companion animal to a trainer, a dog owner may refer to their Alsatian's anxious and aggressive nature. In visiting a captive group of baboons, a zoo veterinarian may be only too aware of which individual animals are more relaxed and which are unpredictable. The notion of personality also features in the ways in which we compare different species; most people would use very different personality terms to describe a panda and a tarantula, even if they had never personally encountered an individual of either species.

Traditional ethological and behavioral approaches have assumed that such characterizations reflect anthropomorphic assumptions, that they project human attributes and qualities onto animals in an inappropriate way. Yet assumptions about the evolutionary relationships between species suggest a continuity of emotion, thought, and behavior. In his studies of emotion in different species, Darwin was one of the first to suggest that emotional expression was closely tied to variations in the functioning of the nervous system. The link between biology and personality was echoed almost seventy years later by Robert Yerkes, who drew parallels between personality traits of humans and chimpanzees, suggesting that personality was "the unit of social organization" in chimpanzees. Work on animals by behaviorists such as Ivan Pavlov was fundamental in the development of later human personality models, such as that developed by Hans Eysenck in the 1960s. Pavlov had suggested that differences in temperament in dogs resulted from variation in nervous system functioning; Eysenck's later explanations of personality focused on how traits such as extraversion and neuroticism could be explained in terms of nervous system arousal.

In more recent years, the formal study of animal personality has gained impetus, and a wide range of animal species have been studied across a number of disciplines, from cats, dogs, and piglets to hyenas, octopuses, and gorillas (see Gosling, 2001, for a full review of this work). Studies of animal personality have been used to explore important and difficult issues arising from the field of human personality, such as the genetic and environmental bases of personality, the reliability and validity of personality ratings, and the relationship of ratings to behavior. In its consideration of the psychological processes of personality attribution and the underlying meaning of personality descriptions, animal personality work also provides a window to the dynamics of the human-animal relationship.

Although the assumption that human personality descriptors can be applied to animals seems widespread, researchers in animal personality have struggled with the question of what such descriptions actually mean. Because personality studies of nonhumans cannot make use of self-reports, they rely on observer ratings. Typically, animals are rated on a number of personality-trait terms (e.g., sociability, anxiety, confidence) and behavioral characteristics (e.g., aggression, grooming, play). Ratings are often analyzed using statistical measures that reduce large numbers of traits or behaviors down to a smaller number of underlying dimensions. For example, Thomas Draper analyzed ratings of dogs collected by Hart and his colleagues. Veterinarians, dog handlers, and show judges rated a total of fifty-six breeds of dogs using thirteen behaviors. These included behaviors such as excitability, obedience, playfulness, and destructiveness. Analysis of the ratings produced three underlying personality dimensions: "reactivity-surgency," "aggression-disagreeableness," and "trainability-openness," which seem similar to personality dimensions found in human personality work.

Criticisms of research on human personality in the 1960s and 1970s claimed that personality ratings often reflected the rater's implicit and subjective assumptions about the person being rated rather than serving as an accurate and objective assessment of the person's typical character and behavior. In the absence of self-report data, how do we determine whether personality attributions of animals reflect "real" differences between individuals or reflect the ways in which human observers typically respond to different animals? Although there has been little research in this area, it does seem the case that personality ratings of animals show similar characteristics to human personality ratings. For instance, agreement between different observers rating animals generally tends to be quite high, comparable to the levels obtained in human personality studies. This suggests that people can reliably estimate animal personality. There are some aspects of personality, however, that appear to be more difficult to assess. In both humans and animals, those aspects of personality that are more salient are rated more reliably. It is much easier to rate humans, cats, or chimpanzees on how sociable, confident, or anxious they are than on how deceptive, intelligent, or trusting they are. This suggests the importance of the context of relationship for personality attribution, given that more complex and subtle personality characteristics are probably only apparent upon extended contact. To put it simply: to comment on someone's personality, you have to get to know that individual! Research on humans has found that those raters who were more familiar with the person being rated tended to agree more on personality ratings. The very little work done to date on rater familiarity with animals suggests the same picture. With increased awareness of the complexities of human-animal relationships, this may be an area of research that receives more attention in the future.

If the degree of familiarity with the animal being rated is important, this implies two things. First, the approaches that explore the subjective experiences of those who interact with animals may be more useful at this stage of the research program than more observational or experimental approaches. Approaches that utilize interviews, for instance, such as James Serpell's (1983) study of dog owner's perceptions of canine personality, can be valuable tools to explore in detail the assumptions, emotions, and dynamics of personality attribution within the human-animal relationship. Second, it seems that personality ratings capture very subtle attributions that reflect the rater's attitude toward an animal. Both the type of relationships (e.g., whether the animal being rated is a pet, a working animal, a laboratory subject, or so on) and the level of intimacy of the relationship (e.g., whether the rater responds in an emotionally positive or negative way to the animal) may affect ratings. Closer relationships may not just lead to more positive ratings but may in fact also lead to greater *range* of personality characteristics being applied. This may apply more to those human-animal relationships characterized by close contact, such as relationships with companion animals, than to those human-animal relationships commonly used in animal personality studies, where caretakers or observers rate animals.

However, even in the latter type of relationship, there is a tendency for animals who behave "out of character" or who are hard to decipher or are less expressive to be considered in more negative terms. An early study by Donald Hebb (1946), for instance, measured the personality attributions of a colony of captive chimpanzees. Hebb notes the failure of the colony staff to adequately describe the personality of one chimpanzee, Kambi. Unable to account for the lack of consistency in Kambi's behavior, staff simply labeled her "psychopathic." Instances where personality attribution breaks down provide important clues to the ways in which observers are engaging with animal subjects. Recent work on primate personality suggests that where social groups are undergoing changes in status relationships, and behavior changes very rapidly, personality attributions become more difficult for observers to make. This does not imply that animal personality ratings are inherently unreliable, but rather, it points to the fact that personality attribution is a dynamic, active process whereby observers attempt to categorize and make sense of disparate behaviors.

More recently, some researchers have argued for a "common language" in personality description across species and have urged the use of the human five-factor model (FFM) in rating animals. There seems to be increasing consensus in human personality work as to the existence of five basic personality dimensions: extraversion, agreeableness, neuroticism, openness to experience, and conscientiousness. Personality factors similar to extraversion, neuroticism, and agreeableness appear common across many species, whereas conscientiousness may be restricted to chimpanzees. Using a standard set of personality descriptors would have the advantage of enabling cross-species comparisons of personality structure, including comparisons between humans and animals. For instance, using personality descriptors from the FFM, Gosling and Bonnenburg were able to compare ratings of six species of companion animals (ferrets, dogs, rabbits, cats, horses, and hedgehogs) on the same personality dimensions. They found that dogs, horses, and cats showed similar personality profiles, perhaps because the raters responded to these animals in a similar way.

Cross-species comparisons of personality structure provide important information about perceptions of animals and highlight the central importance of the human-animal relationships as a context within which to consider such measurements. The differences in the relationships, for instance, between dog owners and their dogs and snake owners and their snakes may make the use of a single set of personality descriptors difficult.

Some personality terms simply would not be applicable to some animals or would not feature in some relationships. For instance, it would not be appropriate to talk about the "sociability" and "friendliness" of solitary species. And it may be impossible to discover how "loving" a sheepdog is who is never let into the family home. In other words, aspects such as the type and quality of the human-animal relationship structure how people interpret animal personality and behavior.

There even seems to be some evidence that the personality of owners can influence the personality of companion animals. Valerie O'Farrell (1995) found that dog owners who scored high on a neuroticism scale tended to own dogs who had behavioral problems such as destructive behavior and attention seeking. Later work by Anthony Podberscek and James Serpell (1997), also comparing dog and owner personalities, found that owners who were more tense and less emotionally stable tended to own more aggressive dogs. A simplistic interpretation of this kind of work would suggest that the personalities of animals simply reflect the personalities of those people with whom they interact. But the picture is likely to be much more complex; as in human relationships, animal-human interactions act in a dynamic way to shape the behavior of each respondent. Certainly within more social species, the expression of relatively innate tendencies, as well as the development of new responses, occurs within the context of ever-changing relationships.

Within the field of animal personality, there is increasing interest in how perceptions of animals structure understanding of personality. Human personality research has shown that more attractive people tend to be seen as having more favorable personalities than less attractive individuals. It also seems that we find larger mammals to be more attractive and easier to relate to than smaller and nonmammalian species. In a questionnaire study of attitudes toward animals, Janis Driscoll (1995) found that species such as horses, dogs, chimpanzees, and dolphins were rated as more lovable, intelligent, and responsive than animals such as snakes, slugs, and earthworms. Personality attributions of animals are more than just dispassionate descriptions of animal behavior; they function also as repositories of cultural attitudes, perceptions, and stereotypes.

Our sense of who we are as humans seems intricately tied to our senses of similarity to and difference from other species. The boundaries that we draw between ourselves and other animals often determine the quality of the relationships that we enjoy with them. Perhaps the most valuable application of animal personality work is to enable us to assess those relationships.

Further Resources

De Waal, F. B. M. (1998). *Chimpanzee politics: Power and sex among apes* (Rev. ed.). Baltimore: Johns Hopkins University Press.

Draper, T. W. (1995). Canine analogs of human personality factors. *The Journal of General Psychology, 122*(3), 241–52.

Driscoll, J. W. (1995). Attitudes toward animals: Species ratings. *Society and Animals, 3*(2), 139–50.

Dutton, D., & Andersson, M. (2002). Personality in Royal Pythons and the human-snake relationship. *Anthrozoös, 15*(3), 243–50.

Gosling, S. D. (2001). From mice to men: What can we learn about personality from animal research? *Psychological Bulletin, 127*(1), 45–86.

Gosling, S. D., & Bonnenburg, A. V. (1998). An integrative approach to personality research in anthrozoölogy: Ratings of six species of pets and their owners. *Anthrozoös, 11*(3), 148–56.

Gosling, S. D., & John, O. P. (1999). Personality dimensions in nonhuman animals: A cross-species review. *Current Directions in Psychological Science, 8*, 69–75.

Hart, B. L., & Hart, L. A. (1988). *The perfect puppy: How to choose your dog by its behavior*. New York: Freeman.

Hebb, D. O. (1946). Emotions in man and animals: An analysis of the intuitive process of recognition. *Psychological Review, 53*, 88–106.

King, J. E. (1999). Personality and the happiness of the chimpanzee. In F. L. Dolins (Ed.), *Attitudes to animals: Views in animal welfare*. Cambridge: Cambridge University Press.

O'Farrell, V. (1995). Effects of owner personality and attitudes on dog behaviour. In J. A. Serpell (Ed.), *The domestic dog: Its evolution, behaviour and interactions with people* (pp. 153–58). Cambridge: Cambridge University Press.

Podberscek, A. L., & Gosling, S. D. (2000). Personality research on pets and their owners: Conceptual issues and review. In A. L. Podberscek & J. A. Serpell (Eds.), *Companion animals and us* (pp. 143–67). Cambridge: Cambridge University Press.

Podberscek, A. L., & Serpell, J. A. (1997). Aggressive behaviour in English cocker spaniels and the personality of their owners. *The Veterinary Record, 141*, 73–76.

Serpell, J. A (1983). The personality of the dog and its influence on the pet-owner bond. In A. H. Katcher & A. M. Beck (Eds.), *New perspectives on our lives with companion animals* (pp. 57–65). Philadelphia: University of Pennsylvania Press.

Diane Dutton

Bonding
The Attachment between Humans and Animals

John Bowlby's Attachment Theory

The relationship that some people have with their pets is often described as a strong emotional bond that is unconditional. This relationship is often referred to as the human-animal bond. Interestingly, the behaviors we display toward our pets can be remarkably similar to behaviors directed toward people we love during our life (i.e., family, friends, and romantic partners). The idea that we form strong emotional attachments to people from birth has been theorized and explained by John Bowlby. He wrote a trilogy of books about attachment in the late 1970s to early 1980s. Bowlby utilized ethology, psychology, and other disciplines to develop his principles of attachment theory. His theory states that we are genetically predisposed to display certain behaviors toward a preferred individual (for example, our mother), and as such, that person is regarded as our primary attachment figure (PAF). The lifelong relationships that we form with other people are often termed "relationships of attachment." John Bowlby's attachment theory can be utilized to explain the human-animal bond. To do this, we need to understand the behaviors involved in an attachment relationship.

Bowlby explained four fundamental attachment behaviors (all observable) that infants display toward their primary attachment figure. These behaviors are described as *proximity seeking and maintenance*, *separation distress*, *secure base*, and *safe haven*. These behaviors are best understood in the context of an infant's reaction to his or her mother in an unfamiliar environment. For example, children tend to stay very close to their mother (PAF) and strive to maintain this close proximity (*proximity seeking and maintenance*). If a child is separated from his or her mother for whatever reason, he or she may cry or show signs of distress (*separation distress*). Furthermore, a child tends to feel comfortable exploring, learning, and playing in a new environment when the mother is nearby, thereby using her as a source of security (*safe base*). Finally, if a child is exploring his or her immediate environment, and the child perceives danger and becomes frightened, the child will try to locate and move toward the mother, thereby using her for protection (*safe haven*). Bowlby maintained that these behaviors of attachment could be displayed toward other people we become attached to during our life (e.g., a romantic partner or best friend).

Early research on attachment relationships by John Bowlby and Mary Ainsworth focused on the mother as an infant's only primary caregiver (or AF). However, as infants develop to reach later childhood, they look to other caregivers to provide safety and comfort if the mother is not present. If this happens, the caregiver may now be regarded as a new attachment relationship for the child. Bowlby regarded these new attachment figures as becoming part of a child's "multiple attachments" or "hierarchy" of attachment relationships. As the child reaches adolescence, the child has the opportunity to expand his or her attachment hierarchy outside the immediate family, to include peers and sexual partners. As such, an individual's attachment hierarchy has been shown to include parents, siblings, best friends, and romantic partners.

The relationship that people have with their pets is often referred to as an attachment relationship. The behaviors that some people display toward their pets are similar to those attachment behaviors we display toward people whom we include in our attachment hierarchy. Shana Trinke and Kim Bartholomew (1997) investigated people's attachment hierarchies and stated, "It is likely that repeated contact with an appropriate other at any time during the life cycle could lead to the formation of a new attachment bond." Therefore, the close contact with a pet could result in the formation of an attachment bond consistent with Bowlby's attachment theory.

People sometimes display some of the four attachment behaviors toward their pets when they are strongly bonded with them. In a veterinary office, it is not uncommon to see people hugging, kissing, petting, and stroking a pet and displaying signs of grief when a pet is euthanized or very sick. Bowlby's attachment theory has also been used as a model for pet attachment in children. Gail Melson outlined some variables that would discriminate between children who had close attachment relationships to their pets and children who did not. For example, a child's behavior directed toward his or her pet such as stroking, patting, hugging, and maintaining proximity can be observed and measured and could reflect attachment behaviors as described by John Bowlby. Melson proposed that these variables reflect elements of attachment theory. She outlined and identified a child's attachment behaviors in response to the separation, reunion, and interaction with a pet (separation distress, proximity seeking and maintenance, secure base, safe haven).

However, the surest behavioral sign that a person is attached to a pet is through the display of grief after the death of the pet. The reaction we have to the loss of another human depends on whether we had an attachment to that person. Bowlby found that adults reacted with intense grief to the loss of a parent. He argued that the display of grief is a reflection of separation distress, suggesting that the attachment bonds adults have with their parents have endured. Colin Parkes found that grief and despair were common reactions to the death of a spouse, child, or pet.

The death of a pet and the grief associated with such a loss is an important and real process that affects pet owners and veterinarians. Bereaved pet owners experience the normal signs of grief: painful regret, crying, shock, numbness, deep sorrow, and loneliness. The conclusion of many studies is that the experience of the loss of a companion animal is remarkably similar to the loss of a significant human relationship. From John Bowlby's point of view, the grief people display when they lose a pet is a reflection of the separation distress people experience when they lose an attachment relationship.

A young girl giving kisses to her bunny. Courtesy of Shutterstock.

John Bowlby has introduced to the scientific field a resilient body of work that attempts to explain "relationships of attachment" during our life. The attachment behaviors he proposed as part of his attachment theory (i.e., proximity seeking and maintenance, separation distress, safe haven, and secure base) can be used to explain the human-animal bond or special relationship that can exist between pet owners and their pets.

Further Resources

Ainsworth, M. D. (1982). Attachment: Retrospect and prospect. In C. M. Parkes & J. S. Hinde (Eds.), *The place of attachment in human behaviour* (pp. 3–30). New York: Basic Books.

———. (1989). Attachments beyond infancy. *American psychologist, 44*(4), 709–16.

Bowlby, J. (1969). *Attachment and loss. Vol. 1: Attachment*. New York: Basic Books.

———. (1970). *Attachment and loss. Vol. 2: Separation: Anxiety and anger.* New York: Basic Books.

———. (1980). *Attachment and loss. Vol. 3: Loss: Sadness and depression*. New York: Basic Books.

Parkes, C. M. (1972). *Bereavement: Studies of grief in adult life*. New York: International University Press.

Trinke, S., & Bartholomew, K. (1997). Hierarchies of attachment relationships in young adulthood. *Journal of Social and Personal Relationships, 14*(5), 603–25.

Weiss, R. S. (1988). Loss and recovery. *Journal of Social Issues, 44*(3), 37–52.

Michael P. Meehan

■ Bonding
Behavior of Animals Influences Human-Animal Interactions

Animal Behavior and Its Attraction

Animal behavior, or what an animal does, has a powerful influence on human-animal interactions. We choose pets that have desirable behaviors. We protect and preserve wild species that possess behaviors we appreciate, and we care less for animals with "undesirable" or misunderstood behaviors. We breed livestock and lab animals to have convenient behaviors. Many involved in animal welfare try to introduce the public to interesting and unique behaviors of animals that would otherwise remain unknown or underappreciated. Finally, television programs and movies containing animals are highly popular fare, indicating a broad attraction of animal behavior to human audiences.

Pet Behavior Examples and What They Mean

Popular pets are those with behaviors to which humans relate. The naturally social and tame aspects of many species make these animals attractive pets. As these species have been tamed, certain predispositions in the species' behavior have been exploited by artificial selection for human purposes. Most pets represent a collection of behavior characteristics that are aesthetic, interesting, or useful to humans.

There are over 400 dog breeds, 152 of which are recognized by the American Kennel Club. The highly variable canine behavioral repertoire has been modified and amplified through artificial selection. The result is breeds that are good with children and breeds that are not, quiet dogs and loud dogs, dogs that require little attention and dogs that need to be walked or worked hard daily. Some breeds make good seeing-eye dogs that guide their owners, or therapy dogs that help with managing chronically ill patients. Amazingly, some dogs are capable of anticipating epileptic seizures long before their owners or medical diagnostic testing can anticipate an imminent episode. Other breeds make excellent search and rescue dogs or guard dogs that protect via barking or aggressive behavior. There are crime-unit dogs to sniff out contraband, hunting dogs to track and recover game animals, fast racing breeds, herding dogs to keep livestock together, and messenger and scout dogs that have served various purposes during war. The hundreds of dog breeds that exist indicate that human preferences have operated for centuries, artificially selecting breeds to fit the fancies and needs of dog owners.

Cats, like dogs, make good pets because of their relative tameness and companionship. Cats show neither the morphological nor behavioral variation evident in dogs. However, many cat owners enjoy the independence of cats in terms of daily needs (e.g., no daily walks and cleaner toiletry, feeding, and drinking behaviors than dogs). Most people are attracted to the soothing purr of the cat and to the way cats play and practice predatory stalking and capture behavior, using a ball of yarn, for instance, as prey.

Most people are attracted to the soothing purr of the cat and to the way cats play and practice predatory stalking and capture behavior using a ball, for instance, as prey. Courtesy of Shutterstock.

Birds are kept as pets because of their plumage and singing. Many breeds also are highly social and will bond tightly to their human owners. Species such as finches engage in captivating mating and parental behaviors. Some species mimic human speech or learn "tricks," and these behaviors enhance their interactions with their human owners.

The most popular fish species kept as pets engage in territorial, schooling, feeding, mating, and parental behaviors. Examples include the nest-building, mating, and egg-caring behaviors of Siamese fighting fish; the courtship, mating, and birthing behavior of live-bearing community fish (e.g., mollies, platys, and swordtails); the territorial, courtship, and birthing behavior of cichlids; the care of the young in discus; and the foraging behavior of catfish, loaches, and algae eaters.

To summarize, it is the behaviors of certain species that play a role in whether they are in demand as pets. Animals that may be cute but are difficult to tame (e.g., deer and foxes) or that are interesting to watch but neither social nor tame (e.g., wolverines and shrews) do not become popular pets.

Game Animals and Their Behavior

Hunting and fishing involve animals with behaviors that attract a particular demographic. Those who fish enjoy the feel of the fish on the line, which really is a way of connecting an animal's behavior to the human tactile sense. As the fish fights to get away, its behavior is amplified through the fishing line and rod. The tug on the rod and the knowledge that the tug is caused by a living wild animal provide excitement. It is interesting that the largest expenditures to endangered or threatened species funded by the U.S. Fish and Wildlife Service have gone to sports fish species. Hundreds of American mammals, birds, invertebrates, reptiles, amphibians, and non-game fish and tens of thousands of species are threatened or endangered worldwide. But the fact that the highest funded in America are sports fish species suggests that humans are more willing to protect species with which they can "interact" behaviorally and therefore with which they can develop a memorable bond. A challenge facing those interested in protecting species and ecosystems is to develop avenues whereby people can develop mutually beneficial bonds with threatened and endangered species (see the subsequent section on zoos).

Zoos: Animal Behavior and Human-Animal Interactions

Zoos preserve rare and endangered species and provide a means whereby the general public can observe species-specific behavior, thereby gaining an appreciation for the diversity and richness of animal life. These two purposes are often at odds. For instance, many animals do not interact or behave well in close proximity to humans—for example, animals that may be shy (okapis) or nocturnal (bats). A major challenge for zoos is to develop habitats in which natural behaviors can emerge while providing the public with visual access to animal behaviors. This has two hopeful outcomes. First, behaving animals are interesting to park attendees, who will gain a better appreciation for preserving the species. Through the animal behavior–human observation interaction at the park, humans will develop an appreciation, which will then transfer to useful action (e.g., donations, political action, ecologically centered thinking, and so on). Second, the relatively natural habitat enhances the chances that the animals will successfully breed and rear young. Such action preserves the species for another generation.

"Rural" Wildlife

One attraction of the "rural" or suburban lifestyle is the potential to see "wild" animals, such as skunks, raccoons, deer, foxes, opossums, squirrels, and numerous bird species. Feeding stations and housing for these animals can be placed in the backyard to improve the chances of observing animal behavior. Most suburban wild species are those that readily adapt to human encroachment, whereas more sensitive species are unfortunately forced out of vital habitat by human housing. Additionally, the desirability of the rural lifestyle often brings more people to the suburbs, which reduces the benefits of human-animal interactions as animal habitat becomes scarcer. Fewer animals are observed, and so new suburbs are planned and developed for those humans desirous of living closer to nature. The paradox is that the desire to be near nature first drives the growth of human communities further out into wilder areas and then changes the ecological nature of those wild areas, which then drives away the animals.

On the other hand, the behaviors of many wild species may be considered nuisances by some. Examples include squirrels that take food from bird feeders, rodents and deer that eat plant bulbs and buds, mice that nest in our homes, raccoons and bears that forage in trash cans, skunks that nest under porches, animal latrines and scent markings in inconvenient places, and even sometimes the noise created by calls and songs. These "problem behaviors" have led to an industry of pest control and management that is designed either to remove animals from the local environment or to prevent animal access to specific areas. However, rather than battling these wild species' behaviors with pest control, it may be wiser to develop policies and lifestyles of cohabitation because these battles usually cannot be won without dire consequences—for example, annihilation of ecosystems and loss of appreciation for the rural lifestyles (see the subsequent section on amphibians).

Birding and Wild Avian Behavior

Avian behaviors play an enormous role in birding's popularity. Birders enjoy listening to and identifying birds by their songs and calls. Many calls and songs are transient in an area, as migrating birds pass through and signal seasonal changes. Nesting and foraging behaviors are enjoyable aspects of birding because these behaviors vary widely among avian species. Species differ in terms of where they can be found (e.g., on the ground, in bushes, in the tops of deciduous or evergreen trees) and how they forage (e.g., catching insects on the wing, diving for fish, eating insects off the ground, pecking seeds against tree branches, impaling lizards on thorns, dropping rocks on eggs, placing nuts in the middle of roads for cars to run over and crack, searching for grubs in tree bark). This diversity helps catch the interest of birders, many of whom travel great distances to experience new species, songs, and behaviors.

Amphibians

Scientific evidence indicates that amphibians are very sensitive to environmental changes; therefore, they are good "early warning" detectors of such harmful changes. About one-third of America's amphibian species are in significant decline. In an attempt to protect these species, some herpetologists have launched survey programs using the argument that amphibians, like birds, have songs and calls that are interesting to hear. These herpetologists argue that amphibians are interesting in the same way birds are interesting, and therefore, people can grow to appreciate amphibians and subsequently act to preserve and protect habitat for them.

Livestock and Laboratory Animals

To many people, livestock and lab animals have utilitarian purposes only. Consequently, these animals have been bred to have the most docile, manageable behavior possible. Examples include cows, pigs, sheep, and chickens as livestock and guinea pigs, rats, rabbits, and mice as laboratory animals. None of these species is aggressive; each can easily be handled or manipulated by humans with little threat to human safety. These species probably became livestock and lab animals of choice because their naturally docile predispositions were recognized and then amplified through artificial selection. Consider how manageable cows and pigs are in comparison with the water buffalo or wild boar. It is these management differences that are responsible, in part, for the selection of species as livestock.

Laboratory animals are selected because they manifest traits that somehow provide insights into the human condition. Often, however, medical science is not interested in the animal's behavior. The animal's physiology, genes, and tissues are of prime importance. This leads to a behaviorally impoverished interaction, which may have unfortunate consequences. For instance, many medical treatments given to humans are greatly enhanced if a patient has strong positive support from, and interacts with, caregivers and family. By the same token, it has recently been "rediscovered" that handling lab rats also improves the animals' health and well-being. But it is unlikely that these findings will change the way rats are handled by laboratory scientists. One must wonder how many potentially advantageous treatments for diseases have been overlooked in laboratory animal tests because the positive support that works so well on humans was not used with the laboratory animals.

Scientific Behavioral Observations

Many scientists thoroughly enjoy interacting with animal subjects, and for these scientists, behavior forms an integral part of their work. Studying animal behavior scientifically is challenging. It is relatively easy to measure morphology (e.g., the length of a tail, the weight of an animal, and so on), but it is more difficult to measure behavior. How does one, for instance, measure a rat's exploration, and how would exploration be described in scientific writing? What should be measured to determine whether the behavior is affected by age, gender, or drug treatments, keeping in mind that the measurement and results need to be clear to other readers who may wish to repeat the study? Can behavior intensity, vigor, emotional content, attentiveness, and such be given a number value that scientists could agree on? Scientists who are enthusiastic about animal behavior see these questions as exciting challenges, and it is the close observation of animal behavior that leads these scientists to develop highly detailed understandings of and appreciation for their animal subjects' rich behaviors.

The fact that animal behaviorists are enthusiastic about their subjects is often lost on the public because in most scientific writing, action verbs are replaced by passive or mechanical verbs, and this means that the behavior description loses some of its impact. The technical vocabulary of scientific writing dominates, and the reader cannot generate visual imagery. As a result, the animal's behavior has less capacity to influence readers.

By contrast, the naturalist writer's portrayal of animal behavior is more image-filled. The animal's behavior comes alive, and the reader is able to visualize the animal's behavior and to appreciate the animal more fully. As a result, naturalist writing maintains popularity among those interested in animals and animal behavior, whereas scientific writing on animal behavior does not. A good example of a naturalist writer was Loren Eiseley (see Further Resources). Marc Bekoff also does a good job of translating modern animal behavioral sciences into "public form."

Descriptions of animal behavior can influence and reinforce the human-animal interaction. Scientific, naturalist, and popular descriptions attract different types of readers and reinforce a certain type of interaction with animals.

The shape and appearance of the eagle's head is easily mistaken for behavioral predispositions, such as nobility and stern integrity. Courtesy of Shutterstock.

Mistaken Identities in Animal Behavior

For interpersonal human relationships, facial expression and gestures are critical. Two-thirds of our communication with other people occurs through nonverbal means, which may involve the expressive human face. We try to read others' emotions, mood, and even morality based on expressions. But these critical nonverbal communication skills can be problematic when we extrapolate or project these interpretations onto animals. Most animal species lack the capacity to express themselves through their faces, or if they can, the animals do not attach the same emotional states that humans would to these expressions. As a result, an animal's morphology may influence how we interpret its behavior.

For instance, some people interpret the strong brow ridge of an eagle as a sign of a noble, stern, and austere animal. Because humans have mistaken this morphological trait for a behavioral trait, the eagle has become the national symbol for many countries. Other examples include the slender, slightly upturned snout of camelids, which make these animals appear aloof. Snakes' lack of eyelids and their flicking, sharp, forked tongue give them an ominous appearance and add to their negative image. By contrast, polar bears look soft and cuddly, but they are extremely dangerous to humans.

Humans are "programmed" to feel nurturing toward babies partly because of the baby's flat, rounded face, and large eyes. Animals that have similar features are likewise considered sweet (e.g., pandas, cats, and many baby animals). By contrast, animals that lack the baby-like features or that possess morphologically "opposite" traits are often considered the antithesis of cute (e.g., weasels and sharks). Very much like human stereotypes, animals are more complicated and different in terms of their actual behaviors, and it remains an important challenge to any person struggling with a more enlightened view of animals to see through the stereotypes created by our own human evolutionary histories and preconceptions in an attempt to see the actual richness and

Animals that have features similar to the human baby are likewise considered sweet. Courtesy of Shutterstock.

Humans are "programmed" to feel nurturing toward babies, partly because of the baby's flat, rounded face and large eyes. Courtesy of Shutterstock.

diversity in animal behavior. The promise of the successful struggle is to develop a richer appreciation for and therefore a richer relationship with nonhuman species.

See also

Human Perceptions of Animals

Further Resources

Bekoff, M. (2002). *Minding animals. Awareness, emotions, and heart.* New York: Oxford Press.

———. (Ed.). (2004). *Encyclopedia of animal behavior, Volumes 1, 2, 3.* Westport, CT: Greenwood Press.

Crist, E. (1999). *Images of animals: Anthropomorphism and animal mind.* Philadelphia: Temple University Press.

Eiseley, L. (1979). *The star thrower.* San Diego: Harcourt, Brace.

Geoffrey E. Gerstner and Sarah Najor

Bonding
Behavior of Pets and the Human-Animal Bond

Animals have been a part of human civilization for thousands of years and have been used for a number of different reasons, ranging from food production to helpers in the hunt to human companions. Although physical stature is often grounds for their use in our society, their behavior is a major reason that we keep them for companionship. Their behavior can also be a reason for fracturing the human-animal bond, leading to abandonment and euthanasia.

There are numerous examples of humans utilizing animals in society. Cattle are domesticated for food production, as well as for use as beasts of burden. Horses are domesticated for use as beasts of burden, as well as for companionship. Dogs are domesticated for hunting, protection, and herding, as well as purely for companionship.

Take cattle, for example. Certainly, body size and physical composition play an important role in determining what breed a farmer will raise. However, it is necessary to understand just *how* these animals became domesticated. Domestication is the process by which an animal's behavior is changed through the process of selective breeding. Cattle have to be amenable to handling, and certain space and production restrictions limit farmers to specific breeds of cattle, often based on their behavior.

The same can be said for dogs. Border collies have historically been bred for herding sheep—their behavior defining who they are. In addition to their selected behavioral ability to herd, they must also be responsive to their human handlers for guidance. Physical conformation plays a role in the dog's ability to perform its job, but without the herding behavior, these dogs would be of little use to a shepherd. On the flipside, the herding behavior can become a problem for an owner who adopts a border collie for companionship in an urban environment.

There have been other uses for working dogs in more recent times. Dogs have been used extensively for military and police work, in both protection and guard work, as well as for detection work (drugs and bombs, for example). Behavioral traits are the driving force for their selection and training. These dogs need to have the determination to work, a stable and confident personality, and the ability to follow commands and must be able to pay heed to their handlers in challenging situations.

Certain physically and emotionally challenged people benefit greatly from seeing-eye dogs and other types of service dogs. Physical conformation, obviously, plays a role in their jobs, especially because seeing-eye dogs need to be of a certain size to adequately lead their humans, but their behavioral characteristics need to be of a certain type—calm, confident, and easily trained.

But what is more important to the average pet owner is the companionship aspect. We have selectively bred dogs and cats to be a part of our lives and households. Different owners have different expectations of behavioral characteristics. Some owners want a dog that will play with their children and have a high threshold for aggression-provoking stimuli. These owners should not choose a breed with a higher tendency to display territorial or owner-directed aggression. Other owners want a cat that is quiet and relaxed. These owners should not choose a breed with a higher tendency to be vocal or active. And yet other owners want a dog that will compete in agility or other canine sports. These owners should not choose a breed that is overly anxious or easily distracted. These are broad generalizations, for certain, but Drs. Ben and Lynette Hart (1998) have systematically studied breed tendencies, and these need to be taken into consideration before selecting a pet.

Not only is innate or "natural" behavior important, but "learned" behavior is also important. Currently, there is not a way to measure which has more influence on the outcome—it is the old "nature versus nurture" debate, but we must understand that no animal lives in a vacuum. Certain breeds may be more predisposed to certain behaviors, but it is necessary to take the learned component into consideration.

Animal behavior, learning, and training have been studied for years. In contemporary times, the research of people such as B. F. Skinner and Ivan Pavlov has paved the way to study how animals (and humans) learn, via classical and operant conditioning.

As an overview, classical conditioning, the emotional aspect of learning and behavior, is essentially bringing internal reflexes under the control of a previously unconditioned stimulus. Most people are familiar with Pavlov's dogs (the bell is linked to food, and the dogs then respond to a bell by salivating). But classical conditioning happens with animals all of the time, especially when taking fear responses into consideration. For example, the average dog has not been conditioned to "enjoy" going to the veterinarian's office. No matter how kind or patient the veterinarian is, the dog is unable to cognitively understand that he or she is there to maintain a good health status and that the veterinarian has the dog's best interest in mind. Dogs learn that the veterinarian causes them discomfort with needles, ear cleanings, and anal gland expressions. The dogs then become classically conditioned to associate the veterinarian's office or white lab coat with fear.

Operant conditioning is the process by which the likelihood of a specific behavior is increased or decreased through reinforcement or punishment each time the behavior is exhibited, so that the animal associates the pleasure (or displeasure) with the behavior. B. F. Skinner performed the seminal research, with his development of the Skinner Box, where a rat learned that food was released when it pressed down on a lever. With this simple, yet elegant, study, Skinner was able to expand with more studies evaluating shaping of behaviors, different models of pairing unconditioned stimuli with reinforcements, and different intervals of reinforcement. A current example of operant conditioning in action is clicker training. The dog is conditioned (via classical conditioning) that the sound of the click means that a piece of food is coming. When the owner observes a wanted behavior, such as when trying to teach the dog to sit, the owner "clicks," marking the wanted behavior exactly (operant conditioning). The dog will then be more likely to offer the sit. Eventually the owner pairs a command with the action, and now the dog has learned a new trick.

Even in the face of these early proven methods of modulating animals' behavior, there are people who still heavily rely on more punishment-based methods. Although these methods have been used successfully, such as with emotionally stable military dogs and their well-trained handlers, the average pet owner can have a difficult time using these methods. Owners who choose to use these methods need to understand the principles of punishment. For the animal to be punished appropriately, certain rules must be followed: the punishment has to happen immediately (within a few seconds); it cannot be too aversive (to avoid causing fear and anxiety); it should not directly relate to the owner (to avoid a classically conditioned aversion to the person); and it should work after a few tries.

So, with all of these methods in their back pocket, owners can, and do, affect an animal's good and not-so-good behavior in many ways, and this also affects the human-animal bond. For example, a family may reinforce a dog begging from a table, such as by periodically feeding it from the table, but when the dog is climbing up onto the chair for food, they get frustrated that the dog will not listen to them. In essence, the dog has been listening very well. If the dog performs a certain behavior, whether it is sitting looking cute or jumping on the chair, it gets reinforced with food that is much tastier than its dry kibble.

Another unwanted behavior that some cat owners reinforce is the cat waking them up in the middle of the night, by meowing, pawing at the door, or pawing at their face. The owners may feel guilty—perhaps they forgot to feed the cat that evening? Maybe they did not give the cat enough food? So understandably, they get up and feed the cat. The cat just learned a valuable lesson! The owners continue on with this night-feeding until they are sleep-deprived, even as they remember that they fed the cat a whole bowl of food that evening. The same manners in which these animals learned the inappropriate behaviors can be used to train them to learn appropriate behaviors as well.

It has been shown that behavior problems are a primary reason for euthanasia or relinquishment to animal shelters. Most of the problems can, and should, be addressed in a proactive manner, lest they break the human-animal bond. House soiling, unruliness, and aggression are common reasons for relinquishment of dogs to shelters, and inappropriate elimination and aggression are common reasons for relinquishment of cats. These are not only reasons used for relinquishment or euthanasia, but also reasons for a diminished human-animal bond. Perhaps the cat that urinated outside of its litter box is now an outside cat, with an increased chance of being severely injured. Although, in a case such as this, the primary reason for potentially euthanizing a cat that was hit by a car may be its injuries, if the veterinarian digs deeper, he or she may find that the real reason for euthanasia was that the cat was made to be an outside cat because it had urinated outside of the litter box.

But as we look to solving these problems, we use the behavior-modification methods of classical and operant conditioning. In more recent times, there has been an effort to bring this information to owners, shelters, and other organizations, such as those training working dogs. The American College of Veterinary Behavior is a recognized specialty of the American Veterinary Medical Association, consisting of veterinarians fulfilling special postgraduate education and research responsibilities. Another organization is the American Veterinary Society of Animal Behavior, a group of veterinarians who share an interest in understanding, teaching, and treating behavior problems in animals. The Animal Behavior Society promotes the study of animal behavior and also has a certifying arm for people who have reached a certain level of academic training. There are other organizations related to dog and horse training, as well as organizations concerned with the welfare and training of captive and laboratory animals. With these examples, there are a good number of resources for owners to seek help with their pets.

With help from properly trained people, owners can help change the behavior of their pets. What owners would not want a dog to sit when they came home, instead of jumping up on them? Or a bird who did not scream when they left the room? Or a cat that did not scratch the furniture? If an owner provides opportunities for an animal to perform the correct behavior, rewards such behavior, and properly uses humane punishment techniques, the animal stops performing the inappropriate behavior and, subsequently, performs the appropriate behavior. The methods of classical and operant conditioning apply to everyday life, not just in the laboratory. We, as humans, are often focused on *stopping* a behavior, but we do not focus on what the animal *does* do correctly and on rewarding that behavior.

There is obviously a close relationship between an animal's behavior and the human-animal bond. Without selecting for wanted behaviors, whether for work or companionship, we would not have the domesticated animals we have today. Owners need to understand their pets' behavior, how they influence it, and how they can change it for the better, in order to decrease relinquishment and euthanasia for problem behaviors. And humans need to appreciate the uniqueness and wonder of animal companions, for without them, what would many of us do?

Further Resources

Hart, B. L., & Hart, L. A. (1998). *Perfect puppy: How to choose your dog by its behavior.* New York: Freeman.

Landsberg, G., Hunthausen W., & Ackerman L. (2003). *Handbook of behavior problems of the dog and cat.* New York: Saunders.

Lindsay, S. R. (2000). *Handbook of applied dog behavior and training, Vol. 1: Adaptation and learning.* Ames: Iowa State University Press.

Serpell, J. (1995). *The domestic dog: Its evolution, behaviour, and interactions with people.* Cambridge: Cambridge University Press.

Melissa Bain

Bonding
Chimpanzee and Human Relationships

Western culture frequently sees chimpanzees as intelligent creatures, capable of being trained, learning sign language, and forming bonds with each other and even with humans. People visit zoos and watch with amazement as chimpanzees exhibit human-like characteristics such as making faces, using tools, becoming excited, angry, and scared, and caring for their babies. Their size, mannerisms, eyes, and hands remind us so much of ourselves as we stare in awe and wonder what they are thinking. Could they be thinking the same thoughts about us? It is not a coincidence, after all, that we see such similarity in them. They are our closest relatives in the entire animal kingdom; chimpanzees share 99 percent of the same DNA with humans.

Most people are appalled when made aware of the strong possibility that chimpanzees may become extinct in the wild in our lifetimes. At the same time, we accept the use of chimpanzees for medical research because of their genetic similarity to humans (Goodall, 1995). So close are the two species that chimpanzees are used in infectious disease laboratory research because chimpanzees can catch or be infected with all known human infectious diseases (with the exception of cholera) (Jane Goodall Institute, n.d.).

This western view of chimpanzees as sentient beings worthy of protection and conservation, however, is not unanimously shared. In their natal African countries, chimpanzees and humans coexist in very different ways. Chimpanzees are often seen as distant, neutral, forest-dwelling animals that are not problematic until human "space"—as defined by people—is entered. Perhaps this is the same way that Americans might view mountain lions or birds of prey, or the way that Australians might view kangaroos. When interactions between chimpanzees and humans in Africa become too antagonistic, for example, when the animals raid crops for food, chimps are thought of as a pest species (Hill, 2002; Naughton-Treves et al., 1998). In other circumstances chimpanzees have an important nutritional value and serve as a vital source of protein. In still other circumstances, eating chimpanzee meat is strictly taboo, although such a taboo may not preclude the hunting of chimpanzees if they are considered pests.

Chimpanzees: Imminent Extinction?

Chimpanzees exhibit a number of inherent biological, ecological, and behavioral characteristics that make them especially susceptible to extinction. Whereas other wildlife species may be able to tolerate hunting and habitat degradation to a certain extent, chimpanzees cannot. There are only about 150,000 chimpanzees spread across 21 African nations. Their reproductive and life history parameters, extensive habitat and diet requirements, and genetic similarity to humans suggest that conservation efforts should pay particular attention to these factors when devising protection strategies.

Relative to other mammals, all primates have very slow individual growth rates, birth rates, and death rates; these rates are one-fourth to one-half those of other mammals of similar body size (Charnov & Berrigan, 1992). Chimpanzees, in particular, have an infant stage that spans 60 months, with the average weaning age at 48 months. Chimpanzee females do not give birth until they are 14 years old; even then, females have only a 41 percent chance of surviving to that age (Hill et al., 2001). Once conception is achieved, gestation lasts for 240 days. Female chimpanzees can conceive until about age 43 (Ross, 1991). With the long infancy period and delayed weaning, the birth interval is long, approximately five years. Even if the infant dies (40–60% of all infants die before age 5) the mother does not conceive again for almost two years on average (Tai data: Boesch & Boesch, 2000). In an extensive study of five chimpanzee study populations, Hill et al. (2001) found that wild chimpanzees have a reproductive rate of 0.8—well below replacement capacity.

These parameters tell us that chimpanzees are long-lived species who depend on both immediate family and group members for survival. Like humans, chimp babies rely on their mothers for years, and bonds between family members are strong. Another important aspect of chimpanzee behavior is the fission-fusion aspect of social organization. Chimpanzees may associate with large groups, but they may also regularly split off into subgroups, or parties. Parties primarily consist of males; females transfer from their natal group to another group and thus do not maintain strong ties with their female family members (Williams et al., 2002). Females also frequently spend time foraging alone.

Chimpanzees have a fairly large home range of approximately 1,250 hectares (Wrangham, 1977), but this is actually quite varied. For example, populations in densely wooded regions use 5–38 square kilometers (500–3,800 ha), compared to populations in sparsely wooded areas, who use 25–560 square kilometers (2,500–56,000 ha) (Nishida & Hiraiwa-Hasegawa, 1987). Ensuring the availability of adequate home range areas ensures that there will be enough genetic flow between groups to prevent inbreeding.

These aspects of the chimpanzee social system and range requirements have important implications for long-term field research and conservation efforts. Temporary absences by individuals may not be accounted for in cross-sectional studies commonly used for conservation assessment. Aspects of chimpanzee habitat use and range size are more accurately determined through longitudinal studies (e.g., Gombe, Mahale, & Kibale), whereby the flexibility of group size and composition can be monitored and recorded (Boesch & Boesch, 2000).

In light of these innate biological and ecological aspects that render chimpanzees particularly susceptible to extinction, conservation research needs to explore human perceptions of and behavior toward chimpanzees in their natural habitats in Africa and how they affect chimpanzee survival. If international conservation programs are going to succeed in protecting chimpanzees in the wild, they must take into account the relative value systems and perceptions of those human populations living around and sharing resources with chimpanzees, as well as the resulting behaviors that are detrimental to chimpanzees. Perceptions and behavior toward chimpanzees, just as toward any other wildlife species, depends on the costs and benefits of interactions between the species as well as the level of competition for resources (Weladji et al., 2003; Gillingham & Lee, 1999).

Ethnoprimatology

Ethnoprimatology is the study of "intimate" relationships between humans and primates in the primates' natal habitat (Fuentes & Wolfe, 2002). Intimate does not necessarily mean friendly, or even neutral. Antagonistic, yet close, relationships between humans and primates (and any other animal, for that matter) are found whenever there is conflict over space or resources.

Examples of intimate, noncontentious relationships between humans and primates abound. Usually, there is a value tied to the primate; the value may be religious, ecological, or economic, or the primate may even be given an intrinsic value in their own right. In these situations, there is some valuable benefit from associating with, and effectively protecting, the primate. For example, Cormier (2002) found that among the Guaja people of Brazil, monkeys form not only a significant part of their diet during the rainy season, but also part of their cultural identity. Monkeys are kept as pets and serve as a fertility symbol for the women who care for them. Pet monkeys are also valuable tools for adolescent boys, who learn the monkey calls and become better hunters when they get older. Cormier suggests that Guaja both eat and keep monkeys, illustrating a unique cultural cosmology that emphasizes common kinship and a "like eats like" philosophy. This philosophy ensures the survival of the primate by balancing dietary and cultural needs.

Quite a different example of an intimate relationship between humans and primates comes from Sicotte and Uwengeli's 2002 study of mountain gorillas in the Virunga Mountains of Rwanda. In an attempt to inform gorilla conservation programs, Sicotte and Uwengeli assert that it is necessary to understand the Rwandan view of nature in terms of the forest ecosystem and its inhabitants. They found that people were actually fearful of the forest and, thus, traditionally had very minimal contact with the gorillas. Gorillas were found to be absent from traditional stories, as well as from the people's diet. Any primate meat, in fact, was said to be "dirty." People perceived gorillas as smart, self-controlled, quiet animals. Gorillas were not seen as vermin because they did not raid agricultural crops. Sicotte and Uwengeli's study does not illustrate an antagonistic relationship but, rather, a neutral, somewhat disaffected relationship.

These two examples illustrate the ways in which humans can and do coexist with primates. For the Guaja, the primates are entwined in cultural ideologies as well as in their diet. For the people near the Virunga's of Rwanda, primates occupy a more distant space in their culture. The gorillas are quiet, nonintrusive, and thus generally absent from cultural and nutritional aspects of human life. Here it is clear that there are different costs and benefits of interacting with primates in each setting. What, then, are the costs and benefits associated with human-chimpanzee relationships? Research reveals that relationships are different in West and Central Africa than they are in East Africa. This stems from a different utility of chimpanzees in each geographic area.

Humans and Chimpanzees in Africa

There are a few records of intimate, positive relationships between humans and wild chimpanzees in Africa. Peterson and Goodall (1993a) describe a traditional story in the Ivory Coast in which a chimpanzee saved a lost child in the forest, which then rendered the chimpanzee sanctified. As a result, in some villages chimpanzees are a sacred totem and are thus not eaten. Chimpanzees are hunted, however, in other parts of the Ivory Coast. Peterson and Goodall (1993a) also assert that when the Batoro tribe migrated to areas near Kibale Forest, Uganda, chimpanzee hunting decreased, although previously it had been a subsistence activity. The Batoro do not eat chimpanzee meat and apparently will not even eat from the same plate or drink from the same cup of a person who does. Even with this cultural aversion to eating chimpanzees, the Batoro may inadvertently harm the chimpanzees with their nets and snares used to catch other types of wildlife for food.

Although few traditional chimpanzee stories are known, there is substantial local human knowledge of chimpanzee ecological and economical value. The difficulty lies in discerning the costs and benefits of maintaining a relationship with chimpanzees that either supports their ecological survival and value or places the species' economical value so high that pressure is put upon their ecological value. Examples of regional disparity illustrate the complexity of interactions between chimpanzees and humans and how the species' value is diminishing as human needs are being unmet.

The Bushmeat Crisis in West and Central Africa

Wild forest animals that are hunted and eaten are referred to as "bushmeat." The "bushmeat crisis" refers to the illegal and unsustainable hunting of many of these forest animals and is considered the most significant threat to central and West African wildlife (Bushmeat Crisis Task Force, 2000). Primates make up 15 percent of the bushmeat trade. Great apes make up approximately 1 percent, although this may be underestimated because meat is cut up into indeterminable pieces at the market (Bushmeat Crisis Task Force, 2000).

Peterson and Goodall (1993a, p. 58) recall an anecdote told by a Gambian conservationist:

> Before the Europeans came, Africans traditionally believed that at one time chimps had been human. But then a group of people were cursed by Allah for fishing on a sacred day and banished to the forest—they became chimpanzees. Africans didn't hunt chimps until the Europeans came and taught them that they could make money by killing chimpanzee mothers and selling their babies.

In Gambia and other west African countries, hunting chimpanzees *for profit* began and exponentially increased with the introduction of European market pressures. But it is important to understand that the hunting of bushmeat is not inherently wrong. As Peterson and Goodall (1993a) explain, "eating meat is as old as Africa." A clear illustration of this is the Bantu word *eyama* and variations of this word (yama, ama), which means both wildlife and meat (Quamman, 2003). Eating meat derived from hunting wild animals was normal (perhaps even more normal than eating domestic meat), culturally accepted, and a relevant subsistence activity in many African cultures; it is the *unsustainable* nature of the activity, with increasing populations, growing demands, and more efficient hunting methods, as well as the *risk of extinction* for many wild animals, that has created a crisis situation. Now there is no place in Africa where the hunting of chimpanzees is sustainable, whether for profit or sustenance.

The causes of such unsustainability are complex, but one major factor is the increase in commercial logging throughout West and Central Africa, whereby roads are cut through previously intact forests, and trucks and workers pass through areas full of protein-rich animals. Hidden from forest guards, these workers capitalize on the dietary opportunity to hunt wild animals. Unfortunately, such reliance on bushmeat for protein has created major pressure on the animal populations. Whereas the loggers still need a source of protein, the animal sources they have come to depend on are no longer substantial enough to support the activity. In this case, chimpanzees are seen as an important, perhaps vital, source of food.

In areas where hunting bushmeat has become an industry for profit, not just sustenance, it is highly probable that people are desperate for food and are willing to hunt *and buy* bushmeat, regardless of the fact that the system is unsustainable and depletive. Trefon (1998) states that urban populations in Central and West Africa are growing at 2–4 percent per year; yet, many urban families view livestock, such as cattle, goats, and chickens, as savings and assets rather than a food source. Given this preference for nonlivestock protein, as long as there are chimpanzees in the forest and inadequate laws to protect them, they will remain in the bushmeat market.

Although chimpanzee conservation is intrinsically entwined with the larger issue of poverty alleviation, research shows that increases in wealth continue to be associated with increases in the demand for wildlife. Yet there are ways, however daunting, that the bushmeat crisis can be alleviated. The extinction of chimpanzees and other vulnerable bushmeat animals is directly related to expanding human populations, poverty, and food scarcity. Rowcliffe (2002) argues that entirely preventing hunting is not feasible or ethically acceptable, because entire populations of humans rely on the activity for food. Therefore, it is essential to address human conditions and needs as part of a plan to protect the conditions and needs of chimpanzees.

Ling et al. (2002) suggest that solutions must involve a complex reorganization of conservation policy to include enforcement and human livelihood development; this will involve biology, economics, anthropology, and other disciplines that understand the age-old dilemma of common-pool resources and the tenuous interaction between humans and their environment. Robinson and Bennett (2002, p. 332) argue that "the only way out of this crisis will be offered by long-term, integrated efforts that provide alternative sources of protein and income for the rural poor, curtail the commercial trade in wildlife, secure wildlife populations in protected areas, educate hunters and buyers, and involve government, the not-for-profit and the private sectors." In other words, people have to *perceive* that they have easier, cheaper, and better options for protein consumption and income generation.

Crop Raiding and Habitat Overlap in East Africa

There is yet another chimpanzee "crisis" developing in East Africa. In Tanzania, Kenya, and Uganda, chimpanzees face habitat destruction and human conflict on the borders of agricultural lands and forest borders. Although chimpanzees are not hunted and sold in markets with the regularity found in West and central Africa, the conflict between humans and chimpanzees can be just as intense because they involve natural resource competition and scarcity. Interestingly, Boesch and Boesch (2000) suggest that one of the reasons why the hunting and eating of chimpanzees has remained minimal in East African countries is that there is heavier Islamic religious influence in these areas than there is in West and Central Africa. According to the Islamic tradition, killing animals for food requires special hygienic measures, which may be too inconvenient for hunters to conduct.

Peterson and Goodall (1993a) argue that East African chimpanzees, like chimpanzees in West and Central Africa, are susceptible to shamanistic purposes. They explain an incident in Uganda where crop-raiding chimpanzees were killed and then, with the exception of their brains, fed to dogs. The chimpanzee brains were sold to a shaman who believed the brains would speed the healing of broken bones.

A well-known example of chimpanzee-human interactions is found in and around Budongo Forest, Uganda. This large forest is home to 600 chimpanzees, 40 of which have been closely studied for the last 15 years (Reynolds, 2005). Villages around Budongo are primarily inhabited by subsistence farmers. Their crops, however, often border the forest, creating a tempting new food source for chimpanzees, baboons, and other primates. Watkins (2006) found that people living in Nyakafunjo village actually had a positive perception of chimpanzees, primarily because the people identified with chimpanzee behavior; chimpanzees share their food with each other, nurture their babies, groom, fight, scream, and sleep—all behaviors found in people. Not coincidentally, crop raiding

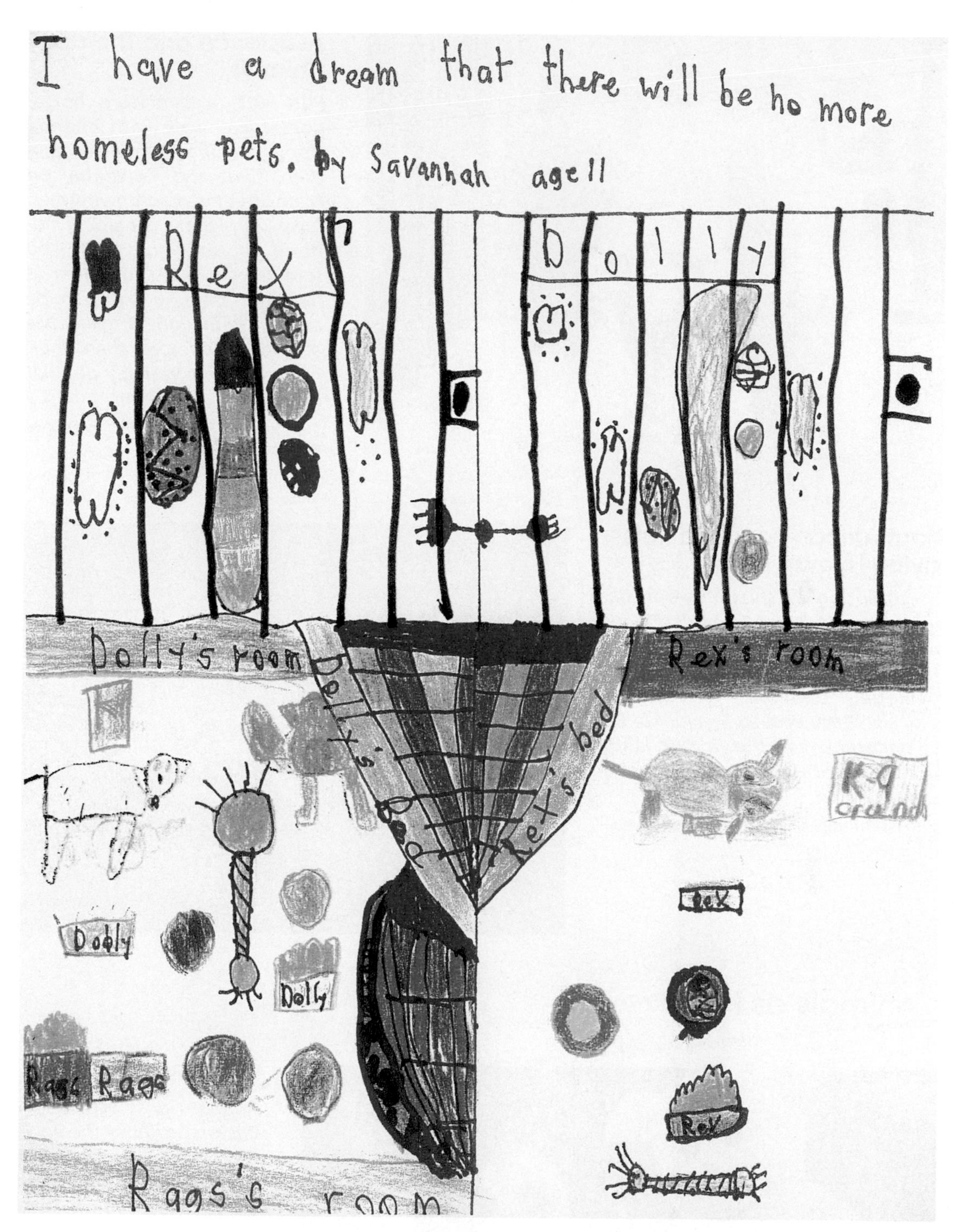

As part of a worldwide project that Marc Bekoff is conducting with Jane Goodall (Roots & Shoots program, http://www.rootsandshoots.org) called "We have a dream," we ask children to fill in the blank in the statement "I have a dream that ____" and then draw a picture. All of the pictures we have collected show deep concern for the well-being of animals and the close relationship children have with their pets and other animals. This drawing shows that Savannah Nystrom (age 11) is very concerned about homeless pets: she portrays them in cages (above) and then in their rooms in an imagined home (below).

Animal Assistance to Humans

Assistance and Therapy Animals

Putt-Putt, a miniature horse, goes for a walk with Rhonda Applegate (back) and her children, Clint and Samantha, on Main Street in downtown Ames, Iowa. Applegate is helping the horse adjust to the sights and sounds of a populated area as part of its training to be a guide animal. With higher intelligence and longer lifespan than dogs, ponies are increasingly popular as guides for the blind.

Giant African Pouched Rats Saving Human Lives

The attention of this mine-detecting rat is triggered by a buried land mine in Morogoro, Tanzania. The giant African pouched rat's highly developed olfactory nerve and small size have proved invaluable in the dangerous tasks of finding land mines and locating trapped earthquake victims.

Photo by Christophe Cox.. Used by permission.

Animals as Food

Veganism

Animal rights activists in Boston promote the environmental benefits of a vegan diet. Many vegans argue that raising livestock drains environmental resources such as water and fossil fuels, pollutes nearby waterways with animal waste, and uses potentially productive farm land for grazing.

Stock Dogs and Livestock

A border collie "works" a group of sheep on a Scotland farm. Kelpies, border collies, and huntaways are genetically predisposed to the strenuous but controlled task of herding animals. Stock dogs and owners have the opportunity to show off their skills at various sheep and cattle trials held throughout the United Kingdom.

Courtesy of Shutterstock.

Archaeology

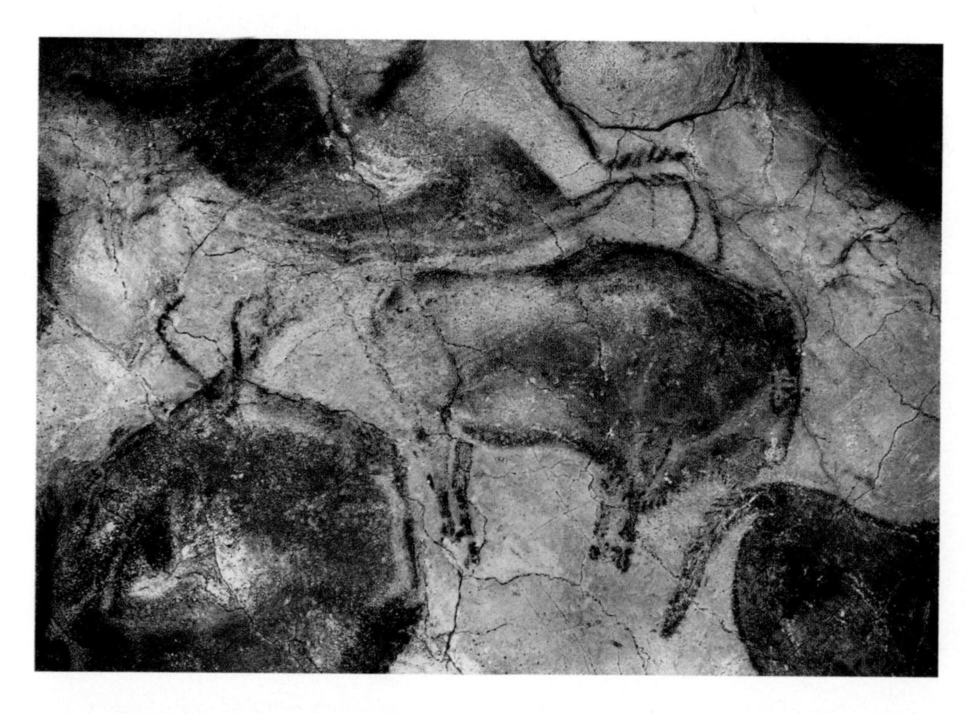

Archaeology and Animals

Cave art featuring animals can be found throughout the world, illustrating human beings' universal respect for them. The bull is frequently depicted as a symbol of power. Shown here is the Hall of the Bisons, a Paleolithic cave painting found in Altamira, Spain.

©Giraudon/Art Resource, NY.

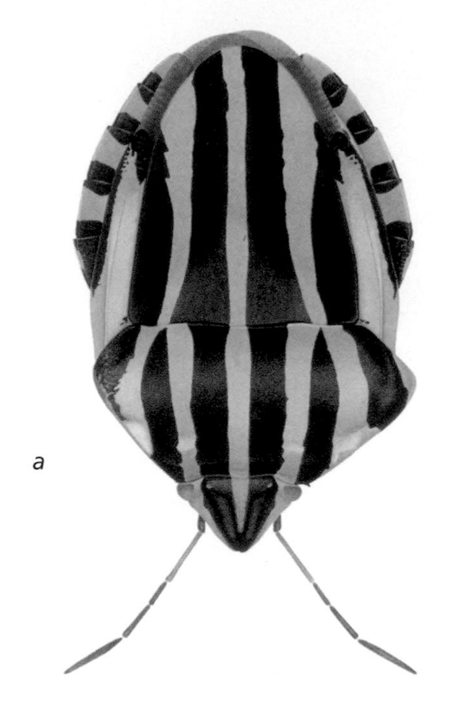

a

b

Insects in Art

Insects have helped inspire environmental conservation movements, and art by (a) Cornelia Hesse-Honegger and (b) Andy Warhol instills awareness of human-induced environmental damage. Hesse-Honegger paints insects with mutations that she has found near Chernobyl, Three Mile Island, and other nuclear installations.

Art and design often incorporate nature. Karen Anne Klein augments still life with insects. Pictured here, details from Entomological Cabinet of Curiosity.

Bonding

Attachment between Humans and Animals

A little girl snuggles with her pet cat.

Courtesy of Shutterstock.

Chimpanzee and Human Realtionships

Like humans, chimpanzee babies rely on their mothers for years, and bonds between family members are strong.

©Jeryltan/Dreamstime.com.

Dogs as Social Catalysts

Research has proven that dog walkers have significantly more chance encounters and longer conversations with complete strangers whenever they are accompanied by their dog, than when walking exactly the same route without an animal.

Courtesy of Shutterstock.

Communication and Language

Birdsong and Human Speech

A limited definition of bird song includes only attracting mates and competing with members of the same sex for territory, but it is clear the issue is more complicated.

© Farmer/Dreamstime.com.

Great Apes and Language Research

Talbot J. Taylor, a professor of English and linguistics at the College of William and Mary, is coauthor of Apes, Language and the Human Mind, *which contends that apes have the ability to use language to communicate with humans. Behind him is a Lexigram keyboard, a diagram of symbols through which some apes have been trained to communicate.*

©AP Photo/Daily Press, Adrin Snider.

Arctic Char, Zander Fish, and Global Warming

The Arctic char, a loser in global warming. Native to Finland's cold, shallow inland lakes, char are being forced north in search of colder climates as rising temperatures heat the shallow water.

Courtesy of Juha Jurvelius.

Birds and Recreationists

While species such as the great crested grebe have adapted well to human disturbance, instincts such as the need to defend their territory remains strong.

Courtesy of Shutterstock.

Conservation Medicine Links Human and Animal Health with the Environment

We often attribute species decline to habitat destruction. What is alarming is that many amphibians occupying undisturbed wilderness habitats are also disappearing at a previously unseen rate.

Courtesy of Shutterstock.

Exotic Species

Humans frequently import wild exotic species as household pets and sometimes lose or abandon them, such as this parrot in San Francisco, who likely belongs to one of the colonies established in the urban jungles of cities such as San Francisco, New Orleans, and even New York.

Courtesy of Shutterstock.

Ecosystems and a Keystone Species, the Plateau Pika

Plateau pikas are vital to the health of the Tibetan plateau's ecosystem. Besides providing a staple food source for predators, the plateau pika digs extensive networks of burrows that provide shelter for birds and lizards and encourage the growth of several types of plants.

Courtesy of Andrew Smith.

by chimpanzees is minimal in this village. Kiwede (2001) found that subsistence farmers in the area also viewed chimpanzees as "well-behaved," but farmers growing sugarcane as a cash crop viewed chimpanzees as serious vermin. For these farmers, the costs of allowing chimpanzees to eat the crop are simply too high; eaten sugarcane represents a real monetary loss. This represents a growing problem for the chimpanzees of Budongo and surrounding forests because the sugarcane industry is lucrative and growing fast (Reynolds et al., 2003; Reynolds, 2005).

Dunbar and Barrett (2000) argue that apes are often perceived as being aggressive simply because of their large size and are killed out of fear in order to get rid of what people perceive to be a dangerous enemy. This fear is unfortunately becoming a reality in a small forest fragment near Budongo. Although the habitat of Kasokwa Forest Reserve is quite different than the intact, dense Budongo Forest—it is a mere 73 hectare riverine fragment—the 13 Kasokwa chimpanzees exhibit the same humanlike behaviors as any other group of chimpanzees. However, crop raiding and face-to-face interactions with humans are much more of a problem in Kasokwa than in Budongo. Kasokwa is one of a number of forest fragments that are separated by degraded forest edges, village settlements, and subsistence crops. The borders of such fragments are becoming increasingly pressured by sugarcane crops. This, in turn, has promoted an increase in chimpanzee crop-raiding, as well as incidences where chimpanzees have seized human infants who have been left unattended near the borders of subsistence crops and the forest edge (Reynolds et al., 2003; Reynolds, 2005).

In these cases, the perception of chimpanzees varies with the extent of negative interaction. The solution to the problem of chimpanzee endangerment in these areas is similar to those in bushmeat-ridden countries: Reduce the costs and increase the benefits of coexisting with chimpanzees. Perhaps this means buffering cropland with plants distasteful to chimpanzees; perhaps it means mandating a large enough "no-crop zone" around the perimeter of the chimpanzee's forest habitat. For this to happen, however, local people must be able to acquire wood and water, now obtained from the forest, from another source. Many of Robinson and Bennett's (2002) solutions also apply, such as providing alternative sources of resources and income, stabilizing the commercial trade of forest resources, educating local people, and involving all levels of support, including the government, not-for-profit, and private sectors.

Saving Chimpanzees

People who do not live in Africa might wonder why it is important to care about the survival of chimpanzees. There are three reasons. One is the human dimension. When animals are threatened, it is almost always linked to a human problem. Endangered animals can be ecological signals that humans are behaving in unsustainable, environmentally and ecologically detrimental ways.

The second reason to care about chimpanzee survival is that they are humans' closest living relatives. The impact of their survival spans the globe in universities and research facilities where behavioral and evolutionary research is being conducted. We can learn an immense amount about ourselves through the study of chimpanzees. As western zoo visitors and native Africans alike have seen, chimpanzees exhibit physical and behavioral traits that illustrate just how close they really are to humans; tool use, culture, and warfare are prominent features of chimpanzee society. Losing wild chimpanzees would be a huge loss to many scientific disciplines.

The third reason is for the sake of the chimpanzees themselves. If one subscribes to the idea that humans are indeed connected to the rest of the natural world, it is hard to ignore the demise of any species. Peterson and Goodall (1993b) argue that "changed attitudes can lead not only to changes in our personal lives but also to policy changes at the corporate and governmental level." Support for both formal and environmental education, poverty alleviation programs, and environmental protection initiatives are all crucial for the survival of chimpanzees.

Unfortunately, Stanford (2004) correctly argues that the bushmeat crisis has become only one of a myriad of environmental and social causes connected to the survival of the chimpanzee. With so many causes vying for attention and support, chimpanzee conservation must rely on the fact that the plight of the species is deeply intertwined with the plight of humans in Africa. Although Quamman (2003) argues that this is inherently an "African issue," many believe that people in the international community have an obligation to inform themselves about impending wildlife extinctions and their root causes. From many conservationists' points of view, optimism not pessimism, realism not despair, and global-mindedness rather than narrow-mindedness, are crucial to the continued search for successful chimpanzee conservation.

See also

Animals as Food—*Global Diversity and Bushmeat*
Bonding—*Animals Beings I Have Known*
Communication and Language—*Great Apes and Language Research*

Further Resources

Boesch, C., & Boesch, H. (2000). *The chimpanzees of the Tai forest: Behavioral ecology and evolution.* Oxford: Oxford University Press.

Bushmeat Crisis Task Force. (2000). *Bushmeat. A wildlife crisis in west and central Africa and around the world.* Silver Spring, MD: BCTF.

Charnov, E., & Berrigan, D. (1992). Why do primates have such long life spans and so few babies? *Evolutionary Anthropology, 1,* 191–94.

Cormier, L. (2002). Monkey as food, monkey as child: Guaja symbolic cannibalism. In A. Fuentes & L. Wolfe (Eds.), *Primates face to face: The conservation implications of human-nonhuman primate interconnections* (pp. 63–84). Cambridge: Cambridge University Press.

Dunbar, R., & Barrett, L. (2000). *Cousins, our primate relatives.* London: BBC Worldwide Ltd.

Fuentes, A., & Wolfe, L. (Eds.). (2002). *Primates face to face: The conservation implications of human-nonhuman primate interconnections.* Cambridge: Cambridge University Press.

Gillingham, S., & Lee, P. C. (1999). The impact of wildlife-related benefits on the conservation attitudes of local people around the Selous Game Reserve, Tanzania. *Environmental Conservation, 26*(3), 218–28.

Goodall, J. (1995). Why is it unethical to use chimpanzees in the laboratory? *Atla-Alternatives to Laboratory Animals,* 23(5), 615–20.

Hill, C. M. (2002). Primate conservation and local communities—Ethical issues and debates. *American Anthropologist, 104*(4), 1184–94.

Hill, K., Boesch, C., et al. (2001). Mortality rates among wild chimpanzees. *Journal of Human Evolution 40,* 437–50.

Jane Goodall Institute. http://www.janegoodall.org/

Kiwede, T. Z. (2001). Attitudes of local people towards crop raiding with particular reference to chimpanzees: A case study of Nyabyeya Parish. Dissertation for Uganda National Certificate in Forestry, Nyabyeya Forestry College.

Ling, S., Kumpel, N., & Albrechtsen, L. (2002). No new recipes for bushmeat. *Oryx, 36*(4), 330.

Naughton-Treves, L., Treves, A., Chapman, C., & Wrangham, R. (1998). Temporal patterns of crop-raiding by primates: Linking food availability in croplands and adjacent forest. *Journal of Applied Ecology, 35*(4), 596–606.

Nishida, T., & Hiraiwa-Hasegawa, M. (1987). Chimpanzees and bonobos: Cooperative relationships among males. In B. Smuts, D. Cheney, R. Seyfarth, R. Wrangham, & T. Struhsaker (Eds.), *Primate societies* (pp. 462–74). Chicago: University of Chicago Press.

Peterson, D., & Goodall, J. (1993a). *Visions of Caliban: On chimpanzees and people*. New York: Houghton Mifflin Company.

———. (1993b). Towards kinship. *The Animals' Agenda, 13*(4), 44.

Quammen, D. (2003, June 15). Almost cannibalism: The chances of keeping Africans from eating all their apes don't look good. *New York Times Book Review*, p. 16.

Reynolds, V. (2005). *The Chimpanzees of Budongo Forest: Ecology, behavior and conservation*. Oxford: Oxford University Press.

Reynolds, V, Wallis, J., et al. (2003). Fragments, sugar and chimpanzees in Masindi District, Western Uganda. In L. Marsh (Ed.), *Primates in fragments: Ecology and conservation* (pp. 309–20). New York: Kluwer Academic/Plenum Publishers.

Robinson, J. G., & Bennett, E. L. (2002). Will alleviating poverty solve the bushmeat crisis? *Oryx, 36*(4), 332.

Ross, C. (1991). Life history pattern of New World monkeys. *International Journal of Primatology, 12*(5), 481–502.

Rowcliffe, J. M. (2002). Bushmeat and the biology of conservation. *Oryx, 36*(4), 331.

Sicotte, P., & Uwengeli, P. (2002). Reflections on the concept of nature and gorillas in Rwanda: Implications for conservation. In A. Fuentes, & L. Wolfe (Eds.), *Primates face to face: The conservation implications of human-nonhuman primate interconnections* (pp. 163–82). Cambridge: Cambridge University Press.

Stanford, C. (2004). Eating apes (book review). *The Quarterly Review of Biology, 79*(1), 113.

Trefon, T. (1998). Urban threats to biodiversity in the Congo Basin. In C. Besselink & P. Sips (Eds.), *The Congo Basin: Human and natural resources* (pp. 89–99). Amsterdam: IUCN.

Watkins, C. (2006). Local ecological perceptions of chimpanzees and forest resources: A case study near Budongo forest. In N. E. Newton-Fisher, H. Notman, J. D. Paterson, & V. Reynolds (Eds.), *Primates of Western Uganda* (pp. 423–37). New York: Springer.

Weladji, R. B., Moe, S. R. et al. (2003). Stakeholder attitudes towards wildlife policy and the Benoue Wildlife Conservation Area, North Cameroon. *Environmental Conservation, 30*(4), 334–43.

Williams, J. M., Pusey, A. E., Carlis, J. V., Farm, B. P., & Goodall, J. (2002). Female competition and male territorial behaviour influence female chimpanzees' ranging patterns. *Animal Behaviour, 63*, 347–60.

Wrangham, R. W. (1977). Feeding behaviour of chimpanzees in Gombe National Park, Tanzania. In T. H. Clutton-Brock (Ed.), *Primate Ecology* (pp.503–38). London: Academic Press.

Cristy Watkins and John Bock

Bonding
Companion Animals

In the not-so-distant past the notion of studying social relationships between humans and other animals was seen as heresy. In Europe, until the early modern period, animals were viewed as irrational beings put on this planet for the economic benefit of humans. Most scholars of the time insisted that affectionate relationships between people and animals were not only distasteful but also depraved. Happily, attitudes toward animals have changed, and during the past three decades, the subject of relations between people and other animals has become a respectable area of research.

Though this area of inquiry has gained respect, questions still remain: Why study relationships between animals and people to begin with? What purpose does it serve? One answer would be that it allows us to "create theoretical and conceptual bridges that not only link together widely separated disciplines but also span the gulf between the world of humans and the life of the rest of the planet" (Podberscek, Paul, & Serpell, 2000, p. 2). Scientists studying animal relationships have noted that we owe animals a debt of gratitude because we are who we are "as much because of our relationships with non-human animals as because of human ones, and we do ourselves a great disservice—and probably great harm—by denying or ignoring this" (Podberscek, Paul, & Serpell, p. 2). Many believe that understanding our connection to nonhuman animals, and companion animals in particular, widens our perspective of the world both emotionally and sociologically.

According to James Serpell and Elizabeth Paul (1994, p. 129), "the word 'pet' is generally applied to animals that are kept primarily for social or emotional reasons rather than for economic purposes. Another salient characteristic of pets, as opposed to domesticated animals, is their not having a direct commercial utilization. The worth of most domesticated animals is measured by the practical services and economic resources they provide. The value of owning a pet, however, appears to come from the relationship itself. That is, people love and cherish their pets not necessarily because they are useful but because they fulfill certain social and emotional needs.

The current use of the term "companion animal" as an alternative to "pet" further emphasizes the difference between humans' relationships with domesticated animals and the animals they choose to take into their homes and hearts. Rather than simply looking after a pet, many owners of companion animals appear to enter into some form of relationship with them that is analogous to a human-human relationship. It has been suggested that the most common role a companion animal can fulfill is that of a friend to its owner. Companion animals are also frequently described as family members, or even as surrogate siblings or children. Finally, companion animals assist in the achievement of trust, autonomy, responsibility, competence, and empathy toward others (Beck & Katcher, 1983).

The literature about companion animals is extensive. A great deal of the information breaks down into four major categories: sociohistoric and economic aspects of domestication and pet-keeping, physical benefits of pet ownership, psychological benefits of pet guardianship, and the case against pet guardianship. Also significant are the ways in which pet guardianship affects notions of human identity.

Sociohistoric and Economic Aspects of Domestication and Pet-Keeping

According to Roger Caras (2002), the goat may have been the first animal humans ever domesticated. It came into the lives of humans in the preagricultural stage of our development, during the Middle Stone Age, between twelve thousand and fifteen thousand years ago at the end of the last glacial period. The only other preagricultural domestic animals that humans made a part of their lives were dogs, reindeer, and sheep. The goat probably did come first, judging from the bones that have been found and their demonstrable age.

The dog in Japanese lore is considered one of the five most prominent animals. In fact, a story is told of a dog that was traveling with his mistress and a silkworm. The dog got hungry and "ate the silkworm and he later produced a large amount of silk in a single strand coming from one of his nostrils. When the dog died, his mistress buried him under a mulberry tree and prayed to Buddha to thank him for such a wonderful dog. Instantly the mulberry tree was covered with silkworms and the woman became rich" (Caras, 2002, p. 75).

From preagricultural societies forward dogs have been present. The earliest fossil remains, dating from about 12,000 years ago, come from Iraq and Israel. Later finds have been discovered in the United States from about 10,000 years ago, Denmark and the United Kingdom from about 9,000 years ago, and from China about 7,000 years ago (Olsen & Olsen, 1977). The evidence available suggests that the earliest domestication of the dog took place in the Near East during the preagricultural Mesolithic period of human culture, which followed the end of the last global ice age (Messent & Serpell, 1991).

Evidence shows that the human-companion animal bond has existed for many thousands of years. It appears that the human capacity to relate on a social level to wild animals of other species lies at the heart of the domestication process. With regard to dogs, useful canine skills, such as scavenging, hunting, and territorial barking, may have helped to solidify the attachment but were probably not necessary for its original formation. The initial switch from pet keeping to domestication was most likely precipitated by sudden and favorable changes in the environment in specific geographical locations.

The rich, the poor, and the middle class have all contributed to the nurturing of pets over the last several thousand years. In the year 1062, for example, toward the end of the Han dynasty, the Chinese Emperor Ling became so "infatuated with his dogs that he invested them all with the rank of senior court officials. This entitled them to the finest food available, sumptuous oriental rugs to sleep on and a personal bodyguard of hand-picked soldiers" (Serpell, 1986, p.34). When the Hans were displaced by foreign dynasties, the new Emperors followed Ling's example by displaying a great love of dogs, especially the modern Pekingese.

In Europe, Greece and Rome especially, class distinctions related to pet guardianship were less extreme. The early Greek inhabitants of the city of Sybaris were avid pet lovers, their favorites being long-haired Maltese lap dogs. These dogs were paraded about the streets and taken everywhere, including to the public baths and wrestling schools. They were also encouraged to share the beds of their owners at night. The well-known Roman authors Ovid and Catullus wrote poems to commemorate the deaths of their mistress's pet birds, and the Emperor Hadrian insisted that "monumental tombstones be erected over the graves of his favorite dogs" (Serpell, 1986, p. 37).

A more careful consideration of the factors that have contributed to the popularity of dogs and cats as pets can provide additional insights into the relationship between humans and companion animals. For example, why were the "Asian subspecies of wolves and wildcats, rather than other available wild species, the only ones initially chosen for domestication?" (Messent & Serpell, 1991, pp. 11–12). One reason may be that, once the role of companion animal had been filled by dogs and cats, there was no longer a need to search for and adopt alternative species (Messent & Serpell, 1991).

Pets provide a sense of security in fragmented and confusing times. That plus the fact that many families and individuals now have more disposable income, larger homes, and more leisure time have all contributed to an increase in pet-keeping. In 2001, "58.3% of all households in the United States owned at least one pet at some time during the year. Pet ownership conveys considerable benefit to family members. Pets were considered family members by 46.9% of all pet-owning households. A shade over half, 50.9% of all households considered the animals in their homes to be pets or companions. Only

2.2% of pet owners thought of their pets as 'property under their care'" (American Veterinary Medical Association, 2002, p. 6).

The person with primary responsibility for the care of a pet is overwhelmingly more likely to be female than male. Approximately two-thirds, 72.8%, of pet owners that have primary responsibility for pet care are female (American Veterinary Medical Association, 2002).

Households comprising parents and children and those comprising nonrelated roommates were most likely to own pets. Households of "one person were the least likely. Only 39.5% of single households owned pets compared with 68.9% of households with parents and children. Among singles, pet ownership decreased with age: 50.2% of young singles owned pets compared with 44.4% of middle-age singles and 29.7% of older singles" (American Veterinary Medical Association, 2002, p. 45).

The likelihood of owning a pet increases with household income. About "half of all households with incomes less than $20,000 own pets. In contrast, 62.0% of households with income greater then $54,999 own pets. Type of residence is also a factor in pet ownership. Homeowners were more likely to own pets than those who rented, 58.7% of homeowners owned a pet compared with 46.7% of renters" (American Veterinary Medical Association, 2002, pp. 47–48). Households most likely to own pets were located in places with a population of less than 100,000. Overall, pet ownership rates declined as community size increased (American Veterinary Medical Association, 2002).

Physical Benefits of Pet Guardianship

A common theme in the literature on companion animals is that of the physical benefits humans experience when in the company of their pets or of companion animals in general. For example, Aaron Katcher, professor of psychiatry at the University of Pennsylvania, Philadelphia, has found that pets may in fact "have important effects on the lives of adults that are independent of and supplementary to human contact" (p. 50). This prompts him to suggest that pets may serve not as substitutes for human contact but offer instead a special relationship that other human beings cannot supply. Evidence of this is present in Katcher's work on the physiological stress of talk and the effect of companion animals. In his study respondent's were first isolated from their animals and then measured for blood pressure and heart rates using an automated blood pressure monitor. Respondents were "observed at rest and when they were reading from a section of uninteresting text. We [the researchers] then brought in the dog and asked the [custodian] to pet the animal. During the greeting, blood pressure and heart rates were significantly lower than during any time in which the subject was reading aloud" (p. 56). Katcher's findings indicate the profound influence pets can have.

Pet ownership may also affect risk factors associated with cardiovascular disease. Garry Jennings, Christopher Reid, Irene Christy, Janis Jennings, Warwick Anderson, and Anthony Dart studied some 5,741 participants receiving free risk profile screenings at the Baker Medical Research Institute. Of these individuals 784 were pet owners and 4,957 were non-owners. Pet owners had significantly lower systolic blood pressure and plasma triglycerides than non-owners. Male "pet owners had significantly lower systolic but not diastolic blood pressure than non-owners and significantly lower plasma triglyceride levels and plasma cholesterol levels" (Jennings et al., 1998, p. 163). Systolic but not diastolic pressure was significantly lower in female pet owners older than 40 years, and plasma triglycerides tended to be lower. Pet owners and non-pet owners, however, did not differ in body mass index, socioeconomic indicators, or smoking habits. Pet owners exercised significantly more than non-owners (Jennings et al., 1998).

These findings suggest the possibility that pet ownership reduces cardiovascular risk factors. On the other hand, these findings may be a marker of other undefined characteristics of those who owned pets that led to lower levels of blood pressure and lipids, and the association was not causal.

Social support, pet ownership, and one-year survival after acute myocardial infarction (severe heart attack) is the focus of a related study by Erika Friedmann and Sue A. Thomas. Here 369 participants in the CAST (Cardiac Arrhythmia Suppression Trial) were followed for at least one year. For these individuals "dog ownership and social support made significant independent contributions to survival beyond the effects of the physiologic measures of the severity of the cardiovascular disease" (Friedmann and Thomas, 1998, p. 198). Dog ownership and the amount of social support were independent predictors of survival. Although owning a dog had a positive influence on health, that effect was complementary to, rather than a substitute for, other sources of social support. In contrast, cat ownership was not related to survival and was not independent of social support.

The findings of this study should not lead one to conclude that cat ownership is harmful. It is likely that differences in other characteristics are responsible both for cat ownership and the apparent lower survival of cat owners. For example, a study by Serpell found that cat owners were significantly more likely to be more sedentary than people who owned dogs and those who did not own any pets (Friedmann & Thomas, 1998).

The relation of cat ownership to other factors that might be related to survival is also possible. None of the women who owned cats also owned dogs, whereas 24 of the 39 men who owned cats also owned dogs (Friedmann & Thomas, 1998). Differences between "cat and dog owners were pronounced. A much higher percentage of cat (11.4%) than dog owners (6.9%) were women. Overall, women had higher 1-year mortality than men in the CAST" (Friedmann & Thomas, 1998, p. 198).

Attachment to pets may account for differences in cardiovascular benefits of pets among pet owners. In the CAST study, pet ownership was defined as claiming a pet as one's own rather than just having a pet in the household. Interestingly, those who own both dogs and cats have been reported to have less attachment to their cats than their dogs. This difference does not appear to be responsible for the observed differences in mortality between dog and cat owners, however. Among men in the current study who owned only cats, "the mortality rate was significantly higher (2 of 15) than among those who owned dogs with or without cats (0 of 81). There were not large differences in mortality among women (1 of 5, 1 of 6, respectively)" (Friedmann & Thomas, 1998, p. 199).

The effect of companion animals on Alzheimer's patients also appears in the literature on pets. Long-term association of companion animals with persons with Alzheimer's living in the home has been associated with fewer episodes of verbal aggression and anxiety compared with those not exposed to companion animals (Fritz, Farver, Kass, & Hart, 1995). Individuals with Alzheimer's who were attached to their pets also had fewer reported mood disorders. However, when the effects of companion animals on the caregivers of persons with Alzheimer's were studied, few benefits were found.

Psychological Benefits of Pet Guardianship

The connection between humans and companion animals is beneficial in other ways too. For instance, animals play a vital role in promoting emotional well being. Descriptions of pet ownership often highlight emotional and esteem support as elements of the relationship. It is plausible then that "these aspects of perceived support from a pet may have greater stability than similar elements of support from a human companion. That pets are not human may be advantageous because there is no fear that the relationship will be damaged by displays of weakness, emotion, or by excessive demands" (Collis & McNicholas, 1998, p. 116). The opportunity to nurture a living being is also suggested as a form of support. In certain circumstances pets may provide instrumental support (e.g., service animals such as guide dogs and assistance dogs). In addition, individuals' social networks may be provided or enhanced by pets through their role as social catalysts facilitating person-to-person contacts.

Many hypotheses attempt to explain how pets function as providers of social support. Pets are perceived as always there, predictable in their responses, and nonjudgmental. They enhance the self-esteem of their pet custodians in that they are perceived as caring about them and needing them, regardless of the owner's status as perceived by self or others. Companion animals "can also provide tactile comfort and recreational distraction from worries" (Collis & McNicholas, 1998, p. 117). Pets are less likely to experience provider burnout. As a result they may be a consistent source of support when human support is scarce or lacking altogether. No social skills are required to elicit attention from pets so there may be a reduced likelihood of mismatches between required and received support. Pets may also provide relief from the strains of human interactions, allowing a freedom from the pretenses that must be erected between giver and receiver of support in order to protect the relationship.

Social workers and psychologists who bring companion animals into nursing homes as part of their work with the elderly report that nursing home residents who refuse to interact with any of the people around them spontaneously and immediately open up to the animals. And, in other settings, horses (among other animals), for example, seem to know what people really need. They ignore the outward form and respond, instead, to the person's inner substance. When in the presence of the mentally ill, who often see the world as full of frightening people, a horse, sensing this vulnerability instinctively "slows down and softens its movements so as not to cause alarm" (McCormick & McCormick, 1997, p. 56).

Kidd, Kelley, and Kidd (1984) found interesting differences in the personalities of owners of horses, birds, snakes, and turtles. Male "owners of horses were found to be aggressive and dominant, while females were easy-going and non-aggressive. Bird owners were socially outgoing and expressive while snake owners were relaxed, unconventional and novelty-seeking, and turtle owners were hard-working, reliable, and upwardly mobile" (p. 203). Other studies indicate that personality differences exist among owners of dogs belonging to different breed groups. For example, researchers have found that

> [T]oy breed owners were the most nurturing and least dominant of all the dog owners. Herding breed owners were the most aggressive and orderly, sporting breed owners were the least orderly, non-sporting breed owners were the least nurturing, working breed owners were the most dominant, hound owners were the friendliest, and terrier owners were the least aggressive but the most dependent on others for emotional support. (Katz, Sanders, Parente, & Figler, 1994, p. 22)

In addition to revealing personality traits, pets may possess the ability to ground and balance us as humans. Animals stay connected to the world around them. They do not block their awareness; they stay in touch with the energy at the core of themselves and others (Lasher, 1996). Animal companions encourage us to get back in touch with ourselves. With animals, humans do not have to hide or hold back aspects of the self. It has been found that, if we are afraid, our dog picks up and resonates the fear, enabling us to experience the fear more clearly (Lasher, 1996). In addition, animals expand our ability to experience life, enhancing feelings such as certainty and happiness. When we arrive home and the dog leaps and trembles with joy, we too experience joy in a relatively pure form. This daily ritual can open humans to experiences of joy at other times (Lasher, 1998). Animals can create for us the sense of trust in another creature. When an animal trusts in us we are able to re-connect to a sense of our own trust. Because animal companions are relatively free of the cultural labels assigned in human relationships, our connections with them have the potential to free us from the limitation of this kind of thinking. On occasion roles are reversed and our animals exhibit the greater amount of knowledge. For example, Lasher (1998) insists that "although I may be dominant in the house, when I go with my dog into the woods he becomes dominant" (132).

In recent years, a growing interest in human relationships with other species has developed. Martin maintains that such interest is probably inspired by "the considerable amount of boundary crossing going on in the contemporary world, not just between humans and animals, but between humans and machines and society and nature" (1991, p. 269). It is not only the crossing of boundaries but also the way they are subject to continual redefinition and conflict that is of interest. Once it was common to assume that some sort of conceptual boundary between human and animals, like that between culture and nature, was universal. As categories, both animals and nature are now more likely to be "described as culturally or historically specific, with a number of scholars insisting that in many non-Western societies, nature is not a category that ordinarily can be opposed to culture or society" (Willis, 1990, pp. 6–8). Even in societies that do share human-animal oppositions, these often seem not to involve a hierarchy of value; boundaries between human and animal are fluid, with animals thought of as persons (or capable of personhood) and humans thought capable of being reincarnated as animals and vice versa.

The Case against Pets

Despite the enormous numbers of pets in the Western world and the vast emotional and financial investment that these animals represent, there have been few serious attempts to explain why pet keeping exists or what purpose it serves. It is only within the past two decades that the subject has attracted any significant scientific interest. Though actual research conducted on people and pets has only scratched the surface, theories on the nature of this relationship abound. One of the most widespread and popular of these theories is the belief or suspicion that pets are no more than substitutes for so-called "normal" human relationships. The view of pet-keeping as a "gratuitous perversion" of natural behavior has been reiterated time and time again throughout history and is today most often expressed by a general tendency to regard people's relationships with their animal companions as absurd, sentimental, and somewhat pathetic. When it comes to pets, these darker aspects of the self can take many forms. For example, some believe that pet-owners are socially inadequate and that they use their pets in much the same way that drug-users use heroin, as artificial substitutes for reality. Others see the relationship between pets and people as an excuse for playful domination. Further, this relationship is thought by some to be essentially sexual in nature. Finally there are those who regard the relationship between pets and people as a phenomenon that consumes people's positive emotions and therefore contributes indirectly to the oppression and physical or psychological annihilation of human beings (Serpell, 1986, p. 20).

More damning evidence against pet-ownership comes in the form of two studies published in 1966 and 1972. The authors of these studies claim that "pet-owners did not like people as much as non-owners, that they did not feel liked by others, that they liked their pets more than they liked people, and that urban pet-owners tended to have weak egos" (Cameron, Conrad, Kirkpatrick, & Bateen, 1966, p. 884; Cameron & Mattson, 1972, p. 286). There are, however, some problems with this study. For instance, the instrument employed for assessing relative liking for people and pets was distinctly crude, and the statistical methods used were highly questionable (Serpell, 1986, p. 29). Despite these flaws in design, the authors asserted that "pet-owners are less psychologically healthy than non-owners" and that "psychologically the pet seems to function as a detriment to effective social relationship and consequently to the person's mental health" (Cameron et al., 1966, p. 886; Cameron & Mattson, 1972, p. 299).

In addition to being considered substitutes for human companions, pets are also seen as frivolous and silly, as instruments of folly. Time, energy, and money spent on the physical comfort of pets are often seen as wasteful and foolish. For example, the pet boutiques of Los Angeles can, for a price, "supply over-indulgent owners with custom-made water beds, gold plated choke-chains and personalized leather-covered dining suites for the pet who has everything. There are also backpacks, bow-ties, plaid coats, and even real or imitation fur stoles. Pet boutiques also offer dog raincoats, frilly dresses, underwear, and cosmetics, such as nail polish, available in a wide range of colors. Some services will plan pet birthday parties, complete with cakes in dog-attractive shapes, such as bones.

Upwardly mobile pet owners can enroll their companion animals in daytime "play-groups" and when they go on vacation can make use of dog hotels and summer camps with swimming pools and organized romping. Cedar chip, mint-scented, and mashed corn-husk cat litters are some of the recent luxuries created for cats (Phillips, 1992).

The desire to save and care for pets, to indulge and to honor them, if taken to the extreme, can lead to unhealthy behavior. When help is extended to more animals than can reasonably be cared for, hoarding or animal collecting begins to take root. Hoarding is defined as the accumulation of possessions that are useless and that interfere with the ability to function. In the case of animal hoarders, specifically, people often live with dozens to hundreds of animals both alive and dead in apartments, trailers, and single-family homes.

Arluke, Frost, Luke, Messner, Nathanson, Patronek, Papazian, and Steketee, using case reports from 28 states and 1 Canadian province, found that nearly three-quarters of the 71 hoarders they studied were single, widowed, or divorced. In more than half of the cases other individuals were living in the home, including children, bedridden or dependent elderly people, and disabled people. Of the people for whom employment status was provided, most were described as unemployed, retired, or disabled (2002). Though the hoarding of animals is often seen among socially isolated individuals, animal hoarding has also been found among physicians, veterinarians, bankers, nurses, teachers, and college professors. Hoarders holding jobs "appeared to be able to live a double life, with co-workers never suspecting the true conditions in their homes until animal rescue authorities were called to investigate" (Frost, Krause, & Steketee, 1996, p. 116).

Hoarding of animals and/or possessions occurs in 20 to 30 percent of people with Obsessive Compulsive Disorder and is also observed in people with anorexia nervosa (Frankenburg, 1984), psychotic disorders, depression, and other mental disorders. Some research suggests that people who hoard animals may suffer more severe impairment than people who hoard only possessions (Patronek, 1999).

Hoarders' reasons for collecting and living with animals were many. Love for animals was, of course, a common theme as was the role of animals as children or surrogate family members. Collectors accumulated animals through a variety of means. Accidental breeding of "animals was the single most common reason, ranked first or second in 56%

of cases. In addition, active solicitation from the public was ranked first or second in 46% of cases" (Arluke et al., 2002, p. 4).

Pet Guardianship and Its Effect upon Self-Definition

Though relationships with companion animals can be fulfilling, humans don't all relate to pets in the same way. According to Michael Fox there are four categories that describe owner-pet relationships. The first of these categories is the "object-oriented" relationship. Pet custodianship may be considered object-oriented when a family has one or more dogs and cats simply because they believe in the old adage that a house is not a home without a pet. Like the puppy or kitten given as a Christmas present, interest in the pet may wane with time as its novelty fades and her/his more appealing traits are replaced by less endearing behavior and temperament. Humans may also keep pets for purely decorative purposes, regarding the animal with concern and empathy in proportion only to its monetary worth or uniqueness (1981, p. 31).

A second category of relationship between pet and human is "exploitative/utilitarian." Here the animal is used, trained, manipulated, or exploited to varying degrees for the sole benefit of humans. This category includes "the use of animals in all forms of biomedical research, in military work and in agriculture as food converters for human consumption" (1981, p. 31). Possession of an animal for any utilitarian function, such as in the case of guard dogs, bird dogs, guides for the blind, and dogs for show and breeding purposes, involves this kind of relationship to differing degrees. Further, a pet that is kept as a "learning experience" is being exploited if the learning is not combined with empathy for the animal.

The "need-dependency" relationship is a third way of relating to animals. This type of need-dependency bond is often a major reason why people have pets, especially cats and dogs. Here the pet satisfies needs for children or adults by serving as a companion, a confidant, a link with nature, and a break from more superficial, impersonal, and often dehumanizing human interactions. The deeper the emotional needs of the person the more important the pet becomes in this type of relationship. Companion animals are gaining recognition and respect as therapeutic assistants, creating an emotional bridge between the emotionally withdrawn patient and the therapist. In "normal" settings, too, pets facilitate and enhance interactions between family members (Fox, 1981).

A deaf woman gets emotional as well as physical support from her companion dog. Courtesy of Shutterstock.

Fox's final category is known as the "actualizing" relationship. In this type of exchange a pet is related to essentially as a respected significant other. The pet is appreciated for itself as opposed to loving it for reasons of status, utility, or emotional substance. For this relationship to succeed a person's perception and understanding of their pet must shift from one of dependency to one that is less egocentric and more all-encompassing. For many it is easier to forge such a relationship with an animal than with a person because human insecurities and ego defenses can create barriers to deep and meaningful interaction (1981).

Conclusion

Both the domestication of animals and the subsequent practice of pet-keeping have existed for thousands of years. From preagricultural times to the present, animals as helpmates and companions have played a vital role. Goats and sheep provided early humans with milk and wool and allowed them to travel from place to place with a source of food and clothing in tow. Today animals are kept for a variety of reasons. One important aspect of modern pet-keeping centers on the physical benefits that companion animals can provide. Lower blood pressure, quicker recovery from illness, and a longer life are just some of the positive aspects of owning a pet. The psychological benefits of having a companion animal are also numerous. Pets are nonjudgmental, can provide tactile comfort, and often help to distract us from worries. They may also serve to ground and balance us. There are, however, certain downsides to the keeping of pets. Some pet owners may overindulge their animals, failing to recognize that they are dealing with feeling and thinking beings, not toys to be dressed up and shown off. Other pet owners hoard or collect more pets than they can reasonably care for. This practice can place the lives of both the owners and the animals in jeopardy. Although most people do not hoard or collect pets, we all relate to our companion animals differently. Some of us see our pets as objects and others view them as entities to be manipulated and trained. The best of all possible worlds involves seeing our companion animals as equals.

See also

Animal Assistance to Humans
Bonding
Culture, Religion, and Belief Systems—*Taiwan and Companion Animals*
Enrichment for Animals—*Pets and Environmental Enrichment*
Ethics and Animal Protection—*Hoarding Animals*
Living with Animals

Further Resources

American Veterinary Medical Association. (2002). *U. S. pet ownership and demographics sourcebook*. Schaumberg, IL: Center for Information Management of the American Veterinary Medical Association.

Arluke, A., Frost, R., Steketee, G., Patronek, G., Luke, C., Messner, E., Nathanson, J., & Papazian, M. Press reports of animal hoarding. *Society and Animals, 10,* 1–23.

Ash, E. C. (1927). *Dogs: Their history and development*. London: Ernest Benn.

Basso, C. B. (1973). *The Kalapalo Indians of central Brazil*. New York: Holt, Rinehart & Winston.

Beck, A. M., & Katcher, A. H. (1983). *Between pets and people: The importance of animal companionship*. New York: G. P. Putnam.

Blumer, H. (1969). *Symbolic interactionism: Perspective and method*. Berkeley: University of California Press.

Bodson, L. (2000). Motivations for pet-keeping in ancient Greece and Rome: A preliminary survey. In A. Podberscek, E. Paul, & J. Serpell (Eds.), *Companion animals and us: Exploring relationships between people and pets* (pp. 27–41). Cambridge: Cambridge University Press.

Bonas, S., McNicholas, J., & Collis, G. M. (2000). Pets in the network of family relationships: An empirical study. In A. Podberscek, E. Paul, & J. Serpell (Eds.), *Companion animals and us: Exploring relationships between people and pets* (pp. 209–36). Cambridge: Cambridge University Press.

Cameron, P., Conrad, C., Kirkpatrick, D. D., & Bateen, R. J. (1966). Pet ownership and sex as determinants of stated affect toward others and estimates of others' regard of self. *Psychological Reports, 19,* 884–86.

Cameron, P., & Mattson, M. (1972). Psychological correlates of pet ownership. *Psychological Reports, 30,* 286.

Caras, R. (2002). *A perfect harmony: The intertwining lives of animals and humans throughout history*. New York: Simon & Schuster.

Clutton-Brock, J. (1981). *Domesticated animals from early times*. Austin: University of Texas Press.

Collis, G. M., & McNicholas, J. (1998). A theoretical basis for health benefits of pet ownership: Attachment versus psychological support. In C. C. Wilson & D. C. Turner (Eds.), *Companion animals in human health* (pp. 105–22). London: Sage.

Council for Science and Society. (1988). *Companion animals in society*. Oxford: Oxford University Press.

Elmendorf, W. W., & Kroeber, K. L. (1960). The structure of Twana culture with comparative notes on the structure of Yurok culture. *Washington University Research Studies,* 2(28), 114.

Fox, M. (1981). *Returning to Eden: Animal rights and human responsibility*. New York: Viking Press.

Frankenburg, F. R. (1984) Hoarding in anorexia nervosa. *British Journal of Medical Psychology,* 57(1), 57–60.

Friedmann, E., & Thomas, S. (1998). Animal companions and one year survival of patients after discharge from a coronary care unit. *Public Health Reports, 95,* 307–12.

Fritz. C. L., Farver, T. B., Kass, P. H., & Hart, L. A. (1995). Association with companion animals and the expression of noncognitive symptoms in Alzheimer's patients. *Journal of Nervous and Mental Disease, 183*(7), 459–63.

Frost, R. O., Krause, M. S., & Steketee, G. (1996). Hoarding and obsessive-compulsive symptoms. *Behavior Modification, 20,* 116–32.

Galton, F. (1883). *Inquiry into human faculty and its development*. London: Macmillan.

Goffman, E. (1959). *The presentation of self in everyday life*. Garden City, NY: Doubleday Anchor.

Halliday, W. R. (1922). Animal pets in ancient Greece. *Discovery, 3,* 151–54.

Irvine, L. (2004). *If you tame me*. Philadelphia: Temple University Press.

Jennings, G., Reid, C., Christy, I., Jennings, J., Anderson, W., & Dart, A. (1998). Prospects for the non-pharmacological control of hypertension. In J. K. McNeil, R. W. F. King, G. L. Jennings, & J. W. Powles (Eds.), *A textbook of preventive medicine* (pp. 133–45). Melbourne: Edward Arnold.

Katcher, A. H. (1981). Interactions between people and their pets: Form and function. In B. Fogle (Ed.), *Interrelations between people and pets* (pp. 41–67). Springfield, IL: Charles C. Thomas.

Katz, E., Sanders, C., Parente, J., & Figler, K. (1994). *Nature as subject: Human obligation and natural community*. Lanham, MD: Rowman & Littlefield.

Katz, J. (2003). *A dog year: Twelve months, four dogs, and me*. New York: Random House.

Kete, K. (1994). *The beast in the boudoir: Petkeeping in nineteenth-century Paris*. Berkeley: University of California Press.

Kidd, A. H., & Kidd, R. M. (1980). Personality characteristics and preferences in pet ownership. *Psychological Reports, 46,* 939–49.

Knapp, C. (1998). *Pack of two: The intricate bond between people and dogs*. New York: Delta.

Lasher, M. (1996). A relational approach to the human-animal bond. *Anthrozoos, 11,* 130–33.

Malek, J. (1993). *The cat in ancient Egypt*. London: British Museum Press.
Marsh, F. O. (1994). International pet food market. *Anthrozoos, 8*, 55–57.
Martin, M. (1991). *Animal models in psychiatry II*. New York: Humana Press. Health Communications, Inc.
Masson, J. (2002). *The nine emotional lives of cats: A journey into the feline heart*. New York: Ballantine Books.
McCormick, M., & McCormick, A. (1997). *Horse sense and the human heart*. Deerfield, FL: HCI.
McNicholas, J., Collis, G. M., Morley, I. E., & Lane, D. R. (1993). Social communication through a companion animal: The dog as a social catalyst. In M. Nichelmann, H. K. Wierenga, & S. Braun (Eds.), *Proceedings of the International Congress on Applied Ethology* (pp. 368–70). Berlin: Humboldt University.
Melson, G. F. (2001). *Why the wild things are: Animals in the lives of children*. Cambridge, MA: Harvard University Press.
Messent, P. R. (1983). A review of recent developments in human-companion animal studies. *California Veterinarian, 5*, 26–50.
Messent, P. R., & Serpell, J. (1991). An historical and biological view of the pet-owner bond. In B. Fogle (Ed.), *Interrelations between people and pets* (pp. 5–22). Springfield, IL: Charles C. Thomas.
Nagle, T. (1974). What is it like to be a bat? *Philosophical Review, 83*, 435–50.
Phillips, M. (1992). Savages, drunks and lab animals: The researcher's perception of pain. *Society and Animals, 1*, 61–82.
Podberscek, A., Paul, E., & Serpell, J. (2000). *Companion animals and us: Exploring relationships between people and pets*. Cambridge, MA: Cambridge University Press.
Serpell, J. (1986). *In the company of animals: A study of human-animal relationships*. Oxford: Basil Blackwell Ltd.
Serpell, J., & Paul, E. (1994). Pets and the development of positive attitudes to animals. In A. Manning & J. Serpell (Eds.), *Animals and human society: changing perspectives* (pp. 127–44). London: Routledge.
Thomas, E. (1993). *The hidden life of dogs*. New York: Pocket Books.
Voith, V. L. (1985). Behavioral disorders. In S. J. Ettinger (Ed.), *Textbook of veterinary internal medicine*, Vol. 1 (pp. 208–27). Philadelphia: W.B. Saunders Co.
Weiss, R.S. (1974). The provisions of social relationships. In Z. Rubin (Ed.), *Doing unto others* (pp. 17–26). Englewood Cliffs, NJ: Prentice Hall.
Willis, R. (1990). *Signifying animals*. New York: Routledge.
Zasloff, R. H., & Kidd, A. H. (1994). Loneliness and pet ownership among single women. *Psychological Reports, 75*, 747–52.

Lisa Sarmicanic

Bonding
Discrimination between Humans by Cockroaches

When most people think of meaningful interactions with animals, they don't have Madagascar hissing cockroaches in mind. Nevertheless, some people keep insects as pets, and there is scientific evidence that a variety of insect species can be "tamed" following prolonged handling by humans. The real question is whether such taming is a generalized response to all humans, or whether it might go further and reflect the insect's ability to discriminate a particular individual. For the past ten years, my students and I have been looking at just this question with a variety of animal species, although none more exotic than hissing cockroaches.

The Madagascar hissing cockroach (*Gromphadorhina portentosa*) is a large (70–80 mm), docile arthropod that has become increasingly popular in "petting zoos" as well as teaching laboratories. Many people find them extremely creepy, because of both their size as well as the highly audible hissing response they make when disturbed or threatened. In order to determine whether this species could be "tamed," we investigated whether hissing would show habituation over the course of exposure to a human handler and, further, whether such habituation might be specific to contact with a particular person.

We began with an initial sample of twenty adult insects taken from a large, experimentally naive colony but soon found that only twelve were reliable hissers, which made them ideal subjects for our research. Subjects were removed from the colony by a technician and placed in individual plastic containers. Hissing was recorded during each trial by suspending a microphone approximately 30 centimeters over the work area. It is known that the hissing response is created by arching the body and expelling air from modified abdominal spiracles. Pilot studies confirmed that most hisses were approximately one-second in duration. Their occurrence was easily detected and reliably counted by independent observers.

Two female experimenters, similar in age, height and weight, served as handlers. Neither person had prior experience with hissing cockroaches, and both people underwent a two-week familiarization period with nonexperimental animals. Both handlers maintained a consistent program of personal hygiene and diet during the experimental period and never used heavily scented soaps or perfumes. Prior to testing, each handler was randomly assigned to work with half the subjects. Thus, both handlers served as the Familiar person for one group of subjects, and the Novel person for the remaining subjects. For each trial, the handler started the tape recorder, removed the lid of the container, and lifted the cockroach and placed it in the palm of her hand. The subject was allowed to explore the hand and arm of the handler for two minutes and then returned to the container. The handler stroked the subject continuously during the two-minute session. Stroking consisted of exerting gentle downward pressure on the dorsal surface of the subject, slowly running the index finger from the anterior to posterior of the shell. To control for the effects of possible scent marking by the subject, handlers rinsed their hands and arms thoroughly with water between trials, and dried them with paper towels. In addition, the sequence in which subjects were tested was varied randomly to minimize the effects of lingering cues from previously tested animals.

An A_1–B–A_2 design was used for each subject; A_1 consisted of exposure to the Familiar handler for a minimum of 17 sessions until six consecutive hiss-free sessions occurred. When this criterion for habituation was met, subjects began the B phase, during which they were exposed to the Novel handler for a single session. In the final Phase A_2, subjects were re-exposed to the Familiar handler for one additional session. Thus, we could plot the course of habituation and determine whether it was specific to the Familiar person.

Ten of the twelve cockroaches showed habituation of the hissing response during repeated contact with a human handler in Phase A_1. In four of these cases, testing in Phase B revealed that habituation was specific to the Familiar person; that is, hissing habituated to the Familiar person in Phase A_1, was reinstated during exposure to a Novel handler in Phase B, and did not occur again during reexposure to the Familiar person in Phase A_2. In addition to hissing at the Novel handler, one of these four subjects also regurgitated on the person's arm during the two-minute test in Phase B, a form of behavior in Madagascar hissing cockroaches that has been associated with "extreme distress."

In six other subjects, habituation of the hissing response occurred but appeared to be a generalized response to human contact; that is, once habituated in Phase A_1, hissing did not reappear during exposure to the Novel handler in Phase B. Two remaining subjects did not habituate over trials and are notable exceptions. In both cases, there was visible damage to the insect's body (sustained during communal living prior to testing). One subject had lost a portion of her antenna; the other was missing a leg. Both of these subjects were noticeably more irritable and hissed consistently across all handling sessions.

Of the ten subjects that showed habituation of the hissing response, we found evidence of both general "taming" as well as what appears to be discrimination between humans. The generalized response (i.e., once subjects were "tamed," it did not matter who held or touched them) may be newsworthy to those who do not hold insects in general or cockroaches in particular in very high regard. In truth, habituation has actually been widely documented in species ranging from *Aplysia* to *Homo sapiens* and needs little in the way of explanation. By far the more interesting response is the evidence from four subjects that habituation may be confined to a particular human. In these cases, contact with a Novel person (or persons, as further post-experimental testing revealed) caused the subjects to hiss, and reintroduction of the Familiar person reduced hissing to zero. Such human discrimination by cockroaches may not surprise those who keep them as pets, but our research is the first attempt to move beyond anecdotal reports to the realm of controlled experimental research. Person-specific habituation is of considerable interest. From an evolutionary point of view, it is a more conservative strategy and requires the animal to discriminate between "safe" vs. unknown or "dangerous" individuals with whom it has contact. Obviously this ability in cockroaches has not evolved to support discrimination between humans. However, there is evidence that *G. portentosa* are capable of differentiating between individuals of their own species in both mating and the formation of social hierarchies based upon dominance.

Although the presence of four person-specific cases of habituation in our research provides tantalizing evidence of human discrimination in an insect species, the small number of subjects suggests that although such an ability is present, it may not be widespread or occur under all conditions. Although our findings do not address the sensory basis of the discrimination we observed, it is known that *G. portentosa* send and respond to olfactory signals and are sensitive to pheromonal cues. Either of these mechanisms are likely the basis for the discrimination between humans that we observed in our study. For those doing behavioral research with hissing cockroaches, our findings suggest that caution may be in order. The ability to discriminate between human

The first attempt to move beyond anecdotal reports to the realm of controlled experimental research suggests evidence that habituation may be confined to a particular human. Courtesy of Hank Davis.

handlers may have implications for the design and analysis of research. Even with subjects as "simple" as arthropods, routine handling may come at a price. The discrimination between individual humans sets the stage for unanticipated Pavlovian conditioning that could have measurable effects on both physiological and behavioral research.

One of the handlers holds up a Madagascar hissing cockroach used in this habituation experiment. Courtesy of Hank Davis.

Further Resources

Bell, W. J. (1981). *The laboratory cockroach.* New York: Chapman & Hall.

Davis, H. (2002). Prediction and preparation: Pavlovian implications of research animals discriminating among humans. *ILAR Journal, 43,* 19–26.

Kummer, H., Daston, L., Gigerenza, G., & Silk, J. (1997). The social intelligence hypothesis. In P. Weingart, P. Richerson, S. D. Mitchell, & S. Maasen (Eds.), *Human by nature: Between biology and the social sciences* (pp. 157–79). Hillsdale, NJ: Erlbaum.

Nelson, M. C., & Fraser, J. (1980). Sound production in the cockroach *Gromphadorhina portentosa*: Evidence for communication by hissing. *Behavioural Ecology & Sociobiology, 6,* 305–14.

Persoons, C. J., & Ritter, F. J. (1979). Pheromones of cockroaches. In F. J. Ritter (Ed.), *Clinical ecology, odour communication in animals.* Amsterdam: Elsevier Press.

Hank Davis & Emily Heslop

Bonding
*Discrimination between Humans by Emus and Rheas**

During the past two decades, ratites (e.g., ostriches, emus, rheas) have experienced unprecedented growth in economic importance in North America. Virtually every part of these large flightless birds has been exploited, including their meat, oil, feathers, eggs, toenails, and leather from their hides. Such commercial attention has turned ratites from semi-exotic creatures, most often confined to zoos, into a cash crop, living in the ever-widening circle of "ostrich farms." Nevertheless, confinement in zoo settings has remained extensive, despite growth in the agricultural importance of these animals.

*This project was supported by a Grant from the Natural Sciences and Engineering Research Council of Canada (NSERC). These data were presented in part at the 1999 annual AALAS meeting and the 2000 annual meeting of the Society for Comparative Cognition. The author thanks Allison Taylor for her support and insights, as well as Susan Boehnke, Christina Norris, Meredith Smith, and Lill Svendsen for their technical assistance.

Given their adaptability (emus, for example, can endure temperatures ranging from −12 to +48°C), ratites have been transplanted from their native habitats in Australia, Africa, and South America to geographically diverse regions of the North American continent, from Mexico to Canada. Because ratite farming is still in its infancy in North America, commercial exploitation has raised practical questions about ratite husbandry and welfare. Not surprisingly, most of the recent literature on ostriches, emus, and rheas has focused on issues such as nutrition and disease control, as well as minimizing the stress of egg collection and transportation.

However, increased contact between humans and ratites has raised other questions as well. Because ratites are large and potentially dangerous animals, a particular focus of many manuals has been on optimal handling techniques. Anecdotal evidence suggests that ratites fare best with "slow and easy" treatment, stressing consistency in both handling routines and handlers. Familiarity is the keynote of such techniques. For this reason, we explored the ability of emus and rheas to discriminate between novel and familiar persons with whom they have contact. Like many other animals and birds (including penguins and domestic fowl), could emus and rheas make discriminations between individual humans?

We studied a total of eight emus and twelve rheas. Animals were housed (and tested) in groups of various sizes as follows: three pens of emus consisting of a Male-Female pair of mature birds; a pen of five juveniles (M and F); and a single mature M bird. Rheas were penned and tested in two groups of six juveniles (3 M and 3 F in each). All birds were housed in indoor pens (approximately 2 m by 2.5 m) at African Lion Safari in Rockton, Ontario. Birds were tested during the winter months when the park was closed to visitors and the birds were confined indoors.

Each pen of subjects was exposed to the same Familiar handler for ten 10-minute sessions conducted twice daily (4 hr apart) on five consecutive days. On the sixth day, subjects received two 10-minute test sessions, also conducted four hours apart. An A–B–A design was used in which A = exposure to the familiar (F) handler and B = exposure to a novel (N) person, matched for gender, approximate age, and clothing (barn overalls). Three different persons were used in the role of Familiar handler.

During each of the ten initial sessions, the F person entered the pen and stood midway against the rear wall with a covered food bowl containing pieces of freshly cut apple in her left hand, which was positioned by the side of her body. A single piece of apple was held in her right hand, which was extended from her body. The F handler remained motionless and talked softly to the birds, allowing them to explore her person freely. When food was taken from her hand, she slowly replenished the piece from the bowl. The Novel (N) person behaved in an identical way during the single B-test session. In order to be certain that N persons were "comfortable" during their interaction with subjects, each N person also served in the capacity of Familiar person for a different group of subjects in the experiment. A video recorder located in the front corner of the pen, approximately 3 meters from the handler, remained in the pen during all sessions. Data were recorded during the first 5 minutes of the final two pretest A-sessions, the single B-session, and the final A-session, during which subjects were re-exposed to the Familiar person.

A decision was made to treat birds within a pen as a unit rather than to subject individuals to the stress of separate testing. This resulted in considerable variability between pens in the birds' behavior. For that reason, a number of different response measures were recorded for analysis. These included the number of pecks directed to the handler, latency to first peck, latency to take food from the handler, and duration of visual inspection of the handler's body and face. Video records were scored by two independent raters who had no knowledge of whether the person was Familiar or Novel to the birds.

Each pen of birds showed evidence that they differentiated between familiar (F) and novel (N) persons, although there was variability in the behavioral indicators of this ability. The male-female pair of adult emus spent considerably more time in overt inspection of the N person while virtually ignoring the F person in their presence. This result is consistent with previous findings with seals, who similarly directed increased attention to a novel person in their midst while virtually ignoring the person with whom they had previously interacted. Similarly, the pen of five juvenile emus as well as the single adult male emu directed greater attention toward the N individual. In the case of the juveniles, this difference appeared in the number of pecks directed toward the N person; for the adult male, the discrimination was revealed in the latency to the first peck. Group 1 of juvenile rheas responded similarly to their emu counterparts, showing an increase in the number of pecks directed at the Novel handler. The second group of juvenile rheas waited more than twice as long to take food from the outstretched hand of the Novel person.

Despite differences in the manner in which they expressed their discrimination, every pen of ratites we tested made it clear that they knew the difference between a Familiar and a Novel human in their midst. This readily obtained discrimination between humans sets the stage for human-based Pavlovian conditioning (Person as CS) which, in turn, may have a direct bearing on how individual birds or aggregate pens behave in the presence of different persons. Pavlovian conditioning is a highly adaptive response, allowing the animal to anticipate and prepare for what is likely to happen. The implications of such anticipatory conditioning—whether it expresses itself in behavioral or physiological terms—are likely to be considerable for both research and husbandry. Knowing that ratites focus on individual persons, rather than on the general class of "people," should underscore the importance of consistent and positive husbandry, in other words, regularity in both care-giving routines as well as caregivers.

Further Resources

Coody, D. (1987). *Ostriches: Your great opportunity*. Lawton, OK: Coody Publishing.

Davis, H., & Balfour, D. (Eds.). (1992). *The inevitable bond: Examining scientist-animal interactions*. New York: Cambridge University Press.

Eastman, M. (1969). *The life of the emu*. Sydney, Australia: Angus & Robertson, Ltd.

Jensen, J. M., Johnson, J. H., & Weiner, S. T. (1992). *Husbandry and medical management of ostriches, emus and rheas*. College Station, TX: Wildlife and Exotic Animal TeleConsultants.

Taylor, A., Davis, H., & Boyle, G. (1998). Increased vigilance toward unfamiliar humans by harbor (*Phoca vitulina*) and gray (*Halichoerus grypus*) seals. *Marine Mammal Science, 14*, 575–83.

Hank Davis

Bonding
Discrimination between Humans by Gentoo Penguins (Pygosceli papua papua/ellsworthii)

Public consciousness about penguins probably reached an all-time high in 2005, thanks to the critically acclaimed documentary "March of the Penguins." This consciousness was strengthened with the 2006 success of the animated movie "Happy Feet." Examining the lives and breeding cycle of Emperor penguins, "March of the Penguins" offered a sensitive and detailed look at the travails of penguin life in Antarctica. Even before this award-winning film, it was well documented that most penguin species breed in large colonies and interact extensively with other members of their colony. It is folkloric to the point of humor that penguins appear highly similar to most human observers. Indeed, a syndicated comic strip ("Rubes") consistently pokes fun at the difficulties penguins must have discriminating among themselves (Rubin 2000). Nevertheless, successful breeding and maintenance of monogamous pair bonds require a well-evolved ability to differentiate among conspecifics (members of the same species), which penguins plainly have. Evidence suggests that such behavior is based on both visual and auditory cues. The question remains whether the ability to discriminate among individuals extends beyond other penguins to the humans with whom penguins may have contact.

Though somewhat rare, contact between humans and penguins is likely to occur under one of three conditions:

1. unsystematic and largely undocumented occurrences of eco-tourism
2. field studies (i.e., investigations of penguin behavior in natural settings)
3. zoos and aquariums, in which penguin colonies are maintained to promote education as well as conservation. Additionally, public aquariums and zoos have themselves served as settings for the study of penguin behavior, using both ethological and experimental approaches.

A considerable body of anecdotal evidence suggests that penguins in captivity develop considerable bonds with, and direct a range of behavioral responses to, specific human caregivers. In some cases, this bonding is based on imprinting and results in social dysfunction of the individual penguin as well as disruption of nesting pairs in the colony. Of greater relevance to the present study, however, are those cases in which adult birds may have learned to discriminate between novel humans and those with whom they have regular contact. Our research explored this possibility. We studied twenty-seven Gentoo penguins (*Pygoscelis papua papua* and *Pygoscelis papua ellsworthii*), ranging in age from one to twelve years. Subjects were housed in a mixed-species colony of Gentoo and Chinstrap penguins, maintained by the Wildlife Conservation Society's Central Park Wildlife Center in New York. All birds had free range of the enclosed, temperature-controlled indoor habitat twenty-four hours per day. A filtered, 180,000 L, fresh-water pool occupies the front-third of the exhibit. The back two-thirds (on which testing occurred) consists of beach space of approximately 108 square meters with artificial rock of irregular height and length. Except during the two- to three-month nesting season, when the level of husbandry duties increase, keepers enter the penguin habitat four times daily for exhibit maintenance and feedings. On average, the penguins are exposed to caregivers for approximately two hours per day. A glass window separates the enclosed habitat from a public viewing area, and penguins have visual exposure to park visitors approximately six hours per day.

We tested the birds' ability to discriminate between a familiar and novel human during a single session consisting of two 3-minute segments, with a time lapse of 10 seconds in between. The test session was conducted 40 minutes before the first colony feeding of the day. The penguins could approach or avoid, at will, both a familiar (F) and novel (N) human. The F human was a primary penguin caregiver at the zoo since 1988. The N person had no prior exposure with the penguins tested in this experiment. Both persons stood (in order to allow a 360-degree approach) approximately 3 meters apart on an elevated flat rock ledge (30 cm high). The ledge was naturally separated into two distinct beach areas, and the natural boundaries of each beach were used to determine whether a bird had approached either person.

Both persons stood in a designated social area (i.e., removed from the feeding area) during both sessions to signal that this was a social, rather than a feeding, event. The feeding station, where penguins are hand-fed twice daily, was located approximately 12 meters away from this site. This, along with the absence of stainless steel feeding buckets regularly associated with feeding sessions, further underscored the nonfeeding nature of this contact period.

Familiar and Novel persons were matched in physical appearance and behavior. Both persons wore standard caretaker uniforms (blue coats with zippered life vests, green pants, and black rubber boots). Both F and N were female, similar in age, weight, and hair color, length, and style. Both persons stood with minimal movement during the timed trials except for head turns to visually identify the birds around them. Vocalizations directed to the birds were kept to a minimum but were not eliminated since penguins are accustomed to receiving vocal input from their caregivers. F and N attempted to keep their vocalizations reasonably similar, alternating two- to five-word phrases (e.g., "Hey guys, who's coming over?") and repeating each other's phrases when they spoke using similar vocal inflections. Birds' names were not used by either handler. Because F and N persons were not able to enter the exhibit at the same time (one had to follow behind the other), the N person was allowed to enter first during Session 1 to ensure that she would be viewed first by any of the penguins in the area.

The N person entered the exhibit, followed immediately by F. When both persons reached their respective locations 3 meters apart, the timing of the first trial began. Both persons stood at their respective locations for 3 minutes (preexchange), occasionally vocalizing as described above. At the end of 3 minutes, a 10-second interval occurred, during which F and N persons exchanged positions. When the second 3-minute trial (postexchange) was complete, both persons left the exhibit immediately.

The primary data selected for analysis were the number and identity of birds in the immediate location (i.e., a 1 m radius) of the person. Video records were sampled continuously. The appearance of any bird for a minimum of 5 seconds resulted in its presence being recorded for that minute. For example, a frequency count of "nine" during minute 1 indicated that nine different birds were present in the area for at least 5 seconds each. Although there was a decline over both 3-minute test periods, the general pattern of preference for the familiar (F) person is evident. Using a statistical technique called a randomization test, we found a significant difference ($p < 0.02$) in the number of birds per handler across all minutes of the test periods. Plainly, this effect was more pronounced during the initial (preexchange) phase, and most pronounced during the initial two minutes of testing.

A number of more specific observations underscore this pattern of preference for the Familiar person. During the first minute, eight birds chose contact with the F handler exclusively (i.e., they did not shuttle between location or handler at any time). One of these birds was a first-year chick, with the remaining subjects ranging in age between two and twelve years. In contrast, only one bird (a six-year old female) spent time exclusively with the Novel person. Every other bird that appeared in proximity to the Novel person either returned to the Familar person or left the test area. When handlers exchanged locations, one bird (a twelve-year old female) followed the F person to her new station, walking directly behind her. A second bird, which had appeared next to the N person during the final 30 seconds prior to the exchange, simply remained in that location and stayed with F throughout the remaining test. Only one bird followed the N person to her new location and remained exclusively with her for the final 3-minute test.

This subject, a highly social twelve-year old male, began the initial test with the F handler, then crossed over to the N person after 30 seconds, returned to F at 1:30; crossed back to N at 2:30, and spent the balance of the test with N.

Both an overall "body count" as well as tracking individual subjects suggests that penguins prefer contact with the person with whom they were familiar. These data support much anecdotal evidence that animals in captivity are readily able to distinguish among the humans in their presence and, initially at least, exhibit a preference for the person familiar to them. These results are also consistent with experimental findings with rats, which similarly show a preference for humans with whom they had previously interacted, and seals, which show decreased vigilance to familiar persons in their midst. Collectively, these results raise the question whether such preference for familiar humans is based on prior sensory contact, per se, or depends on a history of positive experiences or, at least, the lack of negative experience.

It is worth noting that in the present circumstances, contact with the Familiar person was considerable, extending to the lifespan of many of the birds tested. There is likely a point at which positive contact with a specific person generalizes to all humans and may paradoxically reduce the evidence of preference for a specific person. There may indeed have been such instances in the present experiment, as we have found to be the case with rats. However, in most cases, prior contact yields measurable evidence, even if fleeting, of both discrimination between or among humans as well as preference for the most familiar.

There is an obvious survival advantage in being able to differentiate between familiar and novel aspects of one's environment. If animals are able to recognize the humans

It's known that penguins have a well-evolved ability to differentiate among individuals of their own species, and there's also evidence they show preference for individual human beings. Courtesy of Hank Davis.

with whom they have regular contact, then these same persons can become predictors in a Pavlovian sense of the salient events (e.g., food, pain) in the animals' lives. Such human-based Pavlovian conditioning (Person as CS) is well established and has direct bearing on both animal research and management. Knowing that individual persons, rather than the general class of human beings, are the focus of attention should sound a cautionary note in our varied interactions with the animals under our management. The ability of penguins to discriminate between individual humans underscores the importance of consistent husbandry; that is, regularity in both caregiving routines and caregivers.

Further Resources

Ackerman, C. (1997). Nest-site preferences, pair fidelity and site-fidelity in Gentoo and Chinstrap penguins. *International Zoo News, 44,* 327–33.

Cheney, C. (1999). Ecotourism and penguins: Measuring stress effects in Magellanic Penguins. *Penguin Conservation, 12,* 24–25.

Davis, H., & Balfour, D. (Eds.) (1992). *The inevitable bond: Examining scientist-animal interactions.* New York: Cambridge University Press.

Davis, H., Taylor, A., & Norris, C. (1997). Preference for familiar humans by rats. *Psychonomic Bulletin & Review, 4,* 118–20.

Jouventin, P. (1982). *Visual and vocal signals in penguins: Their evolution and adaptive characters.* Berlin: Parey.

Rubin, L. (2000). Rubes. Available online at http://www.creators.com/comics/rubes.html.

Williams, T. D. (1995). *The Penguins* (Bird Families of the World, Vol. 3). New York: Oxford University Press.

Hank Davis, Celia Ackerman, and Anne Silver

Bonding
*Discrimination between Humans by Honeybees**

Do honeybee colonies react to their keepers differently than to strangers? Beekeepers frequently question whether their bees (*Apis mellifera*) learn to recognize them as individuals. For honeybees to have this capacity requires sensory capabilities not widely recognized in insects. Interspecies recognition of individuals has been demonstrated in a wide variety of mammals (rats, rabbits, cattle, sheep, llamas, and seals) and birds (chickens, emus, rheas, and penguins). However, with the exception of Madagascar hissing cockroaches (see previous), there is presently no evidence that insects have the perceptual ability to make such discriminations.

Individual honeybees (*Apis mellifera* L.) undergo a regular temporal pattern of division of labor, starting with cell-cleaning upon emergence, followed by brood care duties, food handling activities, and finally foraging. Some medium-aged workers perform "guarding" behaviors that involve patrolling near the colony entrance and inspecting

*We thank Paul Kelly for his assistance with the bees and beehives. Funding was provided by the Ontario Ministry of Agriculture, Food and Rural Affairs (G.O.) and Natural Sciences and Engineering Research Council of Canada (H.D.).

bees entering the nest. There is a significant positive correlation between the guard bee population and colony responsiveness to disturbance, suggesting that the number of guard bees reflects the colony's preparedness for defense.

We allowed a colony the opportunity to become habituated to a specific person in the absence of any reward. Following this period of habituation, the colony was subjected to the identical behavior sequence performed first by the Familiar person, then by a Novel person, and finally by the Familiar person again. The evidence for individual recognition would be a stronger reaction (i.e., more bees guarding the hive entrance) to the Novel person.

The beehive we studied had an extended bottom board (41 cm by 56 cm) to facilitate observations. The experimenters provided the only human contact with the colony from the time it was moved into an isolated section until completion of the experiment. A video recorder focused on the entrance of the hive was located inside a shed, 4.5 meters from the hive. Both the Familiar and Novel persons were 22-year-old women of approximately the same height (1.55 m and 1.62 m) and weight (54.5 kg and 52.3 kg). Both persons wore denim coveralls, brown leather boots, and a white plastic hat with a mesh veil to cover the face. Cuffs of the coveralls were taped to prevent bees from climbing up the legs. The Familiar person maintained a constant personal hygiene routine, not changing products (e.g., soap, deodorant) throughout the experiment.

Each session consisted of a 5-minute pretrial during which video recordings were made to establish a baseline for the session, but no experimental manipulations were conducted. After the 5-minute pretrial, a 15-minute trial began. The experimenter walked around the beehive in a clockwise fashion originating from the right side of the hive. She carried a wooden stool in her left hand and a stopwatch in her right hand. After walking around the hive and passing the entrance of the hive twice, the experimenter sat at the left front corner of the hive, approximately 0.5 meter from the entrance, for the duration of the 15-minute trial.

At the end of the 15 minutes, the experimenter arose and walked away from the hive, to the left, not passing the front of the hive again. At this point, a 15-minute posttrial period began. The 15-minute posttrial was similar to the pretrial in that bee counts were taken but there was no experimenter present at the hive. Thus, each session consisted of a 5-minute pretrial, a 15-minute trial, and a 15-minute posttrial period.

Habituation occurred over twelve days, with either one or two sessions as described conducted at variable times each day. Following sixteen sessions of habituation to the Familiar person (during which the bees had the opportunity to learn that the observer posed no threat to the colony), a test of response to the Familiar versus Novel individuals was conducted. The Novel person had experience with bees prior to testing so that her comfort level with the bees resembled, as closely as possible, that of the Familiar person. In addition, the Familiar and Novel person rehearsed their movements at another hive to establish a consistent mode of circling the hive and positioning the stool. All sessions during the test day were conducted in the same manner as during the habituation period. An A_1–B–A_2 design was used: the first session of the test day was with the familiar person (A_1) to whom habituation had occurred, the second session was with the Novel person (B), and the final session with the familiar person (A_2) once again. Video records of the front of the hive were analyzed by counting the number of bees during selected freeze-frame portions of video recordings. The bees on the left half of the video frame were selected for analysis due to the fact that both the Familiar and Novel persons sat on the left side of the hive. Counts were made two minutes into the 5-minute pretrial; 2, 8, and 14 minutes into the 15-minute trial; and at 2, 8, and 14 minutes after the start of the posttrial. Seven measures were thus taken for each session.

The reduction in the number of guard bees present at the front of the hive suggests that the hive habituated to the repeated presence of the Familiar person. The subsequent test for discrimination revealed a significant increase in the number of guard bees when the Novel person was in front of the hive. The same pattern of statistical differences reported for trial data was also observed for post-trial measures. These data reveal that effects of the person's presence on the hive were still apparent for at least 15 minutes after each trial. Details of this analysis are given below.

The honeybee colony we investigated exhibited a decline in guarding behavior with repeated exposure to a consistent nonthreatening person. In short, the bees in the colony became habituated to the experimenter. However, this low level of defensiveness was immediately replaced with a greater level of guarding in response to the Novel person. This was notable because the Novel person was highly similar in appearance to the Familiar person and performed the same behavioral sequence. Given the relatively coarse visual acuity of bees, it seems likely the discrimination was made on the basis of odor differences, although this was not addressed directly in our experimental design.

Beekeepers frequently claim that their bees get to know them as individuals, based on the perception that defensive behavior declines with repeated manipulation of a colony. Our results agree with this beekeeper "lore." Additionally, the heightened number of guard bees that persisted well after the observer had left the vicinity of the hive suggests that researchers should use caution when conducting observations at the entrances of beehives. Anecdotal evidence suggests that minor disturbances of bee colonies, such as standing in front of their entrances for a few minutes, can influence foraging behavior for hours afterward. The results reported here suggest that it is not simply the presence of an observer near the hive entrance, but the specific identity of the individual that elicits defensive behavior of honeybees.

Inter-species recognition of individuals has been demonstrated in a wide variety of mammals and birds, but only recently has this question been turned to insects such as honeybees and cockroaches. Courtesy of Hank Davis.

Data analysis

A Friedman two-way analysis of variance (ANOVA) revealed a significant difference across the three phases of the A_1–B–A_2 discrimination test consisting of (A_1) exposure to the familiar person after habituation, (B) exposure to a novel person, (A_2) reexposure to the familiar person ($x_r^2 = 30$; $p < 0.01$). *Post hoc* binomial paired comparisons revealed significant differences ($p < 0.01$) in the number of bees present between both A_1 versus B trials, as well as B versus A_2 trials. There was no difference between the number of guard bees in the A_1 versus A_2 trials involving the familiar person.

Further Resources

Breed, M. D., Rogers, K. B., Hunley, J. A., & Moore, A. J. (1989). A correlation between guard behaviour and defensive response in the honeybee, *Apis mellifera*. *Animal Behavior, 37*, 515–16.

Breed, M. D., Smith, T. A., & Torres, A. (1992). Role of guard honey bees (*Hymenoptera: Apidae*) in nestmate discrimination and replacement of removed guards. *Annals of the Entomological Society of America, 85*, 633–37.

Moore, A. J., Breed, M. D., & Moor, M. J. (1987). The guard honey bee: Ontogeny and behavioral variability of workers performing a specialized task. *Animal Behavior, 35*, 1159–62.

Moritz, R. F. A., & Southwick, E. E. (1992). *Bees as superorganisms*. New York: Springer Verlag.

Taylor, A., Davis, H., & Boyle, G. (1998). Increased vigilance toward unfamiliar humans by harbor (*Phoca vitulina*) and gray (*Halichoerus grypus*) seals. *Marine Mammal Science, 14*, 575–83.

Hank Davis, Lisa Collis, and Gard W. Otis

Bonding
Dogs as Social Catalysts

A social catalyst is a stimulus that promotes interactions between people. Children tend to be strong social catalysts, particularly babies, who can attract a considerable degree of attention, often from complete strangers. Recently, it has been discovered that dogs, like children, can also serve as social catalysts and, in turn, may indirectly help to promote the psychological well-being of humans.

Domestic dogs have long been noted for their socializing role. Interview studies have repeatedly shown that dog owners believe that their pets can facilitate the development of friendships and help to promote social interactions within the home and community. Observational studies have confirmed these owners' reports. Peter Messent (1983), for instance, discovered that dog walkers have significantly more chance encounters and longer conversations with complete strangers whenever they are accompanied by their dog than when they are walking exactly the same route without the animal. Deborah Wells (2004) found that the presence of a dog is even more likely to attract social attention than the presence of slightly out-of-place objects (e.g., houseplants) or stimuli that normally serve as strong social releasers (e.g., teddy bears). These studies suggest that dogs can somehow act as "ice-breakers," allowing strangers to strike up conversations more easily. Initially, conversations may revolve around the animal itself, but eventually proceed to other topics of mutual interest.

Unfortunately, not all dogs can facilitate social interactions to the same degree. The age of the animal, for example, can contribute to the equation. Deborah Wells (2004) recently discovered that puppies encourage significantly more social acknowledgements than adult dogs of the same breed. Konrad Lorenz (1971) originally suggested that the young of most mammalian species can act as strong social releasers. Their large foreheads, big eyes, short limbs, and clumsy movements are generally perceived to be more endearing or "cute" than those of their adult counterparts, and hence puppies are more likely to act as social facilitators than older animals.

Dog breed may also influence public perceptions and the degree to which an animal can serve as a social facilitator. The Rottweiler, for example, is considerably poorer at encouraging social interactions than the Labrador Retriever. It is still unclear why these breeds differ in their ability to facilitate interactions between people. It may be the case that breeds reputed to be aggressive (e.g., the Rottweiler) do not serve as social catalysts to the same extent as those with a less fierce reputation. Differences in color, as opposed to breed per se, may be an alternative explanation for the differences in social catalysis exerted by the Rottweiler (black and tan) and Labrador (yellow). Indeed, earlier work by Deborah Wells (2004) revealed that people have a much greater preference for blonde, over black, colored dogs, often perceiving the latter to be more unfriendly.

While factors relating to the dog (e.g., breed, age, color) may influence the degree to which the animal serves as a social catalyst, factors relating to the person do not seem to play as important a role. June McNicholas and Glynn Collis (2000) found that the outward appearance of the person walking a dog does not detract greatly from the ability of the animal to facilitate interactions. Thus, scruffily clad individuals are just as likely to reap the social benefits of being accompanied by a dog as more carefully groomed people. Likewise, men are just as likely as women to attract attention from passers-by when they are in the presence of a dog.

The ability of dogs to encourage social interactions has important implications for the psychological health of humans, and hence the human-animal bond. People who have a good network of social contacts tend to experience higher levels of self-esteem, self-worth, and social confidence than those with a smaller number of acquaintances. By promoting social interactions, dogs may thus indirectly contribute to enhanced mental well-being. This may be particularly useful for certain sectors of the community. For example, research has shown that disabled people receive more social acknowledgments when they are accompanied by their service dog than when they are alone. It seems that the presence of the dog can somehow normalize people's reactions to those who may otherwise be overlooked because of their disability. Service dogs can thereby decrease the feelings of isolation that many with physical disabilities are prone to and help to improve social confidence, self-esteem, independence, and social identity.

The elderly in nursing homes and those residing in other types of institutions (e.g., prisons) can also gain enormous psychological benefits from the socializing role of dogs. Various studies have reported increased social interactions in residents and staff in nursing homes following the introduction of a dog. So called "pet-facilitated therapy" schemes involving visits from pets such as dogs are now relatively commonplace in Western society. These schemes have been shown to reduce levels of anxiety and depression and break the vicious cycles of loneliness that many people in institutional settings can experience.

The domestic dog will never have the power of speech. Nonetheless, the findings above suggest that this particular companion animal can act as a strong social catalyst. Like children, dogs can facilitate social interactions, help their owners to make friends, and offer a source of solace to the depressed, lonely, and institutionalized. Further work is now needed to unravel whether the brief interactions that dogs facilitate can lead to more substantial relationships and long-lasting psychological well-being.

See also

Bonding—*Companion Animals*

Further Resources

Lorenz, K. (1971). *Studies in animal and human behaviour* (Vol. II). London: Methuen.

McNicholas, J., & Collis. G. (2000). Dogs as catalysts for social interactions: Robustness of the effect. *British Journal of Psychology, 91,* 61–70.

Messent, P. (1983). Social facilitation of contact with other people by pet dogs. In A. H. Katcher & A. M. Beck (Eds.), *New perspectives in our lives with companion animals* (pp. 37–46). Philadelphia: University of Philadelphia Press.

Wells, D. (2004). The facilitation of social interactions by domestic dogs. *Anthrozoös, 17*(4), 340–52.

Deborah Wells

Bonding
Gorillas and Me: A Personal Journey

I was born with autism. People are surprised when I tell them that because they have in their minds a kind of black picture, a smothering and oppressive picture. Perhaps it is a fragmented picture, an old picture with the corners torn away, a picture with the shadows of the Rain Man, or developed in the dark, positive images of the negatives they have seen, the upsetting black and whites of words on pages.

When most people think of autism they think of violent, unreachable people in worlds of their own making, worlds without keys, feeling no empathy, lacking imagination, and unavailable to the deepest of human needs for contact and love. Having autism is the worst fate parents can imagine befalling their children, and they dread its impact on their families. Experts endlessly speculate on what causes autism and race to find a cure for it. I acknowledge that the phenomena of autism can cause great pain, both to those who have it and those who live with those who have it. But there is beauty in our way of being.

Since I was born I have felt the breath of butterflies and felt the turning of the heavens. I have bent from the weight of the breathed air and covered my eyes to shield myself from the burning beauty of other people's eyes, so full of wonder and regret. I have always felt everything—the too bright sun, the deafening loudness of whispers—I can taste sound and smell colors.

When I grew up and had my first, unhappy job experiences in a big city, I knew I needed to be near some kind of nature, so with one of my first paychecks, I went to the zoo. It was pleasant there, and I enjoyed the trees and the sounds of the living things. But something very primal and basic happened to me when I turned the corner and saw the gorillas sitting there, dark and solid, slow and waiting. I sat down in front of them, and suddenly, the world that had been too loud and bright, too foreign and fast, just slowed down and I felt, for the first time, that I was in the presence of creatures that I could understand and who could understand me. Their stoicism, their predictability, their gentleness, all these let me come home to them. I had always had an interest in the primal, in the prehistoric, and in evolution from the time I was very young. I can remember thinking that it was because of our shared ancient history that I was drawn to them, as if I really belonged to the Gorilla Nation and not to humanity. I felt like my way of being was somehow older, more archaic, than most other people's and that I belonged with gorillas for that reason.

Sometimes I would go each day and sit for hours. Sometimes I could only make it there once a week. But I always went as often as I could. I did this for years before I actually began to volunteer there. After a while, people who worked there began to recognize me, and I believe they were touched by the way I cared about the gorillas and other animals. I was blessed as they began to take an interest in me and helped me to become a part of the zoo. I got a variety of paid and unpaid positions there, and I spent part or all of every day there, sometimes even coming in on weekends.

The gorillas allowed me to enter their world and regain my humanity through their safety. Almost like a baby, I watched them and learned from them. I wasn't overwhelmed when I was with them, so I could pay attention to social cause and effect. Then, with the sense of balance I retained from their strength, I could go out in the world and connect with it. They gave me my personhood. Animals are less complex and deceptive on a social level than are human beings. This isn't to say they are not just as complex emotionally and perhaps even spiritually and that combination is magical for people who can't connect comfortably because they are overwhelmed. I still relate to humans best when I think of them as primates who have, at the end of the day, very simple needs and desires: to be loved, to be appreciated for their special abilities, who want to leave something meaningful behind them. It is the layers on top of those basic things that get autistic people, and maybe all of humanity, into trouble. Perhaps we started to lose sight of our purpose because we have become captive ourselves.

Some captive gorillas are very bitter about their state of captivity. We know from sign language studies that wild-caught gorillas remember the trauma of seeing their families killed and the brutal journey to captivity and that they resent their loss of freedom. But, like human people, some of them find ways to open up to their circumstances and love with a whole heart. Congo, a 500-pound silverback gorilla and my best friend, was just such an animal. He gave everyone a chance and tried to connect with them. Though our relationship was special to me beyond measure, I think the touching thing about Congo is that he loved so many, reached so many. So, his immediate healing toward me was to love me simply and unconditionally, in a way we all want, regardless of our normalcy. I could relax in his protection, be contained by his spirit. In the warm dark of his care I could look up, look around, stretch my sore soul and see the world. Then, I realized if he could do this, love human people in spite of all he had been through, then so could I; indeed I had a responsibility as a person in the Gorilla Nation to do so. I still will think to myself when I am making a decision or doing some thing, "What would Congo think of me?" If I know he would be proud, be honored, then I am on the right path.

Eventually, I became more stable in my personal life, and I was able to begin my university studies. Still finding it difficult to be around many people, I found a private Swiss university—Universitat Herisau—that would allow me to work with mentors and design my program in ways that maximized my ability to learn while giving me the best guidance. Being able to learn long distance, away from the crowds of a campus, without having to go to new places again and again during the day for instruction, and being able to choose and then demonstrate which areas of learning were relevant for me, all of these things made my success possible. I found a home in academia and flourished there; I went on to become an adjunct professor of anthropology at Western Washington University, in the town where I now live.

I am sad to say that Congo has since died. It is a great loss for many. I still use his spirit, his essence, so much like Buddha to me, to guide my way. When I think of him, I remember clearly the day I was sitting with the gorillas and thought, "They have given me everything. Too much to even thank them for because there aren't human words or concepts for what they give. Now I am going to have to leave and make the world a better place." It has only been in the last two years that I lost that sense of needing to run toward the ghost of a future I thought I had lost. Now I feel I am where I am meant to be, but if I look over my shoulder I see the crouched spirits of a thousand people who didn't make it; people like me who were broken by being different.

We are all strange and broken and beautiful in our own ways. We are each so afraid of disconnection, and yet it can't be easily escaped; it is an inevitable state of being and, perhaps, the price of consciousness. That fact makes our connections to other living things all the more important because we are conscious and can make choices about living. There is beauty in our difference and also beauty in our sameness: sameness with other animals, sameness with one another. We feel the loss of so many things: falling forests, disappearing animals, the loss of each other as we move far and fast in our culture. I think the loss has made us tired, maybe so tired we feel like we can't change things. But we can. It will take all of us autistic people, too. We have a lot to bring to the challenge of existence and we are ready to tell our part of the story—our own stories, the stories of things hidden, the stories of the animal nations, the stories of a living world.

Further Resources

Fossey, D. (1983). *Gorillas in the mist.* New York: Houghton Mifflin.

Graham, B. (2000). *Creature comforts: Animals that heal.* Amherst, MA: Prometheus Books.

Grandin, T. (2004). *Animals in translation: Using the mysteries of autism to decode animal behavior.* New York: Scribner.

Prince-Hughes, D. (2001). *Gorillas among us: A primate ethnographer's book of days.* Tucson: University of Arizona Press.

———. (2004). *Songs of the gorilla nation: My journey through autism.* New York: Random House.

Dawn Prince-Hughes

Bonding
Homelessness and Dogs

Pet ownership among the homeless can alleviate the loneliness and isolation the situation brings and is a recognized conversation starter with members of the public. As noted by Alan Beck and Aaron Katcher (1996), animals are indifferent to the various situations a person can find themselves in, offering protection against change and a sense of constancy. These factors often create a strong bond between a homeless owner and his/her dog, whereby the dog is regarded as a trusted and loyal companion.

The study of the relationship between homeless people and their pets has only recently come to life. Heidi Taylor, Pauline Williams, and David Gray (2004) investigated several aspects of this relationship, considering and expanding upon previous research by other authors. The study incorporated the use of two questionnaires, one designed for securely housed participants and one for homeless participants, to compare attachment and empathy between the two populations and between dog owners and non-dog owners within these populations. A study of crime, drug use, health, and medical use among the homeless and public opinion of homeless dog owners was also included.

Those who have no shelter at all, those who reside in hostels, refugees, or live with friends and family can all be termed homeless, making a single universally accepted definition difficult to compose. Some of the commonly recognized causes of homelessness include alcohol and drug abuse, inability to afford to buy or pay rent because of the shortage of affordable homes, ex-prisoners, ex-servicemen, and those discharged from psychiatric units. Homeless people frequently experience damp, overcrowded, cold, and unsanitary environments, which are accompanied by numerous health complaints. The study of animals and their effects on human health has been well documented; Sara Staats, Loretta Pierfelice, Cheongtag Kim, and Riley Crandell (1999) have shown that older pet owners visit the doctor less than their nonpet-owning peers, and many studies have looked into the reduction in cardiac disease-related risk factors in pet owners. One might expect that such positive effects of pet ownership on health would be displayed in homeless populations; however this is rarely the case. Oswin Baker (2001) studied the health of both homeless dog owners and non-dog owners, finding no significant difference between the two. The study of Heidi Taylor and her coauthors (2004) also found the same. Baker (2001) concluded that this was an unsurprising result because of the extreme pressures of homelessness, which destroy health, outweighing any positive effects that a pet may bring. Access to medical treatment can be difficult for someone who has no permanent address, and to complicate things further, those homeless who keep dogs are even less likely to receive medical attention as they cannot take their pet in with them, and many would rather go without treatment than leave their companion outside.

Many homeless, who have no income, take to crime such as shoplifting to support themselves. However, the study by Taylor, Williams, and Gray (2004) found that homeless dog owners are less likely to turn to crime because of fear of separation from their pet due to imprisonment. Also, no difference in drug use between homeless dog owners and non-dog owners was found. This may be explained by the loneliness and anxiety that homelessness may entail; thus one may appreciate the desperation behind turning to drugs.

Laurel Lagoni, Carolyn Butler, and Suzanne Hetts (1994) give a number of occasions when human-animal relationships may be perceived as stronger as and more important than the average human animal relationship, such as the pet getting the owner through a time of difficulty or being a significant source of support, both of which can be true to the homeless person-pet situation. Lynn Rew (2000) speaks of homeless youths who "keep going in life" because they feel that their pet's survival is critical to their existence. Aline and Robert Kidd (1994) compared the attachment to animals between homeless dog owners and homeless nonowners, revealing a higher attachment in those with dogs. Randall Singer, Lynette Hart, and Lee Zasloff (1995) found the securely housed to be less attached to animals than the homeless, and Taylor, Williams, and Gray (2004) repeated such results, finding the difference to be highly significant. The latter study also looked at empathy toward animals, using a scale formerly devised by Elizabeth Paul (2000) to measure empathy toward animals to compare the securely housed and the homeless populations, and found the latter to be highly significantly more empathic. Anecdotal evidence suggests that most homeless, having had many adverse experiences with people, may greatly appreciate the existence of those less judgmental, such as animals.

Taylor, Williams, and Gray investigated public attitudes toward homeless dog owners, revealing that 72 percent of the sample believed that the homeless should be allowed pets if they wish (highly significant); interestingly, the majority of survey respondents were females (74%), tying in with previous research showing females to be more empathic toward animals. Perhaps contradictory, a greater percentage of females than males were more concerned for a homeless dog's welfare than the homeless person's, even though they agreed that the homeless should be able to own pets. The majority of respondents (70%) believed the bond between a securely housed person and his or her dog is equal to that of the homeless person-pet bond; however, research has shown this to be highly unlikely.

Taylor, Williams, and Gray conclude that the homeless person-animal bond could be considered different to the general human-animal bond because of differences in trends in the average population in attributes such as attachment, health, crime, and drug use. It is ironic that a homeless person's pet(s) can be the one factor in their lives that keeps them going; however, it is also the main factor for keeping them on the streets, for many shelters and most housing do not accept pets, and many homeless pet owners interviewed would rather remain homeless than give up their pet.

Further Resources

Beck, A., & Katcher, A. (1996). *Between pets and people. The importance of animal companionship.* West Lafayette, IN: Purdue University Press.

Kidd, A. H., & Kidd, R. M. (1994). Benefits and liabilities of pets for the homeless. *Psychological Reports, 74,* 715–22.

Lagoni, L., Butler, C., & Hetts, S. (1994). *The human-animal bond and grief.* Philadelphia: W. B. Saunders Company.

Paul, E. S. (2000). Empathy with animals and with humans: Are they linked? *Anthrozoös, 13*(4), 194–202.

Rew, L. (2000). Friends and pets as companions: Strategies for coping with loneliness among homeless youth. *Journal of Child and Adolescent Psychiatric Nursing, 13*(3), 125–32.

Singer, R. S., Hart, L. A., & Zasloff, R. L. (1995). Dilemmas associated with rehousing homeless people who have companion animals. *Psychological Reports, 77*(3), 851–57.

Staats, S., Pierfelice, L., Kim, C., & Crandell, R. (1999). Exploring the bond: A theoretical model for human health and the pet connection. *Journal of the American Veterinary Medical Association, 214*(4), 483–87.

Taylor, H., Williams, P., & Gray, D. (2004). Homelessness and dog ownership: An investigation into animal empathy, attachment, crime, drug use, health and public opinion. *Anthrozoös, 17*(4), 353–68.

Heidi Leader and Pauline Williams

Bonding
The Horse "Beautiful Jim" and Dr. William Key: A Personal Essay

As the author of a book about the horse Beautiful Jim, there are certain questions I'm regularly asked, especially how I came to write a book chronicling the strange but true story of Dr. William Key—former slave, Civil War veteran, self-taught veterinarian, and one of the most successful black entrepreneurs at the turn of the twentieth century—who took a crippled foal named Jim, originally bred for the turf, and raised him like his own son, teaching him without force to spell common words and names, to add, subtract, multiply and divide numbers up to thirty, to work a cash register, cite Bible passages, "debate" politics, and more, as the two traveled the nation to become box office stars, winning the hearts of 10 million Americans, ultimately legitimizing the fledgling animal rights movement.

"So, are you a horse person?" is asked so often, by so many self-described "horse people" that I've taken to offering a disclaimer. Am I horse person? If by yes I would have spent my formative years living, breathing, and dreaming horses, or that now I occupy all free time riding, breeding, and racing horses, or that I am a veritable encyclopedia of equine expertise, then the answer would be no, not in that way. Do I love horses? Yes. Have I ridden a horse a time or two, and even thrilled to galloping bareback through the wooded hillsides of east Tennessee? Yep, back in the day. Have I ever looked deep into a horse's eyes and seen a primordial wisdom there and known those eyes were looking back into my own to get into my head too? Yes, without question. Have I always felt a visceral, blood-level connection to horses that tugs on something ancient inside—like DNA? Yes. In that way, I'm a horse person.

If you consider how horses have always been essential to humans, only becoming less so in the last century, and if you look at how horses have been represented across cultures and recorded history—often as supernatural beings, frequently shown in flight, defying laws of gravity—you could say we're *all* horse people.

But there was something more personal than the universal human-horse connection that compelled me to spend two years excavating the lost history of Jim and Doc Key. The real hook was how hearing their saga helped me recover a lost memory—or a series of childhood memories—recalled from long car trips during which my father told epic tales about a made-up character named Blue Horse. Dad, a renowned nuclear chemist and 1960s civil rights activist, didn't fit the conventional definition of a horse person either. Nonetheless, his take on equine intelligence was clearly reflected in his ugly duckling story of a misfit foal who had been born blue and scrawny but who grew up to be the wise leader of a rare tribe of like-minded and like-colored horses capable of high-level human communication. In mythic fashion, the Blue Horses rarely convened with people, except when called upon to rescue them from their all too-human predicaments.

In episodes Dad told, which typically involved me and my siblings in trouble after not heeding parental warnings, Blue Horse was not only linguistically superior, he could also carry three kids on his back, fly, swim, and defeat enemies without a lot of bloodshed. How? By outsmarting and even out-punning them. That's right, Blue Horse was *funny*.

Understandably, I was predisposed to love the real-life version of Blue Horse as embodied by Beautiful Jim Key and to believe that he was as intelligent and witty as claimed. Whatever skepticism I had was erased by a 1902 *Syracuse Post-Standard* article by a reporter who had forgotten to bring the equine star an apple for their interview and who observed that when asked how the meeting had gone, Jim went to his spelling board and pulled down letters that spelled out the word "F-R-U-I-T-L-E-S-S." A groaner pun worthy of Blue Horse, or my dad. If humor wasn't proof of higher, human-like intelligence, what was? Or, if it was a trick, how was it done?

Those questions arose a century ago when a team of Harvard professors set out to study what kind of cuing system Dr. Key used to coach Jim. After sitting through several performances and observing how the Doc stood to the side of the stage as Jim responded to questions posed at random from members of his audience, and after an extensive exam of both horse and trainer, the conclusion, as published in the *Boston Globe,* was that there was no trick and that Jim Key's human-like intellectual abilities were the result of "simply education." Indeed, William Key maintained that while Jim was of above average intelligence, other horses could be similarly educated—with kindness and patience—as long, Key noted, as the horses hadn't been previously abused.

The implications of Key's approach are as relevant now as they were in the late nineteenth century, not only to better understand the intellectual capacities of other species, but also for studying the impact of cruelty on learning and literacy—in both animals and humans. Most educators agree that the more literate and educated a society, the more ethical and less cruel that society will be.

Interestingly enough, at a time when concerns about animal cruelty were largely the domain of an elite fringe of activists, the humane movement's most galvanizing figure—before the Keys came along—was a fictional horse named Black Beauty. On the national stage that the book had set, Jim made his entrance as the real McCoy. Suddenly, the public mind took a giant leap as the reality of a horse that could apparently think, feel, deduce, opine, joke, and, above all, acquire knowledge through learning, made the practice of knowingly harming animals no longer tolerable.

None of this was accomplished without tremendous resistance, forms of which, not surprisingly, remain today—I have been surprised at how intensely the prospect of human-like intelligence in animals continues to threaten a wide range of belief systems.

Historically, the animal rights movement has never been comfortable with animal performers, given the potential for exploitation, and, I am told, the sense that what Jim and other smart animals do in demonstration is the result of learning to perform by rote repetition. Actually, when Jim and his human partners first sought to ally themselves with humane groups, there was little initial interest.

In Jim's day, seeing was believing. For information-age skeptics, however, even alleged "animal people," accounts of a horse who figured out how to correctly file letters—like one addressed to Mr. Smith that belonged in a box marked "R-T"—have the feel of urban myth. Though the science-minded concede that Jim might have been "different" intellectually after spending his first year of life in Dr. Key's house, this group generally stops short of agreeing that animals are capable of cognitive and abstract forms of thinking.

As for bona fide horse people, most believe empirically that horses are more intelligent than we know, while others call them "highly perceptive," even "wily," and others still that they are "basically stupid," either "copycat herd animals" capable of performing by rote for rewards or "too lazy to learn." One breeder swore she had never witnessed a sense of humor in horses, least of all in weanlings, as Doc Key reported.

But no cages appear to be more rattled by dangerous talk about thinking animals than those belonging to people who still take issue with Darwin. In 1897, when a proposal came before the Cincinnati School Board to close schools for a day so that every child could witness a free performance by Jim and Dr. Key, the board voted in unanimous opposition—with one board member declaring that under no circumstances would schools be closed for a "monkey show." The same bias was on display in 1925 when a showdown in Dayton, Tennessee—over the teaching of evolution—became known as the Scopes "Monkey" trial. What would the Keys say about the resurgence of this divisive debate? Maybe they would gently point out that the antievolution arguments based in literal interpretations of the Bible are similar as those used to justify slavery, Jim Crow, opposition to civil rights, segregation, racism, the oppression of women, children, and animals, not to mention the exploitation of our environment that now threatens the ultimate survival of all species. Both Biblical scholars—especially of verses in which horses appear, which Jim learned to recognize by hearing them spoken and to identify the written name of the book and chapter where cited—they might argue that nothing in Darwin directly "denies the story of the Creation of man as taught in the Bible," as detractors claim. Jim might comment that we are not lessened by the admission that we have descended from "a lower order of animals," what Charles Darwin concluded was a "hairy, tailed quadruped, probably arboreal in its habits, and an inhabitant of the Old World." We are, after all, family.

In the end, that was the point of what a horse and a man from Tennessee taught us. Like Blue Horse in Dad's stories, they appeared on the scene when their message was most needed, at a time not so different from this one. They proved how gaps could be bridged between different races and species, between rich and poor, between North and South, between disparate parts of communities, among different religious and political persuasions, even between skeptics and true believers. They taught the power of kindness—both in terms of seeking *kind* ways of treating others and also in terms of connecting different kinds of beings, making each of us, all together, *kindred*.

Further Resources

Rivas, M. E. (2005). *Beautiful Jim Key: The lost history of a horse and a man who changed the world*. New York: William Morrow.

———. (2006). *Beautiful Jim Key: The lost history of the world's smartest horse*. New York: Harper Paperback.

Mim Eichler Rivas

Bonding
Horse Whispering, or Natural Horsemanship

The term *horse whisperer* has become familiar in American culture, thanks largely, but not entirely, to a book and 1998 movie of the same name. Horse whisperers are horsewomen and horsemen who have a seemingly special ability to communicate with horses with very few visible signs of communication. Horse whispering, or "natural horsemanship" as it is generally called, is a practice that endorses humane, nonforceful, and compassionate interactions between humans and horses. It is a style of training and working with horses that is based on the premise that humans can communicate with horses by understanding the horse's thought process and way of being in the world. Natural horsemanship teaches the human to read bodily gestures of the horse to understand if the horse is scared, angry, willing, happy, approachable, and so on. It requires humans to learn to control their emotions of anger and frustration and, instead, learn to empathize with the horse and see the world through the horse's eyes. With this understanding, humans then structure their interactions with horses, not through force, but rather through letting the horse learn what to do through communication.

Long before "horse whispering" was a concept, Native Americans and vaqueros (Mexican cowboys) used similar approaches. Today, two brothers, Tom and Bill Dorrance, are widely credited for the more modern form of natural horsemanship that is so popular today. Tom and Bill grew up in California in the early 1900s and were working ranchers. Though they differed in particulars, they both believed in the horse's intelligence and espoused a philosophy of partnership rather than dominance. Many of the well-known natural horsemanship clinicians (teachers) credit Tom and Bill Dorrance as their mentors.

Though it would be inaccurate to say that there is one form of natural horsemanship, it is a broad umbrella term used to identify this philosophy and its practices. Since the mid-1990s, its techniques have been enjoying tremendous popularity. Across all the riding disciplines, horse people either know about it or practice the techniques. Indeed, natural horsemanship has translated into big business as many clinicians are making thousands of dollars by writing books, creating videos, and giving clinics to promote the concept and techniques of their own brand of natural horsemanship. Many clinicians offer a systematic approach to training horses through natural horsemanship and package their techniques in easy to understand books and videos. Beyond the books and the videos, the most popular way that the philosophy and techniques of natural horsemanship are disseminated is through clinics and demonstrations.

The clinical and demonstration settings have several similarities: the central figure is the horse whisperer (the clinician), who shares the spotlight only with a horse, who is often young or "green." Around them sits an adoring audience watching the clinician work with the horse until it is calm and willing to interact. Once the horse is willing to interact with the clinician, she/he will begin to ask the horse to engage in a variety of exercises. This process builds trust between the human and the horse and becomes a foundation from which humans and horses can learn how to safely interact. Clinicians explain that it is no longer acceptable to "break" a horse; rather, a more acceptable approach of "gentling" or "starting" a horse better represents the heart of natural horsemanship.

Natural horsemanship's popularity is due in part to the changing social role of horses in today's society. Material shifts in human-horse relationships, in part, have fueled its popularity. Because horses are seldom used for labor in the United States, they are becoming a form of recreation more than ever before. This shift has meant a rise in the number of amateurs needing help with horses. It would be fair to say that there has never been a time in history when more people owned horses who knew less about them. Historically, the horse has been used as a worker, a form of transportation, and an implement of war. Beginning with the invention of the combustion engine and followed by the increased modernization of farming and factory work, by the early to mid-twentieth century, "jobs" for horses quickly began to dwindle. As horses' utilitarian value decreased, their value as a form of recreation increased.

Today, the percentage of horses used for labor is minuscule when compared with the percentage of horses enjoyed as a form of recreation. According to the American Horse Council, national statistics show that over 5 million horses are used for show and recreational purposes, and approximately 1 million account for other activities such as farm and ranch work, rodeo, polo, and police work combined. Together, all these changes mean that more people than ever before are struggling with their horses and looking for guidance. This created an enormous market for natural horsemanship.

These conditions were well-suited for horse training systems that could help anyone learn to work with horses by following a series of simple steps. Therefore, part of the financial success of some of the most popular teachers of natural horsemanship is that they offer people a step-by-step method for working with horses. This reflects a much bigger trend in American culture, in which people are drawn to programs and products that promise simple techniques to master complex problems. Witness the popularity of twelve-step programs and scores of books geared at "dummies." It would not be fair to say that all natural horsemanship clinicians promote a simplistic system. In fact, there are horsewomen and horsemen in the business of natural horsemanship who argue against approaching horse training in a simplistic, systematic, linear way. Nonetheless, many of the most successful and most well-known "horse whisperers" offer a systematic approach to horsemanship. For these clinicians, they present horsemanship a bit like a math problem; if you do A, your horse will do B, and the end result will be C. Making horsemanship systematic in many ways erases the individuality of horses, but on the flip-side, it creates a structure that is easy for people to follow. More importantly, step-by-step approaches are easy to package and sell in the form of books and videos. Yet some might rightly argue that thinking of horsemanship in a linear fashion could be dangerous to both humans and horses. A mechanical, systematic approach can offer people false hope: if they just learn the mechanics of method and follow the right steps, they too will become skilled at the art of horse whispering. The skills necessary for training and working with horses take many years to develop, and when people do not achieve with their horses what they see clinicians achieving, this often leads people to feel frustrated and deflated. As a result, this frustration can be directed at the horse as if they are being stubborn or defiant, when in actuality there is a miscommunication and lack of understanding between the human and the horse.

The popularity of natural horsemanship parallels a larger trend of thinking about nonhuman animals as companions who share meaningful relationships with their human partners. Though the commodification and popularity of horse whispering has not been without its problems, there is no doubt that its general philosophy of compassionate and effective horsemanship has been a good thing for horses. Indeed, for centuries horses have been dominated and abused in an effort to make them "suitable" for humans. Thus, the popularization of humane, nonforceful training techniques is a positive step toward ensuring the overall well-being of the equine species.

See also

Bonding—*Agility Training*
Communication and Language—*Human-Horse Communication*

Further Resources

Dorrance, B., & Desmond, L. (1999). *True horsemanship through feel.* Guilford, CT: The Lyons Press.

Dorrance, T., & Porter, H. M. (1994). *True unity: Willing communication between horse and human.* CA: World Dancer Press.

MacLeay, J. M. (2000). *Smart horse: Training your horse with the science of natural horsemanship.* Lexington, KY: Eclipse Press.

Keri Brandt

Bonding
Infant Signals and Adopting Behavior

Forming Bonds with Humans

If anybody is able to distinguish at first sight a nestling bird or an infant mammal from an adult of the same species, it depends partly on our ability to recognize the distinctive cues associated to that species age class. This concept was described for the first time by the Nobel Prize winner Konrad Lorenz in 1943. In his paper he suggested that altricial broods, those fully dependent on parents for survival, of many endotherm species (mammals and birds) are equipped with some common features, particularly located in the head, forming what he named "baby schema." Such a similarity is assumed to be indicative of an evolutionary convergent phenomenon between mammals and birds, allowing an infant to be recognized such not only by its parents but also by strange adults of its own or different species as well.

We assign the status of infant when we recognize the presence of certain characteristics, largely undefined. This means that we classify an animal as infant only if its behavior and appearance coincide with what German speaking people would call infant *Gestalt*. This is not a clear-cut, unitary image, but rather is a combination of broad characteristics, which are otherwise almost undefined if considered singly, applicable to a large number of mammal and bird species. Our use of the label "infant" or "immature" is then the result of a decision taken only after these characters have been recorded. Human infant gestalt is rather similar to that of many animal species. Presumably, this is the reason why humans, after becoming hunters, did not kill indiscriminately all animals they met, but now and then spared and even adopted some, then starting the process of domestication.

The origin and evolution of the infant form appears evident when we consider the natural context in which it is most evident: reproduction. An adult bird or mammal producing offspring will recognize his progeny provided that it bears specific characteristics. The infant must use signals to communicate with its parents, and in turn, the parents must be able to somehow receive its signals. It is important to consider that these signals involve the whole range of sensorial capabilities: vision, hearing, smell, taste, and likely touch as well. They are then called infant signals. When they fade out or disappear, the parents lose interest in their offspring, and it is possible that mammal weaning is partly rooted on a mechanism of this sort.

Lorenz himself described the list of characteristics that can be grouped as infant signals. They include a large head in proportion to the body, large eyes on the midline of the face, round cheeks, short limbs, rounded body, and soft body surface. In mammals, particularly canids, we might add floppy ears. Everybody can check that these characteristics are usually borne by infants of any bird or mammal species, humans included, apart from some obvious exception, such as, for instance, the lack of ears in birds and pinnipeds. The constant presence of these signals together in infants led Lorenz to name them cumulatively "baby schema."

An important, common feature of infant signals is that they have the ability to induce labeling the producer as infant and consequently block the spontaneous aggression that any adult animal has against a strange individual, which in this case would be unprotected from adults. The infant signals emitted by the young have then two important evolutionary functions: to increase the survival probability of the bearer in the case it occurs to be inadvertently separated from parents and to increase the possibility that the progeny of parents having lost their young still can reach the adult age and spread their own genes to the following generation (fitness increase).

If we look at the adult of a number of bird and mammal species we obviously note many differences. However, if we look at the offspring of the same species, particularly their head, we note striking similarities. We can ask ourselves why different, evolutionary far species evolved similarities in the key stimuli associated with their infants. The answer is that an evolutionary convergence in infant signals increased the species' genetic benefit in terms of fitness. In fact, if the infant signals are similar and the infant can be recognized as such, it can then prevent the aggression and trigger parental care in the strange adult it might meet. In order to increase their efficacy, infant signals should work even beyond the species boundary, should then be adapted to a broad range of animal groups. If they have to be recognized anyway by several species they also have to be somewhat indefinite if considered singly, but working when perceived all together, just a gestalt.

Besides body size, other types of signals may release parental attention and even adoption by strange adults. Gannets (*Morus bassanus*), a colonial fully white marine bird, have been reported easily adopting guillemot (*Uria aalge*) chicks, regardless they have a contrasting black and white plumage. It is suggested that it is right such contrasting plumage that acts as a releaser signal for adoption in gannets. Conversely, the white plumage of gannet chicks would be a substimulus, since they are attacked by adult gannets and even by their own parents if they wander away from their natal nest. Color is important in domestic sheep too: Merino ewes, having white pelage, accept their own lamb after birth, provided that they possess specific visual cues. In particular, ewes accept a lamb even if it is partially or almost fully colored black, but only if the head has not changed color. Primates too are inclined in adopting strange young, but a similarity in body coloration between infant and adult is important in eliciting parental care.

The evolution of detection of infant signals within the same species was followed by adoption between different species, which in turn led to the rise of the remarkable mechanism of brood parasitism. Following a coevolutionary path, cuckoos developed eggs closely resembling the host's own eggs, but having an accelerated incubation period. After hatching, the parasite cuckoo nestling exploits exaggerated (called supernormal) releasing signals located in the mouth mucosa, which is lined by brilliant red, more marked than that of stepsiblings. This visual signal "fascinates" the foster parents, which care for their foster hatching and continue restless to feed it even though it soon becomes much greater than themselves. Although being nest parasites themselves, the African widowbirds chose the mimicry evolutionary path: instead of developing supernormal releasers, as cuckoos, they evolved chicks reproducing perfect copies of the typical brightly colored "mouth buttons," again located in the oral mucosa, of their host's hatchlings. As result, the host parents cannot discriminate their own nestlings from the widowbird chick intruders and, therefore, they feed equally offspring and parasites.

Sound signals are used mainly in rodents, whose pups produce typical infant ultrasounds in the range 50–70 kHz during the early days of life. These high-frequency calls are produced when the pups are in distress and have the effect of stimulating and maintaining the dam's hormonal state, whereas they do not affect other typical parental activities, such as nest building. Sometimes it is the visual effect acting as an infant signal. An interesting example has been reported in the mute swan (*Cygnus olor*), whose young are usually gray. In a few cases, however, the latter exhibit the white color typical of adults. This "abnormal" coloration evidently prevents recognizing the young as infant, as territorial pairs attack white young more often than gray ones.

Behavioral patterns typical of young individuals are found in some adult displays named "infantilisms." These behaviors are found in sociosexual contexts and are useful to maintain group cohesion or pair bonding. Everybody certainly experienced a domestic dog crouching on its back when meeting the owner. This behavior is usually displayed by dogs and their wild ancestors, the wolves, to appease one group member when wishing to signal its subordinate rank after being defeated during a ritualized fight. Such a posture originally belongs to pups, but is maintained even at adult age to indicate submission. When a chimpanzee (*Pan troglodytes*) meets a known species member, as appeasement and greeting gesture it extends its hand, palm up, seeking contact with the other, who in turn extends its hand, palm down. As mutual trust grows, the two shake hands, kiss and embrace one another. When frightened, adult chimps embrace a young one and calm down. In all these situations we clearly note the emergence of the infantile search for body contact.

An additional evolutionary phenomenon related to human sensitivity to infant signals is the domestication of many avian and mammalian species. Domestication of these animals likely started in the Neolithic age, when the early group-living humans hunted animals and occasionally encountered newborns of the hunted species. Sensitivity to infant signals blocked hunting motivation in humans and enhanced adopting behavior. On the other hand, this phenomenon continues even nowadays, occurring every time a lonely pup or kitten is encountered and adopted within a human family.

Furthermore, imprinting-like processes led to the formation of social bonds, with consequent stable integration of the animal into the human context. A clear example is provided by the domestic dog, whose domestication was evidently facilitated by infant signals. Man likely preferred, more or less consciously, keeping adult subjects bearing infant traits. Even today, many strains exhibit these characters and pet dogs are a true concentration of infant patterns: large head with floppy ears, elastic and soft body surface, short and thick paws. In contrast, breeds selected to arouse fear or respect, as watchdogs, lack infant characters. Besides, in some cases, such as in the Doberman or Boxer, the owner removes surgically the infantile characters, as the floppy ears, not yet eliminated by genetic selection.

Humans' attraction to infant signals has also been directed toward their own species as well. Infant signals are known to be a strong vehicle to the exploitation of several activities, first of all infant toys and doll industry. Cartoon characters are another clear example; in fact, we can easily note that positive (pleasant) characters are a "concentration" of infant signals, while negative (unpleasant) ones lack them almost completely. The late evolutionist Stephen Jay Gould noted interestingly that Disney's well-known Mickey Mouse became more youthful over the years. In fact, three types of body measurements (eye size, head length, and cranial vault) show that the artists evolved Mickey's appearance toward a progressively more infant-like aspect, likely to increase its acceptance by the public.

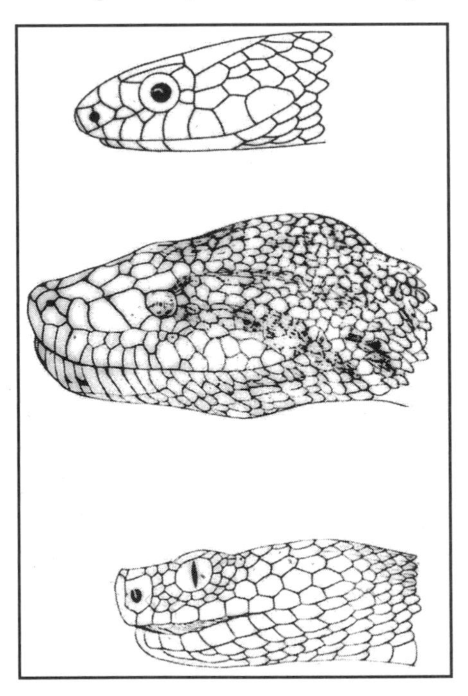

Infant signals are borne by mammals and birds only. Consequently, snakes do not induce our tendency to care for them, at least not because of infant signals. However, everyone can attest that people consider the upper head as the most preferred (or least unpleasant). In fact, it is the one with larger eyes, which in mammals and birds are a typical, strong infant signal.

Lorenz suggested that infant gestalt perception acts as releasing mechanism to protective behavior and this seems to be true in our species too. In fact, adult humans have been demonstrated to prefer slides depicting infants rather than adults. Such a preference develops at puberty, but differs between genders, rising earlier and stronger in women than in men. However, it is possible that social factors became superimposed upon an innate predisposition.

Infantilisms are found in humans too. Similar to other primates, bodily contact has soothing effects after a shock or during courtship interactions. Kissing behavior is usually performed by men rather than by women and clearly derived from parental trophallaxis (oral feeding between two subjects), observed in a number of other primates and still practiced in some human populations.

Finally, it is worth mentioning the possibility that animal species perceive infant signals released by human infants themselves. For instance, domestic dogs are known to discriminate adults from children and behave differently in front of them. It is common experience to observe not only pet dogs but also watchdogs be very patient victims of children's long-lasting play or to observe cats that usually do not tolerate manipulations by adult humans accept being stretched and "tortured" by babies without any even vague retaliation attempt.

Although specific scientific studies are still unavailable, there is a great deal of anecdotic reports about "wild" children—that is very young children abandoned in the wild and adopted and reared by animal species. Baby adoption of this sort is very likely released by the same mechanism underlying retrieving behavior observed in several mammal species. The first and maybe most famous of those children was Victor de l'Aveyron, who was found in 1799 in a wood in southern France and later studied by the French physician Jean-Marc Gaspard Itard. Victor was eleven or twelve years old and was filthy, naked, and covered with scars. Local residents began reporting that a young, naked boy had been seen in the woods five years earlier, presumably living alone for many years and surviving by eating whatever he could find or catch.

There are also two more recent and better documented cases. They concern Kamala and Amala, two Indian girls about eight and one-and-a-half years old, respectively, found together in the 1930s in a wolf den. Their appearance and behavior could hardly be defined as "human": for instance, they walked on legs and arms even when running, ate directly from the dish without using their hands, and lapped liquids. The girls did not speak, but simply grunted when requesting water. They did not accept clothes and rubbed on the ground when wishing to clean their skin, always avoided water, and exposed their teeth and scratched when somebody approached them.

In both the cases of Victor and of Kamala and Amala, it was presumed the children had been abandoned or inadvertently lost contact with their parents when very young and were able to survive because they were adopted by some mammal, which likely lactated them together with their offspring. In the case of Kamala and Amala that animal was likely a lactating wolf female, but other cases are known of gazelles and even a bear as foster parents. This suggests that the tale of Romulus and Remus, the mythic founders of Rome, could have been something real. However, because of the rarity of such episodes, the interpretation of those occurrences is still speculative.

Toys for babies largely use infant signals to make them more appealing. This giraffe is a clear example: note the extraordinarily large eyes, the rounded head, the short and rounded head appendices, and the surprisingly short limbs.

The universally known Barbie doll depicts an adult woman (long neck, well-developed breasts), but with unequivocal and emphasized infant signals (large eyes located on the midline of the face, rounded head, short nose). Courtesy of Davide Csermely.

Further Resources

Gould, S. J. (1979, May). Mickey Mouse meets Konrad Lorenz. *Natural History*, 30–36.

Lorenz, K. (1943). Die angeborenen Formen moeglicher Ehrfahrung. *Zeitschrift Tierpsychologie, 5*, 235–409.

Malson, L., & Itard, J. (1979). *The wild boy of Aveyron*. Cambridge: Harvard University Press.

Davide Csermely

Bonding
Rats and Humans

Rats and humans: it has been a love-hate relationship since Biblical times and probably long before then. By most estimates, there are as many rats on the planet as there are humans. Rats have been disdained and exterminated with a passion bordering on obsession. Yet, by some, they are revered or treated with kindness. In some locales they are worshipped. Rats remain a popular commodity in pet stores and are even more commonplace among university students. The lab rat, by his unwilling participation in medical research, has arguably contributed more to human welfare than rats have cost in other well-publicized ways. As a widely distributed poster from the Foundation for Biomedical Research suggests, "Rats have saved far more lives than [dialing] 9-1-1."

Rats remain a staple of popular culture and often appear in unsympathetic roles in stories such as "The Pied Piper of Hamelin" and successful films such as *Willard* and *Ben*. James Cagney did little for their image when he uttered "You dirty rat," and such expressions as "ratted out," "the rat race," and "rats leaving a sinking ship" do little to convey a positive public image. Ever since the Black Plague killed an estimated 25 million people in Medieval Europe, the rat has been feared as well as hated. In his novel *1984*, George Orwell depicted confinement in a room full of rats as the ultimate in horrific torture. The July, 1977, *National Geographic* continued in that spirit, surveying centuries of battle between humans and rats. Yet, the front page of the March 18, 1984, *Toledo Blade* reported a story (subsequently picked up by the U.S. wire services) titled "Pet Rat Saves Toledo Woman from Fire in Bed."

Love them or hate them, studies have shown that rats appear to enjoy the company of humans with whom they have bonded. Courtesy of Hank Davis.

Can there really be a connection between the affectionate and sociable pet rat and the adaptable wild critter who destroys an estimated fifth of the world's crops each year? Our research has focused on two positive aspects of the human-rat relationship. First, we have reported that rats appear to enjoy the company of humans with whom they have bonded. This is not an empty-headed anthropomorphic judgment. Indeed, Davis and Perusse (1988) demonstrated that rats would actually work in order to be petted by a familiar human. The kind of "work" they engaged in was pressing a lever in a Skinner box, a standard psychology laboratory technique. But instead of being rewarded by food or sweetened liquid, these rats were treated to five seconds of stroking by their favorite human. Once the hand withdrew from the top of their cage, all rats were quick to pounce on the lever again and produce a flurry of responses, which again triggered the human hand from above to pet them. Few pet-rat owners would register surprise at these findings. My own pet rats often came when they were called (perhaps not as reliably as pet dogs, but measurably, in any case) and actively sought out human contact and play intervals, just as other pet mammals are renowned for doing.

That rats really do get to know and prefer contact with individual humans was our second finding. Davis, Taylor, & Norris (1997) gave rats a series of 10-minute daily exposure sessions during which they were petted and talked to by a particular person. Following this experience the rats were allowed to explore two humans who sat blindfolded at opposite ends of a long table. All rats spent the majority of time tentatively exploring the unfamiliar person, but within five minutes each had climbed on to the person with whom they had previously interacted. In some cases they ended up sitting on the person's shoulder. These preferences were maintained for twelve and, in several cases, eighteen months after original exposure—a period constituting nearly the entire lifespan of the rat.

Our findings do not erase the stigma of the Black Plague, nor do they undo the huge economic losses inflicted by this intelligent and highly adaptive rodent. But they do establish the fact that, when not placed in a situation that unleashes a competitive struggle, rats can also be rewarding companions for humans. Words such as "loyal," "friendly," and "loving," often associated with dogs—our favorite pets—are not uncommon among rat owners. As our research suggests, the rat (like the dog) has social needs and social skills and, given the right circumstances, will readily express them in a relationship with a human.

Further Resources

Davis, H., & Perusse, R. (1988). Human-based social interaction can reward a rat's behavior. *Animal Learning & Behavior, 16,* 89–92.

Davis, H., Taylor, A., & Norris, C. (1997). Preference for familiar humans by rats. *Psychonomic Bulletin & Review, 4,* 118–20.

Hank Davis

Cartoons.
See Media and Film

Children
The Appeal of Animals to Children

Animals—both real and representational—are remarkably resonant with children and occupy a central place in their lives as significant others. In support of this natural affinity for animals, the biophilia argument maintains that children, regardless of culture, are "wired" to be drawn to living things, even though they possess a limited knowledge of animals. Evidence for children's natural affinity for animals comes from several sources. Infants have been found to follow strange animals with their eyes and smile at or try to touch them. Nearly all young children want pets or animals in their lives, express love for them by age three, subordinate their interests to them (expressing an especially strong solidarity with dogs up to about ten years of age), watch them with rapt attention (despite distractions or demands), and if denied such companionship or contact, will seek it out by sharing other people's pets. Of course, we should be cautions about over idealizing the natural affinity of children for animals. Wild animals may arouse a different reaction in young children than do domesticated animals.

The social psychological argument for this affinity states that children are drawn to animals because they can serve as the first device for children to construct their identities. Symbols of animals "guide children into deeper understandings of what it means to have a humane self," writes Gail Melson (2001). For this reason, companion animals play a role in identity exploration and achievement through their ability to provide a nonjudgmental "audience," unlike most parents or even friends. In other words, the warm, accepting presence of a nonjudgmental companion animal may offer children a confidant or discussion partner with whom they can explore ideas and feelings freely and safely. Such a context is likely to result in an increased positive effect and is also likely to enhance learning and increased empathy with animals.

Another social psychological argument is that real-life animals, for many children, provide emotional assurance in times of stress because they represent safety, comfort, love, and stability. For example, researchers in the United States have found that 25 percent of pet-owning five-year olds spontaneously reported turning to their pets when they were sad, and 10 percent said that when feeling afraid they would turn to their pet, while 75 percent of children between the ages of ten and fourteen claimed that they turned to their pet when feeling upset. Similarly, researchers in Germany found that fourth graders also reported turning to their pets when they were sad, and almost half preferred the company of their pet to that of other children. Also, in laboratory experiments, children use animals to feel safe and create a sense of intimacy; pairing an animal with a strange human being apparently acts to make that person, or the situation surrounding that person, less threatening to children. In short, preschool to adolescent children may use pets for reassurance when feeling insecure, and these positive feelings may well transfer to animal characters in general.

Children are not only attracted to real-life animals; they are also drawn to images of other species. Animal characters are particularly popular in all cultures as pedagogical tools because of their strong appeal to children. This lure can be seen in computer-based learning, books, television programming, fairy tales, or trickster stories. Given children's comfort with and interest in animals, it is not then surprising that animals are ubiquitous instructional tools. Children actually prefer animal characters to human characters. For example, preschool children's favorite books often include those that feature humanized or fantasy animals as main characters, and elementary school children prefer stories with animals rather than human characters.

This appeal makes animal characters especially popular with marketers, as reflected in their frequent use in print and television advertising. These ads often succeed in getting children's attention and changing their behavior, a point that concerns organizations seeking to protect children from harmful products. For example, in one ad campaign for beer, the American Academy of Pediatrics, the American Public Health Association, and several consumer groups condemned the use of lizards, frogs, and other amphibians because it was an intentional attempt to capture the attention of children. Indeed, a national survey by the *New York Times* (1998) of frequently named commercials reported that children most often named beer ads using frogs and other animals. One advertising agency study showed that more children between the ages of six to seventeen recognized the Budweiser lizards than recognized Barbie. This successful amphibian advertising campaign followed an equally successful marketing strategy based on the Spuds MacKenzie Bud Light dog. Similar concerns were lodged by various organizations against the use of the Joe Camel cigarette campaign that targeted teens as young as thirteen. Reports indicated that Joe Camel ads captured the attention of young smokers. In one brand recognition study, 67 percent of adults knew who Joe Camel was and what he was selling. Among six-year olds, that recognition factor increased to 91 percent, bringing Joe Camel equal to Mickey Mouse in being recognizable to children. The behavior also changed among youthful viewers of these ads, increasing from 1 to 33 percent the number of underage smokers choosing Camel and making it possible for R. J. Reynolds to gain an enormous market share increase. After the Federal Trade Commission (FTC) charged that these ads violated the fair trade practice laws by promoting a lethal and addictive product to children and adolescents, they were discontinued.

Animal characters also have been effective marketing tools to children for less harmful products. In 1995, Nabisco created an Endangered Collection version of their classic Barnum's Animal Crackers, which resulted in a hundred-fold increase in sales during the product's first two weeks on the shelf. More recently, Coke started using an image of a cow to market and sell its new drink, Swerve, to children, much as the camel was used to attract the attention of children. Even positive messages to children have been successfully marketed this way. Animal cartoon characters, such as bears, penguins, and walruses, have been shown to more effectively gain the attention of children than not using animals in antismoking campaigns.

Animal images have such lure because they are enormously flexible symbols to children. Although children recognize that these characters represent the world of make-believe, they easily anthropomorphize them and develop affective reactions to them. Even very young children appear to be comfortable if not eager to think about and reflect on the meaning of these symbols. Children as young as seven are facile in reading, identifying with, and assigning gender to ambiguous animal characters.

Animal characters are flexible symbols to children because they are not saddled with human attributes that could otherwise threaten them. G. Blum (1949) maintains that animal characters "facilitate freedom of personal expression in situations where human figures might provoke an unduly inhibiting resistance." Animal images can do this because they depict characters that possess human qualities in a classless, raceless, ageless, and often sexless way. This symbolic quality of animal characters makes them especially appealing to children, often more than any other type of image.

Animal characters are also highly flexible symbols because they are a medium through which children can express their impulsiveness and socially unacceptable desires, according to the psychoanalytic perspective. What otherwise would be difficult to express is facilitated by these images because they are a highly flexible symbol system that easily becomes any person to whom the child can indirectly and safely target their urges, or become any feeling which the child can safely express without fear of criticism. Support for this argument can be found in children's subconscious image making. Children often dream about animals and commonly express fears of them. Support also comes from children's conscious image making. Psychological testers report that children have a propensity to visualize animals in amorphous Rorschach inkblots and to more readily identify with and express their inner concerns with animal rather than human characters on thematic apperception tests. Children also frequently role-play animal characters and create imaginary animal friends that serve as a source of companionship for those who are lonely or rejected as well as for those who are not. And when drawing, children often spontaneously create animal characters.

See also

Children—*Children and Animals*
Education—*Humane Education*
Education—*Nurturing Empathy in Children*
Literature—*Children's Literature: Beatrix Potter*
Literature—*Children's Literature: Cats*
Literature—*Children's Literature: Dogs*
Literature—*Children's Literature: Horses*
Literature—*Children's Literature: Rabbits*

Further Resources

Arluke, A. (2003). Childhood origins of supernurturance: The social context of early humane behavior. *Anthrozoös, 16,* 3–27.

Bettelheim, B. (1977). *The uses of enchantment: The meaning and importance of fairy tales.* New York: Vintage Books.

Blum, G. (1949). A study of the psychoanalytic theory of psychosexual development. *Genetic Psychology Monographs 39,* 3–99.

Kellert, S., & Wilson, E. (Eds.). (1993). *The biophilia hypothesis.* Washington, DC: Island Press.

Melson, G. (2001). *Why the wild things are: Animals in the lives of children.* Cambridge, MA: Harvard University Press.

Arnold Arluke

Children
Children and Animals

Children and animals seem to go together. Pets are common in children's homes, and most children view their resident animals as family members. Children are fascinated by domestic animals and wild animals in parks, zoos, aquariums, and backyards. Books, movies, television, and other media for children are filled with animal characters, real and imaginary.

Not only are animals common in children's environments, but studies find many ways in which animals are important for children's development. Children show strong attachment to their pets and in times of stress may turn to them for emotional support. Pets also provide children opportunities to nurture and to learn about the needs of individuals very different from themselves. In general, learning about animals advances children's understanding of biology and knowledge about the living world, promotes empathy toward others, and stimulates moral reasoning about other species. Therapies and interventions involving animals are increasing in use with children who have special needs.

Animals in Children's Lives

The world of childhood is "peopled" with animals. Recent surveys by the American Veterinary Medical Association and the American Pet Products Manufacturers' Association find that 63 percent of all U.S. households have at least one pet, with dogs (39%) and cats (34%) most common. Most pet owners have multiple pets and often have pets of different species. Pet ownership is growing; the number of U.S. households with pets increased 8 percent from 2002 to 2004. As of 2004, the family form of father, mother, and children accounted for an estimated 76 percent of all pet owners in the United States. While fewer than one-third of U.S. households include children under eighteen years of age, these homes make up the majority of pet owners of fish, small animals such as hamsters and gerbils, and reptiles.

Many parents acquire animals "for the children" in the belief that pets teach lessons of responsibility and nurturing while providing companionship, love, and (for dogs) exercise. Whatever their age or family circumstances, most pet owners identify their pets as family members.

In addition to pets in the home, children encounter animals in their classrooms, especially in the early years of school. A survey by A. J. Rud and Alan Beck (2003) of 431 Indiana public elementary school teachers found that over a quarter had resident pets in their classrooms, while an additional 46 percent had animal "visitors." Like pet-owning parents, teachers who incorporate animals in their classrooms—usually fish, gerbils, hamsters, and other "pocket pets"—believe that they teach children responsibility and caring, provide enjoyment, and lend the classroom a more "homey" atmosphere.

Children's encounters with animals are not limited to those species kept as pets. Although children's everyday contact with wildlife has been shrinking as Western societies become more urbanized, what one might call *intentional wildlife experiences* persist. Families and schools set up birdfeeders, take children to zoos and aquariums, and go on nature walks in parks. The U.S. Fish and Wildlife Service (USFWS) estimates that over 60 million Americans feed wild birds at home. According to the Association of Zoos and Aquariums (AZA; formerly the American Zoo and Aquarium Association), AZA-accredited zoos and aquariums draw more visitors—disproportionately families and groups with children—than the National Football League, National Basketball Association, and Major League Baseball combined. More than half of all zoos have areas where children can touch animals, and these "petting areas" are the most popular aspects of zoos.

The animal world of children extends beyond direct contact with living animals to encompass exposure through print, audio, and visual media such as the cable channel Animal Planet. This involvement at a remove is gaining in importance; for many children, it is the dominant mode of gaining knowledge about wild animals. For example, when Gary Nabhan and Stephen Trimble asked rural eight- to fourteen-year-

A young girl feeds a goat in a petting zoo in New Mexico. ©Rebecca Abell/Dreamstime.com.

olds living next to a national park and protected wilderness about encounters with animals, over 60 percent said they had seen more animals on television and in the movies than in the wild.

Finally, anyone who remembers childhood is likely to recall fanciful animal characters—Peter Rabbit, Barney the dinosaur, Curious George, or Nemo the clownfish. Children's picture books, stories, toys, games, and media are filled with animal symbols, reflecting in part the view in many cultures that animals and children naturally go together. Seven of the top ten all-time best-selling children's books in the United States are about animals. The most widely used third-grade reading texts in 1980 featured stories about children's relationships with animals, usually pets, nearly as often as children's ties with parents, according to a survey by Martin and Penelope Croghan (1980). Thus, when researchers from Sweden (Maj Carlsson and Ingrid Samuelsson), Hungary (Anna Soponyai), and China (Quifang Wen) wanted to find out how six- and seven-year-olds in their cultures made up stories, they asked each child to tell a story about a dog.

Children themselves report strong interest in animals. When third-graders heard stories with animal characters and identical stories with human characters substituted for the animals, three-quarters of the children preferred the animal stories, Nancy Boyd and George Mandler (1955) found. A survey by Patti Valkenburg and Karen Soeters (2001) of eight- to thirteen-year-old Dutch children's Internet use found that "seeking information about animals" was one of the four most common descriptions of positive experiences with the Internet.

Stuffed animals, animal make-believe, animal stories, imaginary animal companions, and animal dreams figure prominently in children's imagination, especially during the preschool and early elementary school years. When the dream researcher David Foulkes

(1999) collected dream narratives from children, he found the same menagerie found in fairy tales, cartoons, and children's stories. "Top ten" lists of fears of U.S. children and adolescents, compiled through periodic surveys during the 1980s and 1990s, regularly included snakes, sharks, and spiders, with spiders often topping the list. In sum, wherever one looks, animals, particularly those species kept as pets, are an integral part of children's lives—at home, at school, at play, even while dreaming. As the children's book writer Rosemary Wells (1990) observed, "Animals live in a world that children seem to climb right into." Since animals are so much part of children's lives, how they influence children's development is an important question.

The Emotional Bond with Animals

Pets can provide company when children feel lonely, amusement when children feel bored, and attention when children feel ignored. Many children form strong attachments to the animals with which they are growing up. They express their attachment by observing, talking to and playing with their pets, sleeping with or near them, giving gifts, celebrating their birthdays, and displaying their pictures. Gail Melson (2001) has found that children often report that their pet is like a best friend. When Brenda Bryant (1990) asked seven- and ten-year-old children in California to name the ten most important individuals in their lives, on average two pets made the list. Interestingly, species differences do not consistently affect level of attachment to pets; children form strong attachments to dogs and cats, the most common types of pets, but also to many other species, including birds, reptiles, horses, fish, and so-called pocket pets, such as hamsters, gerbils, and guinea pigs.

Emotional Support from Pets

The presence of animals in so many households means that when children feel stress, they may turn to their pets for a sense of support. In interviews that Detlef Rost and Anette Hartmann (1994) conducted with German fourth graders, 79 percent said that they sought out their pets when feeling sad. Similarly, a study by Anita Covert (1985) and her colleagues of Michigan youngsters between ten and fifteen years old found that three-quarters reported that when they were upset, they turned to their pets. Gail Melson's (2001) interviews with five-year-olds showed that 42 percent spontaneously mentioned a pet when asked to whom they would turn when feeling sad, angry, happy, or needing to share a secret. Parents of children who turned to their pets rated the children as less anxious and withdrawn than did parents of children who did not derive support from their pets. When elementary school-age children rated all their ties—friends, parents, siblings, and pets—their pets got top prize as most likely to last "no matter what," "even if you get mad at each other," according to a study by Wyndol Furman (1989). Thus, across different ages and backgrounds, seeking emotional support from pets is common, and children feel that it is effective. One reason for this is that children experience pets as available, nonjudgmental, and accepting.

Another reason why pets may be good sources of support for children is a relaxation effect. The presence of a friendly animal, even an unfamiliar one, may relax children under stress. A study by Erika Friedmann and her colleagues (1983) found that children reading poetry aloud (a mildly stressful experience for many children) in the presence of a dog showed lower blood pressure than when reading alone. Similarly, children having a pediatric examination with a friendly but unfamiliar dog present showed lower blood pressure and distress than when the dog was absent.

Learning How to Care

When children have the opportunity to observe and participate in the care of animals such as pets or domestic animals, children may learn valuable lessons in how to care for others appropriately. "Teaching kids responsibility" is a benefit that over a third of all pet-owners identify, according to surveys by the American Pet Products Manufacturers Association. When children grow up with pets, they often participate in caring for the animals. For example, a quarter of the German eight- to ten-year-olds interviewed by Detlef Rost and Anette Hartmann (1994) said that they had sole responsibility for pet care, while half shared that responsibility with other family members. Children with small animals—gerbils, hamsters, rabbits, birds, and fish—took on more care giving than did youngsters with dogs or cats. However, nearly all (92%) of the children considered caring for their pets a "very important" or "important" part of their relationships with the animal. A survey of U.S. children between the ages of five and twelve by Gail Melson (2001) found that older children spent more time caring for pets than did younger children. She also found that when mothers were employed full-time, children helped out more with pet care than when mothers were not employed or only worked part-time.

Because pets and other domestic animals are dependent on human care, children can learn about the needs of creatures very different from themselves when they take care of them. The ability to take the perspective of another, especially when that perspective is different from one's own, is a crucial building block of empathy. There is some evidence that empathy may be related to involvement with pets. Cindee Bailey (1987) found that three- to five-year-olds who had dogs or cats at home were more skilled at predicting the feelings of others—an aspect of empathy—than were same-age children without pets. Similarly, Brenda Bryant (1990) found that seven- and ten-year-olds who were closely bonded with their pets expressed more empathy toward other children. Finally, Gail Melson, Susan Peet, and Cheryl Sparks (1991) documented a link between five- to six-year-olds' attachment to their pets and their levels of empathy.

In addition to empathy, experience caring for animals helps teach children about the characteristics and needs of different species. Of course, children can learn much about animals from reading books, watching nature programs, and observing animals, but actual "hands-on" care giving, under adult supervision, produces the most learning. Two researchers, Giyoo Hatano and Kayoko Inagaki (1993), showed the power of care giving to produce learning in their study of eighteen Japanese five-year-olds who had been caring for goldfish at home for over a year. Compared to their classmates without goldfish-care experience, these children knew more not only about how to care for goldfish, but also about the hidden biology of these animals, for example, answering correctly the question "does a goldfish have a heart?" Moreover, caring for goldfish helped the children reason about unfamiliar animals. When asked, "can we keep a baby frog the same size forever?" one child replied, "No, because the frog will grow bigger as my goldfish got bigger."

A third benefit of caring for animals may be especially relevant for boys. In Western industrialized societies, as well as others, nurturing the young, helpless, and needy generally falls to females. Cultures often buttress this division of labor by associating care

A young boy plays Frisbee with his pet dog in New York. Courtesy of Shutterstock.

giving with being female and feminine. As a result, compared to boys, girls get more encouragement and practice in nurturing others through activities like doll play, babysitting, and helping out at home with younger children. By age five, boys associate baby care with being female and, hence, show less interest in caring for human babies. However, when it comes to caring for pets and other animals, boys and girls are equally involved and interested, according to a study by Gail Melson and Alan Fogel (1996). Melson (2001) has suggested that pet care may be an area of "gender neutral" learning about caring for others, perhaps the only such area available to boys.

The Morality of Relating to Animals

As children grow, they develop ideas about the "rights" and "wrongs" of behaving toward others. This development of moral understanding includes not just relationships between humans, but also between humans and animals. Beginning at a young age, children begin to think about many current issues related to the ethical implications of human treatment of animals—animal rights, use of animals in scientific research, protection of natural habitats, and animal welfare. Childhood experiences with animals can lay the foundation for more positive attitudes toward these issues. Stephen Kellert (1996) found that children who have direct experience with living animals show more appreciation for the ecology of wild species and the need to preserve their habitats, more moral concern for the proper treatment of animals, and more emphasis on the enjoyment of undisturbed wildlife rather than its use for human needs.

One question that remains unclear is whether developing a morality about the treatment of animals relates to a morality about treatment of other humans. Many humane education programs hope that lessons about proper and respectful treatment of animals will pay off in similar treatment of people. Frank Ascione (1997) found that a one-year humane education program increased both humane attitudes toward animals and also empathy toward other children among first and fourth graders. However, other studies have found no link between morality toward animals and toward humans.

Problems in Children's Relationships with Animals

Although animals can have many benefits for children as they grow up, there may be problems as well. Brenda Bryant (1990) has found that children report worries and concerns about their pets. For example, children worry about pet welfare while they are in school or when the animal is sick. The shorter lifespan of most pets means that children will experience the death of a pet. In fact, Mary Stewart (1983), a Scottish veterinarian who has studied children's experiences with pet loss, concludes that over 80 percent of children experience the first loss of an intimate bond through the death of a pet. While dealing with death can be a powerful learning experience, it undoubtedly causes many children emotional pain.

In addition to pet loss, there can be health risks to pet ownership. Children account for over 60 percent of all dog bite victims, and more children are brought to the emergency room because of dog bites than all other injuries combined, according to studies by Jeffrey Sacks and his colleagues (1996) and Thomas Brogan and his colleagues (1995). Even so, Alan Beck and Barbara Jones (1995) argue that many dog bites of children go unreported.

There are also risks from cat scratches and a variety of zoonotic diseases, or diseases transmitted from animals to humans. Children are especially vulnerable to the latest health risk, bird flu, since in many rural parts of the world, children are in contact with domestic fowl and sometimes small wild birds as well.

Finally, involvement with animals may influence children to be *less* caring and *more* cruel or violent. This is because children's fascination with animals can be shaped in positive or negative directions by the adults around them. There is ample evidence that cruelty toward animals may be encouraged in families where child abuse and domestic violence occur. For this reason, animal involvement might better be thought of as a *potential* for children's positive development, a potential that may be realized by supportive, involved, and caring adults.

Animals Helping Children

This potential of animals has led therapists and counselors to incorporate animals as part of their treatments of children with developmental problems or special needs. Some therapists include one or more resident animals in the therapy room as a distraction, a calming influence, and a focus of attention. The psychiatrist Aaron Katcher (2000) developed what he called "the Companionable Zoo" as an animal-based therapy for children with conduct disorder, a psychiatric disorder that includes out-of-control aggression and lack of impulse control. Children at the "Zoo" learn about and care for small animals, such as rabbits or gerbils, and earn the right to adopt a pet. His research has shown that children with conduct disorder become dramatically less aggressive and calmer when at the "Zoo," and over time, this improvement spills over into other settings, such as the school classroom. Residential facilities for troubled children have been developed that make involvement with animals and contact with nature the centerpiece of treatment; Green Chimneys Children's Services in Brewster, New York, is perhaps the most well-known and influential example.

Assistance dogs are well known for their role in helping children and adults with seeing, hearing, and other impairments. Less well known is the way that assistance animals and even pets can help children become more accepted by and involved with other children and adults. For example, Bonnie Mader (1989) and her colleagues observed ten youngsters who were wheelchair bound; half used assistance dogs to get around. The children with dogs got friendly glances, smiles, and conversation from adults and other children who were passing by, while the children without dogs were ignored. This "animal halo" effect has been found in other studies as well; the presence of a friendly animal next to a child (or adult) makes that person seem more likeable and approachable.

In summary, animals are part of children's growing up in many ways. Animals may enrich children's development through the attachment bonds children forge with pets. Animals have the potential to teach children about appropriate care giving and to help them develop empathy. In turn, animals can be a source of emotional support, relaxation, distraction, and just plain fun for children. The positive potentials of animals depend largely on the adults—parents, teachers, and others—who are responsible for children's well-being.

See also

Children—*The Appeal of Animals to Children*
Education—*Humane Education*
Education—*Nurturing Empathy in Children*
Literature—*Children's Literature: Beatrix Potter*
Literature—*Children's Literature: Cats*
Literature—*Children's Literature: Dogs*
Literature—*Children's Literature: Horses*
Literature—*Children's Literature: Rabbits*

Further Resources

American Pet Product Manufacturers Association (2003). *2003–2004 APPMA national pet owners survey*. Greenwich, CT: APPMA.

Ascione, F. R. (1997). Humane education research: Evaluating efforts to encourage children's kindness and caring toward animals. *Genetic, Social, and General Psychology Monographs, 123*(73).

Bailey, C. (1987). *Exposure of preschool children to companion animals: Impact on role-taking skills*. PhD dissertation, Oregon State University.

Beck, A., & Jones, B. A. (1985). Unreported dog bites in children. *Public Health Reports, 100,* 315–21.

Beck, A. M., & Katcher, A. H. (1996). *Between pets and people: The importance of animal companionship*. West Lafayette, IN: Purdue University Press.

Boyd, N. A., & Mandler, G. (1955). Children's responses to human and animal stories and pictures. *Journal of Consulting Psychology, 19,* 367–71.

Brogan, T. V., Bratton, S. L., Dowd, M. D., & Hegenbarth, M. A. (1995). Severe dog bites in children. *Pediatrics, 96,* 947–50.

Bryant, B. (1985). The neighborhood walk: Sources of support in middle childhood. *Monographs of the Society for Research in Child Development, 50,* No. 210.

———. (1990). The richness of the child-pet relationship: A consideration of both benefits and costs of pets to children. *Anthrozoös, 3,* 253–61.

Carlsson, M. A., Samuelsson, I. P., Soponyai, A., & Wen, Q. (2001). The dog's tale: Chinese, Hungarian and Swedish children's narrative conventions. *International Journal of Early Years Education, 9,* 181–91.

Covert, A. M., Whirren, A. P., Keith, J., & Nelson, C. (1985). Pets, early adolescents and families. *Marriage and Family Review, 8,* 95–108.

Croghan, M. J., & Croghan, P. P. (1980). *Role models and readers: A sociological analysis*. Washington, DC: University Press of America.

Fine, A. (Ed.). (2000). *Handbook of animal-assisted therapy*. New York: Academic Press.

Foulkes, D. (1999). *Children's dreaming and the development of consciousness*. Cambridge, MA: Harvard University Press.

Friedmann, E., Katcher, A., Thomas, S., Lynch, J., & Messent, P. (1983). Social interaction and blood pressure: Influence of animal companions. *Journal of Nervous and Mental Disease, 171,* 461–65.

Furman, W. (1989). The development of children's social networks. In D. Bell (Ed.), *Children's social networks and social supports* (pp. 151–72). New York: Wiley.

Hatano, G., & Inagaki, K. (1993). Desituating cognition through the construction of conceptual knowledge. In G. Salomon (Ed.), *Distributed cognitions* (pp. 115–33). New York: Cambridge University Press.

Katcher, A. (2000). Animals in therapeutic education: Guides into the liminal state. In P. H. Kahn, Jr. & S. R. Kellert (Eds.), *Children and nature: Psychological, sociocultural and evolutionary investigations* (pp. 179–98). Cambridge, MA: MIT Press.

Kellert, S. 1996. *The value of life*. Washington, DC: Island Press.

Mader, B., Hart, L. A., & Bergin, B. (1989). Social acknowledgments for children with disabilities: Effects of service dogs. *Child Development, 60,* 1529–34.

Melson, G. F. (1988). Availability and involvement with pets by children: Determinants and correlates. *Anthrozoös, 2,* 45–52.

———. (2001). *Why the wild things are: Animals in the lives of children*. Cambridge, MA: Harvard University Press.

Melson, G. F., & Fogel, A. (1996). Parental perceptions of their children's involvement with household pets: A test of a specificity model of nurturance. *Anthrozoös, 9,* 95–106.

Melson, G. F., Peet, S., & Sparks, C. (1991). Children's attachment to their pets: Links to socioemotional development. *Children's Environments Quarterly, 8,* 55–65.

Myers, G. (1998). *Children and animals: Social development and our connection to other species*. Boulder, CO: Westview Press.

Nabhan, G. P., & Trimble, S. (1994). *The geography of childhood: Why children need wild places*. Boston: Beacon Press.

Nagergost, S. L., Baun, M. M., Megel, M., & Leibowitz, J. M. (1997). The effects of the presence of a companion animal on physiologic arousal and behavioral distress in children during a physical examination. *Journal of Pediatric Nursing, 12,* 323–30.

Rost, D., & Hartmann, A. (1994). Children and their pets. *Anthrozoös, 8,* 199–205.

Rud, A. G., & Beck, A. M. (2003). Companion animals in Indiana elementary schools. *Anthrozoös, 16,* 241–51.

Sacks, J. J., Lockwood, R., Hornreich, J., & Sattin, R. W. (1996). Fatal dog attacks: 1989–1994. *Pediatrics, 97,* 891–95.

Stewart, M. (1983). Loss of a pet—loss of a person: A comparative study of bereavement. In A. Katcher & A. Beck (Eds.), *New perspectives on our lives with companion animals* (pp. 390–404). Philadelphia, PA: University of Pennsylvania Press.

Valkenburg, P. M., & Soeters, K. E. (2001). Children's positive and negative experiences with the Internet: An exploratory survey. *Communication Research, 28,* 652–75.

Wells, R. as quoted in Zinsser, W. (Ed.). (1990). *Worlds of childhood: The art and craft of writing for children* (p. 135). Boston: Houghton Mifflin.

Gail F. Melson

Children
Feral Children

There are more than 100 cases on record of feral children—children who have been raised in the wild, often presumably cared for by animals, and have been discovered at some later date (at which point there is typically an attempted reintroduction into human society). The "wild child" or "l'enfant sauvage" has had a place in the myths, stories, history, anthropology, and philosophy of world culture from ancient to contemporary times. At some moments and in some theories, the feral child is characterized as a "noble savage," a human in its most animalistic form, unspoiled by enculturation, institutions, civilization, and traditions. At other times, the feral child is thought to offer a glimpse into the ways in which humans are different from and superior to lowly animals, capable of living a "wild" life yet always capable of so much more. The feral child, that is, sometimes becomes a symbol of purity and strength, and at other times represents a failure, a denigration. Yet it is always the case that feral children have meant something important and rupturing about the way we categorize the world and especially the way we think of, and organize conceptually, animals and humans.

It must be said at the outset that though stories of feral children often seem romantic as well as scientifically fascinating, they are typically tragic, and much ethical care must be given to such investigations. Often, it is the case that a feral child has become a feral child because of neglect, abandonment, and abuse. And as one looks further into the history of how feral children are all-too-characteristically treated by their "benefactors" once discovered, one finds that it is not uncommon for such abuse to continue in new ways in the name of "civilizing" the child and introducing him or her into human society once again.

It is not always clear which stories of feral children we should take as fact, which are myths, and which are hoaxes. Romulus and Remus, the twins who founded Rome, were—so goes the myth—feral children reared by a she-wolf. Greek mythology is similarly filled with stories of feral children—mostly male. Hercules' son, Telephus, was cared for by deer; Poseidon abandoned a son to be suckled by a mare; Ares' twin sons, Lycastus and Parrhasius, were raised by wolves; and it is even said that Zeus, after one of his many dalliances with a mortal, had a child who was left to be raised by bees. Though clearly fictional, we can learn a great deal from such cases about the complex relationship between humans and animals, and thus the way in which we humans construct our own nature. It is interesting to note, for instance, how these feral children of myth have a divine nature that makes them special and good. It is not merely the fact that they have immortal lineage but also their upbringing in the company of animals that gives these children their power and their status within the story.

Tarzan is, perhaps, the most famous modern mythological feral child—a man cared for by apes who grows to full adulthood and becomes a beacon of physical prowess as well as moral rectitude. *The Jungle Book*'s Mowgli, a wolf-boy, comes a close second. In Tarzan, however, we again have a purity of soul, the feral child presented as superior to the other "normal" humans. No doubt, the story of Tarzan has much to do with constructions of race, colonialism, and patriarchy as well, but it is not surprising, of course, that art reflects (and simultaneously constructs) a culture's values. What is interesting is the way in which feral children are put to that purpose.

Jean-Jacques Rousseau, a French philosopher of the Enlightenment, argued that humans are born free but everywhere they are in chains. Furthermore, according to Rousseau the process of social enculturation is one that, in many ways, corrupts humans, for it is within civilization that we learn of war, immorality, and vice of all kinds. The human raised outside of such influences would be savage, by definition, but pure of heart, and thus Rousseau speaks with admiration of the "noble savage" who runs "wild in the woods." In 1799, hunters in the Caune Woods of Aveyron, France, saw just such a boy running wild. The Savage of Aveyron, or Victor, as he later came to be called, was naked, eating roots and acorns, and generally living the life of an animal of the forest. But it did not take long for him to be captured, imprisoned, and put on public display so that all of France could judge for themselves whether or not Rousseau had been correct. While Victor, who was about thirteen at the time of his capture, was never seen in the company of animals, it was assumed that he could not have taken care of himself in the wild. But regardless of whether or not he had been specifically nurtured by animals, the public saw Victor as a "man-animal." Unable to speak, covered in layers of dirt, at times silent then full of rage, and generally willing to wallow in filth and rock back and forth in place for hours, Victor was not, it seemed to most who visited him, a noble creature. The following year, Dr. Jean-Marc-Gaspar Itard took charge of the boy, caring for him as well as studying him. Itard's express intent was to discover just how much of our humanity is a construct—something that is owed to education and nurturing—rather than a necessary essence that is built into a human body and mind. Along the way, Itard vowed to develop Victor "physically and morally." But there was little success. Over the course of several years, Victor showed some signs of coming to a limited understanding of human language, human emotion, and human expectations about being human, but the changes were generally small and fleeting. Itard would go on to help develop the Braille language and to argue for reforms in institutes for those with sensory and mental problems, but the question of the degree to which humanity is separate from animality because of education and enculturation remained unresolved.

Cases involving children thought to have been raised by animals, and not merely left in the wild, are controversial. The variety of animals is itself intriguing. There are stories reported as fact in which children have been found who have been raised by—or at least are feral and discovered in the company of—dogs, wolves, monkeys, gazelles, ostriches, bears, panthers, sheep, goats, jackals, cows, and even chickens. There are, in fact, numerous stories of wolf-children—especially in India—and such stories perhaps have some greater degree of credibility due to the documentation, photos, numerous eyewitness accounts, and arguments presented. The case of Reverend J. A. L. Singh, for instance, is widely accepted as one of the more plausible and reliable stories of this nature.

In the fall of 1920, Reverend Singh and his wife brought back to their orphanage two young Indian feral girls who had been discovered as apparent members of a wolf pack. Singh's journals tell the detailed story of how the local villagers encountered (and feared) the girls, how Singh and his party of hunters tracked the girls and the wolves to their home in a hollow white-ant mound, and how the men then frightened off the male wolves and slaughtered the mother wolf who was protecting the children. Singh named the girls Kamala and Amala. They seemed to be aged eight and one-and-a-half respectively. Back at the orphanage, the Singhs attempted to "restore the humanity" of the girls, which had been apparently stripped away by spending so much time in the company of animals. The girls, for instance, walked on all fours, sniffed and growled at the other orphans, wanted only milk and raw meat to eat, and prowled around in the moonlight. As the Singhs worked to shape both the children's bodies and their minds to be less wolf-like and more human-like, the girls changed, if only superficially so. Singh took photos with his box camera to document the wolf-girls' behavior and their changes, and he kept detailed records of their activities and education. Regardless, their time at the orphanage—and their time on earth—was relatively short. Amala lived less than a year; Kamala managed only nine.

What shall we take away from such stories? It is well known that there have been human babies lost in India because of the tradition of women who work in the fields leaving their children on the edge of the field while they labor. Wolves, then, run off with the babies, presumably as prey to feed the pack. It has been argued that a she-wolf who has either recently given birth or perhaps recently lost the cubs in her litter to predation might direct her maternal instincts toward the human baby, thus saving the child from becoming food for the other wolves. And so we can imagine Amala and Kamala possibly being suckled, reared, and generally cared for by wolves. But if we are looking for a clear answer about what constitutes our humanity (seen, perhaps mistakenly, in opposition to *animality*), then the lesson is not at all apparent.

On the eighteenth-century, pre-Darwinian classification scheme of life, the Swedish scientist Linnaeus (who invented the genus-species taxonomy used today) saw feral children as so rupturing, so importantly different from both humans and other animals, that he gave them a species all of their own: *Homo ferus*. Indeed, our definitions of "human" often include such notions as tool use, language ability, and culture. And by all such measures, feral children thus do not appear human. That so few—if any—feral children have ever been taught such skills is itself part of the controversy over their nature. Is it the case that living apart from humans for so long has made it impossible for these children to have the mental abilities necessary for tool use, language, and culture? Many feral children have shown signs of mental retardation, autism, and other developmental disabilities. Did such conditions arise as a result of their being abandoned, or were the children originally abandoned precisely because they already showed signs of these conditions? Are we thus to understand our own humanity as something that is apart from animality, superior to animality, inferior to animality, or something else altogether conceptually different?

It is said that in 1211, the Emperor Frederick II desired to know the true "language of God," the language any child would speak if left to his or her own devices. The assumption was that animals could not speak or understand language, but humans innately could. Given that there were so many different human languages around, though, reasoned Frederick, no doubt each language was learned through acquisition and local custom. But unlike an animal, if a human were left alone in silence, the emperor proposed, he or she would eventually come to speak the purest and truest language possible, one unadulterated by culture, custom, and nurture. Frederick imagined that language would be Hebrew, the language of Adam and Eve, but was not completely sure this was the case. And so, he ordered that various children in his territory be raised either in silence or left in complete isolation from human society in a grand experiment. One is wise to remember, however, that even in the myth of Adam and Eve, they not only had each other but also all of the other animals in paradise with which to live in community. None of the children in Frederick's horrific experiment came to speak a word, and each—tragically though perhaps unexpectedly—died in early childhood.

Indian girls Amala and Kamala were known to sleep close to one another on the ground after having lived with wolves for eight years. ©Mary Evans Picture Library/Alamy.

Further Resources

Candland, D. K. (1993). *Feral children and clever animals: Reflections on human nature*. Oxford: Oxford University Press.

Feuerbach, P. J. A. (1996) *The wild child: The unsolved mystery of Kaspar Hauser* (J. M. Masson, Trans.). New York: Free Press.

Gesell, A. (1941). *Wolf child and human child*. New York: Harper.

Itard, J-M-G. (1932). *The wild boy of Aveyron* (G. Humphrey & M. Humphrey, Trans.). New York: Century.
Pies, H. (1833). *Kaspar Hauser* (Lieber, Trans.). London: Simpkin and Marshall.
Singh, J. A. L., & Zingg, R. M. (1939). *Wolf-children and feral man*. New York: Harper.
Ward, A. R. FeralChildren.com. http://www.feralchildren.com/

H. Peter Steeves

Classification
Classification of Animals

In biology, classification is the systematic grouping of organisms into *categories* or *kinds*. Its chief aim is to impose conceptual order over living things, thus making it possible for us to write and talk about them by making general references, which in turn helps us to clearly and usefully record and convey information and to make correct predictions based on our knowledge. This kind of "grouping" is typically done by highlighting various similarities (and choosing to ignore certain dissimilarities), including morphological and evolutionary ones. One fundamental unit or category into which kinds of *animals* are classified is that of a *species*. It is worth noting here that while popular opinion has it that a species is a unit of *evolution*, we were happily classifying animals into different species and subspecies long before anyone developed a theory of evolution. Indeed the father of plant and animal classification, Carl Linnaeus, who classified organisms according to their genus and species (the binomial nomenclature method), died over thirty years before Darwin was born.

The species is a larger unit of classification than the subspecies (each subspecies belonging to a species) but a smaller unit than the genus (each species belonging to a genus), which itself belongs to several larger units. For example the subspecies Sumatran tiger (*Panthera tigris Sumatrae*) belongs to the species of a tiger (*tigris*), which in turn belongs to the genus of a panther (*panthera*), the family of cats (*felidae*), order of carnivores (*carnivora*), class of mammals (*mammalia*), subphylum of vertebrates (*vertebrata*), phylum of chordates (*chordata*), taxon of craniate (*craniata*), and kingdom of animals (*animalia*). We classify it so because it is warm blooded, has well-developed claws, sharp teeth, a skull, a backbone, and so on. Which of the above units of classification we choose to convey information about the Sumatran tiger at any one given time will depend upon our interests and purposes.

There is currently a debate among theorists regarding criteria we appeal to when classifying animals into various categories. According to one school of thought known as *essentialism,* there are sharp distinctions between various *natural kinds* in the world that our methods of classification ought to respect. These natural kinds it is argued (or assumed) exist independently of any interest or purpose we might have; hence the only correct way of classifying organisms is through the employment of scientific knowledge, something that will also help us to refine our discriminatory abilities. In this view, there is only one correct answer to questions such as "are whales fish?", and we must look to science, which will in turn help us to discriminate between different essential kinds, to find that answer. Proponents of essentialism include the philosophers Hilary Putnam and Saul Kripke.

A different school of thought is that of *pluralism* about classification. According to the pluralist, organisms can be classified according to various different organizing principles (some biological, others highlighting nonscientific similarities) none of which need in itself be more legitimate than any of the others, though one may, in any given context, be more *pragmatic*. This is not to adopt the *antirealist* or *nominalist* position that there are no real differences between groups of animals for science to discover, but rather a "promiscuous realism," which rejects the essentialist suggestion that the existence of such differences entails that there is only one correct way of answering questions of biological classification. In this view, defended by John Dupré among others, nothing scientists discover could possible answer a question such as "are whales fish?" because terms like "fish" have both a technical scientific sense (viz. cold-blooded aquatic vertebrate with gills) as well an equally legitimate and realist everyday or "ordinary language" sense (according to which aquatic mammals might count as fish), fixed by conventional use. Thus, whether or not we are to count whales as fish depends on which sense of "fish" we are interested in, much like whether or not we wish to call a tomato a vegetable or a fruit depends on whether we are practicing botany or making a fruit salad (it is interesting to note, in this context, that in 1893 the Supreme Court of the United States ruled that the tomato was a vegetable and, therefore, subject to import taxes). The promiscuous realist is also happy to allow that scientific discovery may well come to change our ordinary (folkbiological) concept of a fish (if it has not, to some extent, done so already), perhaps even making it the case that the two definitions will overlap, but he will insist that there is no reason to think that it *must*, or that until it does (or did) our everyday notion is in some way deficient. Biology cannot tell us what a fish is (what its *essence* amounts to) because "fish" is not a biological category. In this view, nothing in nature can determine whether or not there is such a thing as a "mammalian fish."

The view that how we choose to classify an organism depends on our interests can easily also be applied to the issue of human-animal relationships. If we wish to emphasize the similarities between humans and various (other) animals we may chose to do so by saying that human beings are animals too. If by contrast we wish to highlight general dissimilarities between humans and (other) higher order animals—perhaps while also emphasizing similarities between the latter and lower order animals—we might find it effective to do so by reserving the term "animal" for nonhuman creatures. Yet a person who at one time takes the first approach and at another time the second need not be contradicting herself because it is of interest and also important to come to terms with why both the similarities and differences have evolved.

Further Resources

Dupré, J. (2002). *Humans and other animals*. Oxford: Clarendon Press.

Hacking, I. (1983). *Representing and intervening*. Cambridge: Cambridge University Press.

Mayr, E. (1942/1999). *Systematics and the origin of species from the viewpoint of a zoologist*. Cambridge, MA: Harvard University Press.

———. (1975). *Population, species, and evolution*. Cambridge, MA: Harvard University Press.

Constantine Sandis

Classification
The Scala Naturae

The Origins of the *Scala Naturae*

The *Scala Naturae* ("Natural Scale" or "Great Chain of Being") is a philosophical view of nature attributed to Aristotle from the third century BCE. According to Aristotle, nature could be arranged on a graded scale of complexity, perfection, and value. Inorganic objects, such as rocks, occupy the lowest levels of the scale. Plants lie just above inorganic objects. Thereafter the scale moves up from "lower" animals (invertebrates) to "higher" animals (vertebrates), to humans, who occupy a position above all other life forms. In many versions of the scala naturae metaphysical beings, such as angels, occupy a position above humankind and just below god at the pinnacle.

The scala naturae is not just an organizational scheme of nature. It is also a scale of worth. What is higher on the scale is viewed as more valuable than what is lower because, according to Aristotle, the "principle of form" is more advanced in higher organisms than in lower ones. Describing the life forms as one moves up the scala naturae, Aristotle stated that "one after another shows more possession of life and movement." Aristotle saw the scale as eternally fixed with no organism able to move to another level over time. Therefore, the scala naturae engendered a world view that saw god as perfect and all other creatures, including humans, as progressively less perfect semblances of god. Humans, however, occupied a special status in the hierarchy as the most "advanced" species and therefore closer to god than any others.

The scala naturae view of nature might have remained an historical oddity, much the same as the flat-earth theory, had it not been readily passed down by successive generations of scientists, philosophers, and others. Just before the emergence of Charles Darwin's theory of natural selection, another biologist, Jean Lamarck, put forth the idea that nature represented progressive levels of "perfection" in nervous system organization and that as one moves up the hierarchy, new psychological capacities emerge as nervous systems become more "perfect." Although Lamarck may be best known for his abandoned theory of "inheritance of acquired traits," his scala naturae notions about the brain and intelligence remain present in modern thinking.

The Scala Naturae in Modern Times

The scala naturae became known as the "phylogenetic" or "phyletic scale" in modern post-Darwinian times. The phylogenetic scale has an air of scientific legitimacy because it appears to reflect evolutionary relationships among organisms. Yet, like the scala naturae, the phylogenetic scale is a hierarchical scheme that promotes the idea that organisms on a higher level of the scale than others are more "evolutionarily developed" and that, in general, the organisms on the scale represent an evolutionary line of gradation from less evolved to more evolved forms. The notion of the phyletic scale is based on comparisons across modern species. For example, the phylogenetic scale would classify modern teleost fish "below" modern mammals despite the fact that modern fish and modern mammals do not have an ancestor-descendent relationship. Modern fish and mammals, as is true of all contemporary species, are surviving representatives of specific evolutionary lineages. Therefore, the phylogenetic scale confuses biological *relatedness* with *unilinearity* or *descent*.

By the mid- to late-twentieth century scientists possessed a sophisticated understanding of evolution that incorporated molecular genetics. Evolution came to be commonly expressed as a change in the frequency of gene forms (called *alleles*) over generations. Likewise, definitions of evolution reflected our understanding of the dynamic, rather than fixed, properties of nature. Yet the notion of the phyletic scale continued to have a profound influence in many modern scientific fields such as comparative psychology. Many definitions of animal intelligence, for instance, were entirely consistent with the Lamarckian (and hence Aristotelian) view espoused well over a century earlier. For example, in a 1958 paper on the evolution of learning, the highly influential psychologist Harry Harlow (1958) wrote:

> [S]imple as well as complex learning problems might be arranged into an orderly classification in terms of difficulty, and that the capabilities of animals on these tasks would correspond roughly to their positions on the phylogenetic scale. (283)

In 1964 the authors of a popular comparative psychology textbook characterized the evolution of behavior according to the following scala naturae stance:

> As one climbs the scale from fish to primates the principle seems best stated as follows: The higher the phyletic level the greater the multiple determination of behavior. (Ratner & Denny, 1964, p. 680)

As late as the 1970s the phylogenetic scale was given new vigor by the wildly popular book and television documentary *The Ascent of Man* by Jacob Bronowski. Bronowski's book title, along with chapter titles such as "Lower Than Angels" and "Ladder of Creation," is clearly rooted in scala naturae thinking. Therefore, well into the last decades of the twentieth century the scala naturae continued to be highly influential in shaping scientific thought, educational agendas, and the public understanding of nature and our place in it.

In the twenty-first century our understanding of the nature of biological evolution is elucidated by revolutionary methodologies in genomic research. The present model of biological evolution is that of descent with modification. All modern species—sparrow, human, sponge, etc.—are extant representatives of that process. Ancestor-descendent relationships exist between earlier organisms and later forms, but no modern species are directly ancestral to any other. Our current model of nature rejects the validity of the scala naturae. On an *explicit* or public level this is indeed true. However, on an *implicit* unspoken level, the scala naturae remains a powerful idea that continues to bias our thinking about nature and evolution.

Impact and Implications of the Scala Naturae

The scala naturae continues to shape our deep, implicit assumptions about the relationship between humankind and the rest of nature. One of the forms these assumptions take is that of the teleological view. Teleology is the study of ends, purposes, and goals. Teleological thinking is based on the proposition that humankind is the inevitable goal of the evolutionary process. Accordingly, for all other species, evolution is the process of becoming more and more similar to humankind. Teleological thinking supports the misconception that other species are less perfect, less intelligent, or incomplete versions of humans. This idea is elegantly expressed by the great poet Ralph Waldo Emerson:

> Striving to be man, the worm
> Mounts through all the spires of form.

The scala naturae view promotes the idea that humans occupy not only the highest biological position in the hierarchy but also a unique position that amounts to a discontinuity from the rest of nature. Therefore, according to this view, humans are not only the "highest" beings (save for angels and deities) but they are also of a qualitatively different *nature* than other biological beings. In most scala naturae schemes humans are thought of as part animal and part spiritual. The scala naturae says that humans are different in *nature* than other species despite the evidence that we are *all animal*.

It should be noted that a rejection of the scala naturae is not tantamount to rejecting the notion that there are discontinuities across species. Discontinuities exist across all species and are a natural part of the biological world. For example, dolphins are capable of echolocation, a highly sophisticated use of sound echoes to form mental representations. Humans do not have this sense, plain and simple. This is a discontinuity. But it does not make humans or dolphins different *in nature,* only in some features. Likewise, if humans are the only species to possess a syntactically driven communication system (and we do not know this yet), then there is a discontinuity between humans and other animals in terms of this specific feature but not in the actual nature or character of what a human being or other species is. In terms of anatomy, physiology, and neurobiology there is no evidence that humans are any more distinctive in nature from a lion or an aardvark as a lion and an aardvark are from each other. Yet to this day, the scala naturae influences both our scientific thinking as well as how we think about issues of animal welfare.

Scientific Reasoning

One of the most difficult problems in the field of science has been that of interpreting nonhuman animal behavior. A longstanding stricture in the field of animal behavior testing is to give animals' performance the most parsimonious explanation possible, that is, the simplest explanation that explains the greatest number of observations. This principle is known as Morgan's Canon, which itself has been heavily influenced by scala naturae beliefs. In its pure form Morgan's Canon does not preclude complex explanations for animal behavior when most parsimonious, but in actuality, it is almost universally interpreted as stating that we should not attribute "higher" faculties to an animal if the same behavior can be interpreted as the outcome of a simpler or "lower" level process. Therefore, the scala naturae imposes a unidirectional restriction on Morgan's Canon. But sometimes parsimony as interpreted in this way is false and can lead us astray in terms of our scientific thinking about animal behavior. There are many cases in the comparative psychology literature of humans and nonhumans performing equivalently on various convergent tests of cognition in the realms of self-monitoring, social interactions, learning, and memory. Yet scala naturae beliefs would require two explanations for the same phenomenon—a simpler one for the nonhuman and a more complex one for the human. This invokes two neurobehavioral mechanisms for a single phenomenon when there may be, in fact, a single explanation. For instance, both humans and great apes recognize themselves in mirrors when tested in very similar paradigms. The scala naturae view would have us invoke two neurobiological mechanisms for this single cognitive phenomenon. Morgan's Canon, in its pure form, would suggest that the same behavior under the same circumstances in two phylogenetically closely related species with similar brains is most parsimoniously explained by a shared mechanism. Likewise, many scientists have pointed out that the most parsimonious explanation is often the one that recognizes not only the continuity but the functional equivalence of brain and behavior across humans and nonhumans.

Animal Welfare

In addition to how we reason about scientific phenomena across species, the scala naturae view has profound consequences for how we treat other animals. As long as we view other animals as "less than" or "qualitatively different from" us then we may not feel the same moral responsibility for them as we do to each other. Part of this relates to the conclusions drawn from scientific research as mentioned above. Another part relates to the fact that the scala naturae view sometimes overrides scientific evidence. For instance, despite the fact that most mammals react similarly in similar arousing situations and possess the same neurological structures and neurochemicals underlying emotion, there are still many who question whether animal emotions are as "real" as human emotions or whether other animals have emotions at all. The scala naturae view of these reactions in other species is that they are either *lesser* emotions or just *look* like emotions but are something rather different. Therefore, as long as the scala naturae view holds sway, it provides a justification for treating animals as less valuable than humans. Scientific evidence, however, provides no such justification. Yet, the way in which our society uses other species for food, entertainment, clothing, labor, while providing relatively little protection against neglect and abuse, belies the fact that the scala naturae has been rejected in the modern day.

1579 drawing of the great chain of being from Didacus Valades, Rhetorica Christiana. Courtesy of the Dover Pictorial Archives.

Summary

The scala naturae, or great chain of being—the view that humans sit atop a hierarchy of "lower to higher" organisms—has had a strong and lasting influence on thinking in scientific realms as well as in how we view ourselves in relation to other animals. The scala naturae notion has been buttressed by various scientists and thinkers throughout the centuries despite advancements in our understanding of evolution in the post-Darwinian world and genetics in the twentieth century. Although the scala naturae is today rejected on a public level, it continues in a more insidious, and therefore in a less extractable, form. Evidence for the fact that scala naturae is alive and well today is found in our scientific views of human and animal intelligence and in the myriad of ways we treat and mistreat other species.

A deep paradigm shift would be required to finally end the influence of scala naturae on modern thinking. This shift would result in a truly objective view of ourselves in relation to other species and, despite the differences, a fundamental acceptance of the higher order continuity in the *nature* of humans and other species. When this occurs, the scala naturae will go the way of other strange and misconceived theories from the past.

See also

Bonding—*Chimpanzee and Human Relationships*
Culture, Religion, and Belief Systems—*"Dolphin Mythology"*
Human Perceptions of Animals
Literature—*Human Communication's Effects on Relationships with Other Animals*

Further Resources

Campbell, C. B. G., & Hodos, W. (1991). The Scala Naturae revisited: Evolutionary scales and anagenesis in comparative psychology. *Journal of Comparative Psychology, 105,* 211–21.

Hodos, W., & Campbell, C. B. G. (1969). Scala Naturae: Why there is no theory in comparative psychology. *Psychological Review, 76,* 337–50.

Marino, L. (2003). Has scala naturae thinking come between neuropsychology and comparative neuroscience? *International Journal of Comparative Psychology, 16,* 28–43.

Lori Marino

Classification
Species Concept

A species is a biological classification that is lower than a *genus* and higher than a *variety*. Species, as the name suggests, indicates something *specific,* a group of living beings that possess some set of characteristics in common to make them distinct from other groups. The most popular definition of species today indicates an exclusive interbreeding group, the members of which are capable of passing along their characteristics to their offspring, but there is, generally speaking, no universally accepted definition of the term. Some claim that species are real (in the sense that they are natural groups existing in the real world), while others claim that they are human constructs (in the sense that such groupings are not naturally occurring but are, instead, a result of some human need to order the world). Whether real, constructed, or something else altogether, our concept of species speaks directly to the way in which we see ourselves, animals, and the relationships among us.

It is unclear the degree to which there are universal kinds in nature. Folk taxonomy is of little help. For example, which of the following seems most out of place in relation to the others: a pine tree, an oak tree, a cactus, or a daisy? For a botanist, the answer would be the pine tree, for pine trees are *gymnosperms,* whereas all the others are *angiosperms*. Similarly, science recognizes no such groups as *fish* or *flowers*. Trying to base a natural kind on common sense or on how much certain things look the same or share a similar form (morphology) is unhelpful for a scientist. Species are supposedly based on some deeper connection, some deeper shared essence. But if that is the case, how can such an essence be identified?

Categorization in nature no doubt goes back before the days of ancient Greece and Aristotle, but it is with Aristotle that orderly Western classification more or less begins. Members of a kind, according to Aristotle, share a common essence, or *eidos,* and it is the *eidos* that is responsible for making each member the sort of thing it is. In the Middle Ages this Aristotelian model continued to hold sway, mixed, however, with Christian theology until it became the Great Chain of Being. The Great Chain saw each creature in terms of a fixed essence that placed it on a hierarchy ranging from rocks to plants to animals to humans to angels to God. In the mid-1700s, Swedish taxonomist Linnaeus created the two-part genus-species classification still in use today, but as this was before the time of evolution, Linnaeus' categorization was based on a commitment to Aristotelian essentialism. Gone was the notion of the hierarchical single chain, but still there was the idea that genera and species had been supposedly endowed with essential natures as created by divine will. God, for instance, had created peacocks. Or at least we can say that He created the *eidos* of peacocks (*Pavo cristatus*) by creating the first peacocks, and individual peacocks then went about creating more of themselves. But because all of those descendent birds share the same essence—because there is such a thing as *Pavo cristatus*—we may say that a peacock is a peacock is a peacock. And what it means to be a peacock is thus timeless, necessary, and unchanging. Everything, though, would change during the next century with the advent of Darwin's theories on evolution.

Evolution and the Notion of Species

Why did Darwinian evolution cause such a problem for the notion of a species and such an uproar in general? On the one hand—though Darwin himself played down this fact at first—the idea that different sorts of animals change into other sorts of animals across time indicates the likely possibility that humans also emerged as a group from some other sort of animal. European thinking especially had seen the relationship between humans and animals as a relationship between two very different sorts of beings. To some, the idea that humans were direct cousins to apes seemed to endow humanity with less status, less uniqueness, less divinity of our own. In many ways, then, Darwin seemed—and to some, still seems today—to threaten religion, human-animal relationships, and our self-identity. On the other hand, even without thinking about humans, Darwinian evolution was revolutionary and scandalously challenging because it posited a world in which some degree of chaos rather than stability was the norm. That is, evolution may be following universal rules and laws, but the world in which it operates is one in which there are no clear boundaries among different sorts of animals and, most critically, no possibility of a universal essence for those animals. What does it mean to be a peacock in a world where evolution created peacocks? If there once were no peacocks; if peacocks only slowly became peacocks mutation by mutation; if we look back at the fossil record and see some birds that clearly were not peacocks, some that were peacocks, and a lot in between that were sort-of-kind-of-maybe-a-little-like peacocks; if what a peacock is changes with time, then how could there be anything like a fixed essence for peacocks? A peacock would not be a peacock, for there would be no necessary group essence, fixed and unchanging. There would be no species.

If species rely on fixed essences, then evolution challenges the notion that there is anything like a natural species. Yet the picture is still more complicated, for just as much as Darwin's theories call our conception of species into question, they also make a direct appeal to the real existence of species to explain how evolution itself operates. That is to say that evolution paradoxically (at least at first glance) is evidence that there are no species yet simultaneously needs something like a species to work. This is because there must be something on which evolution is acting—there must be a unit of evolution that is doing the evolving—and this unit is not individuals but rather, most likely, a species.

Say that you are a little white moth living in the English countryside around the time of the start of the Industrial Revolution. Nearly all of the members of your group are white, a trait you seem to have in common and that defines you, in part, as a group. Birds, your most frequent predator, try to snatch you up and eat you. You try not to let this happen. In the meantime, you mate and pass along, by means of your DNA, a set of instructions to make a new moth that has some of your qualities, some of your mate's qualities, and some new qualities due to the random mutations that naturally take place when copying genetic codes from parents to children. Sometimes these mutations will change the size of the wings; sometimes they change the make-up of the sensory organs; other times they change the color of the moth; and so on. Usually, if there is a mutation, it will be a bad mutation, one that hampers the offspring and thus makes it harder for them to reproduce and pass along that new quality to their own children. But every once in a while, one of these random changes proves beneficial. As the pollution from the factories begins to take its toll on the sky and the forests around you, for instance, everything comes to look sooty and gray. One day, one of your children is born with a random mutation that makes her wings gray instead of white. It happens by chance, though the mutation process itself was orderly and rule-governed. The important point, however, is that this little moth will be harder to pick out of the sky and the backdrop of trees. Birds will go for the easier-to-spot white moths, leaving your newly camouflaged gray daughter time to have children of her own and pass along her gray-wing trait to them. After several cycles and generations, the birds will have eaten most of the white members of the group, leaving the gray members reproducing successfully and establishing dominance. Having gray wings now seems clearly to be a trait common to the group. The group has evolved (and notice how it did so in conjunction with changes in its environment and, in this case, changes due to a relationship with humans as well).

The interesting question in this story is *what*, exactly, has evolved? Not you, not your daughter, not your family. What you *are* never changed throughout the story. Similarly, your daughter never changed as well. (Her DNA, and thus her gray-wingedness, was there from the moment she first came into being.) If evolution marks change, then what has changed over time in this story? The best answer to that question is the species as a whole. What it means to be this sort of moth has now become something different. The species is now best described as containing mostly gray members. Species are thus needed in order to explain what the unit of change is. Evolution acts on species. And it is thus troublesome, as we saw above, that evolution also suggests that there is no such thing as a species.

Different Ways to Define Species

There are many different ways to attempt to solve this problem. One is to define "species" in such a way that it does not refer to any fixed essence yet still refers to something concrete at any given moment in time. The most popular definition of this type is what is called the "biological species concept" (BSC), first championed by Ernst Mayr in the mid-twentieth century.

According to BSC, a species is a collection of actually or potentially interbreeding natural groups that are both reproductively isolated from other groups and have a reasonable expectation of creating viable offspring. The moths in the example above would thus constitute a species if they were successfully breeding with each other and each other only. Even as morphological characteristics (such as wing color) changed over time within the group—thus indicating that there was no fixed and eternal essence defining the species—the specific group could be isolated as one specific group because of the way in which its members could and could not mate. When the moths were all white, they were one biological species. When the moths were both white and gray, they were still one biological species. And when all the moths in the group ended up being gray, they were still the same biological species. Changes, in fact, will continue to take place, but unless those changes occur so as to create a new group that is functionally incapable of mating successfully with the original group, the group will continue to constitute one species. Should such a new nonmating group appear (because of, of course, random mutations and natural selection involving individuals from the old group), that new group would then constitute a new species.

One of the drawbacks of BSC is that although it focuses on reproduction—an important element in the engine that drives evolution and thus a good candidate for something that makes a species—it focuses solely on *sexual* reproduction. Consequently, we are forced to conclude that until sexual reproduction evolved on Earth, there were no species present. What, then, was evolution working on for the likely one to two billion years that life was around before sexually reproductive organisms *evolved*? Furthermore, there are problems accounting for creatures that are inherently nonreproductive (such as some worker ants and drone bees) as well as creatures that tend not to mate but can do so under the right circumstances (such as tigers and lions that produce tigrons and ligers). Not everyone accepts BSC as the answer to what constitutes a species.

There are other alternatives, including cladistics (closely related to BSC but arguing that the only natural group is one that includes all the descendants of a common ancestor), genetic species concepts, environmental species concepts, and those theories that argue for the irreality of species. But the important point is this: the species concept is both theoretically and practically important to our understanding of the world and our relationships to other animals. And, unsurprisingly, it speaks to the tension and duality in our conception of those relationships. It at once suggests both how the world is divided, ordered, separated, and compartmentalized, while at the same time how we are all alike, how *Homo sapiens* are simply one species among many others—changing, adapting, always unfinished, and intimately connected with all other life.

Further Resources

Bock, W. J. (2004). Species: The concept, category, and taxon. *Journal of Zoological Systematics and Evolutionary Research, 42,* 178–90.

Claridge, M. (Ed.). (1997). *Species: The units of diversity*. London: Chapman and Hall.

Ereshefsky, M. (2000). *The poverty of linnaean hierarchy: A philosophical study of biological taxonomy*. Cambridge: Cambridge University Press.

Griffiths, P. E. (1994). Cladistic classification and functional explanation. *Philosophy of Science, 61,* 206–27.

Hull, D. (1978). A matter of individuality. *Philosophy of Science, 45,* 335–60.

Mayr, E. (1942). *Systematics and the origin of species from the viewpoint of a zoologist*. New York: Columbia University Press.

Wilkins, John S. (2003). How to be a chaste species pluralist-realist: the origins of species modes and the synapomorphic species concept. *Biology and Philosophy, 18,* 621–38.

Wilson, Robert A. (1999) *Species: New interdisciplinary essays*. Cambridge: MIT Press.

H. Peter Steeves

Communication and Language
Birdsong and Human Speech

Millions of years before humans spoke—or produced music of any kind—birds produced thousands of tuneful whistles and rhythms. We will never know how much these sounds influenced our aesthetic perception of sound, but we can speculate that we might not have produced or appreciated music as much as we do unless birds had done so first. Most of our favorite instruments yield pure tones—as do our singers—but songbirds generate the most varied and conspicuous tonal sounds in nature. We would have had very few models for the development of our own melodies, were it not for the birds.

If this is the case, it is not surprising that in many languages, words originally used for human vocal music were extended to the sounds of certain birds. In English, for instance, the first literary reference we have for the word "song" refers to the human voice (in *Beowulf*), but by 1000 CE, the term was first used with regard to birds in the following stanza (translated from Old English):

> When they can hear the piping choir
> Of other song-birds; then do they send
> Their own notes forth. All together
> The sweet song raise; the wood is ringing.
>
> —Anonymous, *Metres of Boethius* xiii. 47–50

Much more recently, ornithologists discovered that most of the world's best melody-producing birds are close evolutionary relatives of each other. They are classified in the suborder *Oscines* in the order *Passeriformes,* and the term "songbirds" is now reserved for this group alone.

Birdsong has often been related to human speech and expression. Female singers (whose voice pitch is closer to that of birds) are called "songbirds" as a compliment. Birdwatchers use spoken phrases to describe bird songs; Thoreau, for instance, thought the song sparrow's song sounded like, "Maids! Maids! Maids! Hang up your teakettle-ettle-ettle." Poets such as Byron and Keats imagined that the songs of birds convey human sentiments. Less romantically, Hawthorne wrote: "Language—human language—after all, is but little better than the croak and cackle of fowls, and other utterances of brute nature—sometimes not so adequate" (*American Notebooks*, 14 July, 1850).

In the next century and a half, Hawthorne's cynical statement would prove truer in some senses than he could have guessed. Three areas of scientific progress have vastly expanded the relationship between birdsong and human speech: the function of vocal communication, the development of learned birdsong, and the way song changes over time in a population.

What Vocal Communication Is For

First, Darwin's work instigated the realization that humans evolved along with the rest of nature and that our complex traits have been maintained by natural selection. Not surprisingly, then, research investigating the function of birdsong has shown that birds make noises for many of the same reasons that we humans do. Both birds and humans beg for food as young; they coo to maintain pair bonds and soothe agitated offspring; they call to keep track of each other and even (in highly social species) to distinguish between individuals; and they squawk and yell to intimidate competitors, scold intruders, and warn kin of danger. If we limit discussion to what biologists (and the most recent edition of the *Oxford English Dictionary*) refer to as bird*song*, as opposed to birdcalls, two functions are overwhelmingly important: attracting mates and competing with members of the same sex (often by ordering them out of one's territory). In most temperate birds, only the males sing. In species where females sing as well, especially in the tropics, the male and female often cooperate to exclude other males or pairs from the territory. In humans, social life is far more complex than in any bird, and we use our language for a multitude of advanced social functions, such as conveying information about third parties, coordinating exchanges, and facilitating effective networks of cooperation and competition. These functions are probably much less important—and may often be nonexistent—in birdsong. Human language is also far richer in meaning, symbolism, and reference than birdsong, which appears to be primarily "about" the singer. But as Hawthorne's statement above hints, we humans, too, perhaps much more often than we care to admit, use our language for the typical functions of birdsong: envious antagonism, self-aggrandizement, and, particularly, making favorable impressions on members of the opposite sex.

Learning How to Sing or Speak

A second area in which science has found parallels between birdsong and human speech is vocal learning. Most vocal organisms produce sounds without imitating models around them; however, together with hummingbirds, parrots, and a few mammals (including humans), songbirds are vocal learners. Birdcalls, which are usually simpler and are routinely delivered in nonmating contexts, may often be unlearned. Almost all human vocalizations involve significant learning, although laughing, crying, and gasping in surprise may be unlearned (or nearly so).

Linguists and psychologists have argued for decades as to whether human language is acquired primarily by instruction (a model associated with B. F. Skinner) or by the maturation of a core body of knowledge that is *inherited* rather than learned (a position championed by Noam Chomsky). Since the 1960s, experimental studies (by Peter Marler and others) on song learning in birds have shown that neither extreme view captures the complexity of vocal development in birds. These insights inspired human language researchers to think less simplistically about their own subject, and subsequent work has revealed a number of striking parallels between birds and humans in the development, and even the neurobiology, of vocalization.

Both birdsong and human speech are copied from older individuals in the beginning of life, most readily during a short window of time at a certain stage in brain development. In several sparrow species, this window is very strict, and the birds learn very few—or no—new sounds after the first few months of life. In humans, and in some birds such as the zebra finch, learning is fastest and most pronounced in early life, but additional learning can happen later. (It is probably not a coincidence that zebra finches are more social than the sparrows, and that, in several respects, their song-learning process is more like that of humans.) The learning process in both songbirds and humans can be divided into two broad phases: the *perceptual* phase and the *sensorimotor* phase. In the perceptual phase, individuals listen to their elders and store what they hear as a collection of neural representations (which in birdsong is called a "template"). In the sensorimotor phase, individuals gradually develop their ability to produce appropriate sounds based on the templates stored in their brains. The young at first produce gibberish that only vaguely resembles song or speech (called "subsong" in birds, "babble" in humans). The learners then refine their output through a process in which one's own vocal output is repeatedly heard, compared to the template, and modified to achieve increasingly accurate copies. Both birds and humans inherit predispositions to learn only the "correct" sounds—only the sounds of their own species. As individuals develop, these predispositions affect what they hear and how they hear it, and then, later, what they produce. With experience, individuals also whittle away irrelevant parts of their repertoires. Swamp sparrows, for instance, will initially produce a broad variety of songs, but only a fraction of these will be perfected and "crystallized" into final form to be used throughout their lives. Likewise, infant humans are able to learn phonetic distinctions found in any language, but by one year of age they have lost the ability to distinguish sounds that are not contrasted in their native language (such as the Spanish /b/ and /p/

among English speakers and the English /r/ and /l/ among Japanese speakers). Other similarities between song and speech learning are: the surprisingly small number of repetitions that are required to learn species-typical sounds, several features of the elemental structure of songs and sentences, the importance of social input during learning, the preference of learners for familiar over unfamiliar sounds, and the physiological mechanisms underlying the onset and offset of learning. Of course, several aspects of the learning process are very different in the two groups, usually deriving from the greater degree of sociality and complexity that characterizes human psychology and communication.

Consistent with the similarities in vocal development, neurobiological studies have uncovered similar control systems in the brain for vocalization in humans and songbirds. In both groups, two general neural networks have distinct functions, and they are not found in vocal nonlearning species. One, the posterior descending pathway, carries substantive information about song or language; the other, the less-understood anterior forebrain pathway, is important for flexibility or plasticity in learning. So far, at least six specific brain regions have been found in humans and songbirds that appear to have analogous functions in vocal communication.

Several differences in the organization of the vocal learning centers arise because birds do not have the layered cortex of mammals and do not exhibit the degree of hemispheric lateralization (the division of labor between halves of the brain) that humans do; and, of course, humans do not exhibit the sex differences that most birds exhibit in the neural basis for vocal behavior. Nevertheless, the similarities between the two groups are striking, considering that the ancestors of songbirds and humans diverged over 300 million years ago. Vocal learning evolved independently in the two lineages (we are separated in the evolutionary tree by thousands of vocal nonlearners), but many of the same structures evolved in parallel. This situation resembles the evolution of flight in birds and bats, where many of the same structures (pectoral muscles, metabolic rate) changed in similar ways, despite the independent evolution of flight in the two groups.

Dialects and Cultural Change

After function and development, the third scientific achievement that intensified the interaction between birdsong and human speech over the last few decades was the study of cultural traditions and change. Culture is the accumulation of social learning over time. It is a rare phenomenon in nature; most organisms do not learn their parents' solutions to old problems, but have to "reinvent all of the wheels" themselves. Both humans and birds, however, learn their vocalizations with slight imperfections and innovations, and they generally learn from local individuals. This results in gradual local changes that eventually produce geographical differences in both birdsong and human language—differences that are referred to in both groups as "dialects." From extensive work on indigo buntings, Bob Payne found that, as in human languages, song traditions arise, split, influence each other, and go extinct.

Changes in bird songs and human languages do not appear to be progressive, but they *do* open up a world of social cues that is much more diverse than that of vocal-nonlearning species. In some social contexts, these cues can have important consequences; for example, one's repertoire of vocal signals can be a hallmark of experience or ability. In great tits, a male survives better—and is reproductively more successful—the more song themes he knows. Likewise, in humans, a large vocabulary can be perceived as indicative of intelligence. As another example, an individual songbird or human can often be identified by voice and then can be treated as a native or a foreigner. Song sparrow males are more aggressive, and females less sexually receptive, toward foreign song sparrow songs than local song sparrow songs. In humans, a regional dialect can be an avenue to social acceptance among those who speak it, but can seem provincial or carry other negative connotations among nonspeakers.

Conclusion

Because of the many parallels, and because of ethical restrictions on human experimentation, studies on birds have provided some of our most valuable empirical insights into human language and speech development. Recently, clinicians have begun to request bird studies in order to understand aphasia, stuttering, and other speech pathologies. Despite all that is still unknown, birdsong is the best understood communication system in nature. In fact, the relevance of birdsong research extends beyond vocal communication to behavior in general. We understand the complex interactions between "nature" and "nurture" (between unlearned and learned factors) more thoroughly in the songs of birds than in any other behavior in any species—including humans. Evidently, birds can teach us much about ourselves. As Democritus said over two millennia ago, "We are pupils of the animals in the greatest matters; of the spider in spinning and mending, of the swallow in building houses, and of the songbirds, the swan and the nightingale, in singing, by imitation."

See also

Communication and Language—*Similarities in Vocal Learning between Animals and Humans*

Further Resources

Doupe, A. J., & Kuhl, P. K. (1999). Birdsong and human speech: Common themes and mechanisms. *Annual Review of Neurosciences, 22*, 567–631.

Hawthorne, N. (1991 [1850]). *The American notebooks: The centenary edition of the works of Nathaniel Hawthorne, Vol. 8)*. Columbus: Ohio State University Press.

Kuhl, P. K. (2000). A new view of language acquisition. *Proceedings of the National Academy of Sciences of the U.S.A., 97*, 11850–57.

Marler, P. (1997). Three models of song learning: Evidence from behavior. *Journal of Neurobiology, 33*, 501–16.

Nottebohm, F. (2005). The neural basis of birdsong. *PLoS Biology, 3*, 759–61.

Payne, R. B. (1996). Song traditions in indigo buntings: Origin, improvisation, dispersal, and extinction in cultural evolution. In D. E. Kroodsma & E. H. Miller (Eds.), *Ecology and Evolution of Acoustic Communication in Birds* (pp. 198–220). Ithaca, NY: Cornell University Press.

Pinker, S., & Jackendoff, R. (2005). The faculty of language: What's special about it? *Cognition, 95*, 201–36.

David C. Lahti

Communication and Language
Great Apes and Language Research

Language research with nonhuman great apes (chimpanzees, bonobos, gorillas, and orangutans) allows for unique interaction between nonhuman animals and humans. In principle, it offers a distinctive window to the understanding of their mental lives;

however, ape language research is considered, by part of the academic world, to be highly controversial.

From the late-nineteenth century until around the 1950s, several attempts were undertaken to teach nonhuman great apes to talk—all of these yielded very little success. Richard Garner, who, in 1893, spent three months in a cage observing wild chimpanzees and gorillas, claims that he taught a young chimpanzee, Moses, a few words, such as "mama" and the French word *feu* (fire). In the 1930s, the Kelloggs, together with their son, Donald, raised an infant chimpanzee, Gua, but Gua never expressed any words. Toward the end of the 1940s, Catherine and Keith Hayes were slightly more successful with the chimpanzee Viki. After six years of education at their home, Viki could breathe—rather than speak—four words: "mama," "papa," "up," and "cup." The overall lack of success in this area has been explained in terms of anatomical differences in the vocal tracts of nonhuman great apes and humans.

All of this changed in 1966, when Allen and Beatrice Gardner pioneered the teaching of American Sign Language (ASL) to the chimpanzee Washoe. When Washoe was four years old, the Gardners reported that she had reliably acquired at least 132 ASL signs. As they wanted to exclude the risk of inadvertent cueing, the Gardners tested Washoe and other ASL chimpanzees individually, requiring them to name objects shown on slides. Two uninformed observers recorded their signs. The chimpanzees usually provided more than 80 percent of the correct responses, and interobserver agreement was around 90 percent. In the 1970s, "Project Washoe" was taken over from the Gardners by Roger and Debbie Fouts, who (at the time of this publication) still care for Washoe (and several other ASL chimpanzees). Similar ASL projects were started with other great apes—such as the gorilla Koko (by Francine Patterson), the chimpanzee Nim (by Herbert Terrace), and the orangutan Chantek (by Lyn Miles). Different communication methods have been used as well. David and Ann Premack taught the chimpanzee Sarah to communicate by means of plastic symbols, and Sue Savage-Rumbaugh uses a computer console with arbitrarily designed geometric forms (lexigrams) for her research with the bonobo Kanzi and other great apes.

In particular, toward the end of the 1970s, ape language research came under heavy fire from many scientists. The single most important blow was provided by Herbert Terrace, a psychologist at Columbia University. Terrace came to question his former research with the chimpanzee Nim after analyzing video tapes of Nim and his teachers. In an article published in 1979 in *Science*, Terrace and his colleagues wrote that the majority of Nim's utterances (87 percent) immediately followed a human's utterance (so-called "adjacent" utterances). Also, nearly 40 percent of these utterances were classified as (partial) imitations of what the human teacher had signed. However, what remained an unfortunate blind spot in the article was the fact that the majority of Nim's utterances were either spontaneously initiated by Nim (13 percent) or composed of novel signs (40.6 percent)—signs that differed from those used by the human teacher. It is also important to take into account the highly controlled training conditions—in a bare classroom of only six square meters—and the problematic psychological state of Nim, which was due to living with a succession of different caregivers (See accompanying sidebar on Nim Chimpsky.) Around the same time, it was suggested that the language apes were merely performing rigid circus tricks in order to beg for food. Linguist Thomas Sebeok and anthropologist Jean Sebeok described the ape language experiments in terms of "unconscious bias, self-deception, magic, and circus performance." During a press conference, Thomas Sebeok reportedly suggested the involvement of fraud, although no evidence was ever brought forward to support his accusation.

Research with other great apes has resulted in different figures than those of Terrace and his colleagues. The total of spontaneous and novel utterances for the bonobo Kanzi, the gorilla Koko, and the orangutan Chantek range between 50 percent and more than 90 percent. Several of the language-research apes have been reported to engage regularly in spontaneous self-signing, for example, during play; this behavior has been confirmed by independent observers. Jane Goodall describes a visit to the Temerlins, where she "watched as [the chimpanzee] Lucy, looking through her magazine, repeatedly signed to herself as she turned the pages. . . . She was utterly absorbed, paying absolutely no attention to either Jane [Temerlin] or me." Roger and Debbie Fouts state that the chimpanzee Washoe spontaneously taught the use of ASL to her adopted chimpanzee son, Loulis. Not only did she demonstrate to him the correct signs, but on several occasions also molded his hands into the proper signing configuration. For six years, the researchers made only seven signs in Loulis's environment (such as "who" and "where"). Loulis, nevertheless, mastered fifty-five signs by the end of the study period.

The well-known linguist Steven Pinker (who teaches at Harvard University) has suggested that "the apes had not learned *any* true ASL signs." His position is based mainly upon the remarks of a deaf man who testified anonymously in Arden Neisser's *The Other Side of Silence* (1983). This man had worked with chimpanzees that were staying with the Gardners, only a few years after Washoe had left with Roger Fouts. The witness was not on good speaking terms with the Gardners. Also, he accepted fewer of the signs made by the chimpanzees as true ASL signs. What he does not mention in his testimony is that some of the signs accepted by the Gardners are *variations* of the ASL signs used by deaf humans. The Gardners have always been explicit about this—for example, in a 1969 article for *Science*, they clearly describe how some of Washoe's signs exactly differ from default ASL signs. One of those signs—the sign for "more"—was rejected by the deaf man for not being an ASL sign.

It should be mentioned as well that deaf people had to fight a fierce emancipation battle before ASL became recognized as a full language. Several of these people clearly felt deeply humiliated by the ASL research with nonhuman apes. Neisser comments: "The entire issue of chimpanzee sign language is a painful one for the deaf. There is simply nothing in it for them—nothing from which they might be able to take comfort or find dignity, but only the opposite. The image of an ape signing echoes the ancient and familiar charge that their language is only suited for the beasts." Unfortunately, critics like Pinker fail to mention this dimension.

In sharp opposition to the anonymous testimonial referred to by Pinker, it is remarkable that the pioneering ASL authority William Stokoe recognized the ability of nonhuman great apes to master ASL signs. This linguist, who taught at the first college for deaf people in the world (Gallaudet College) and was the first author of *A Dictionary of American Sign Language* (1965), saw how, during a walk, Washoe formed ASL signs such as "cow" (the animals were far away in the fields, barely visible for Stokoe) and "flower" (before she ate it). Stokoe concluded his considerations on the ape language experiments by stating: "I find that the critics who attack the experiments have failed to provide any solid basis for denying what the animals have demonstrated."

Joel Wallman (1992) has written that a distinction needs to be made between making trained gestures to obtain a reward, and symbolic communication. The best criterion in favor of the latter is, according to Wallman, the ability to use "displaced reference"—this is to communicate about things removed in time or space. Multiple instances support the suggestion that nonhuman great apes can meet this criterion. The most convincing example is, perhaps, a systematic research project undertaken by Charles Menzel at the Language Research Center in Georgia. On various occasions, Menzel hid objects under sticks, beyond the reach of the adolescent chimpanzee Panzee. The next day, Panzee spontaneously tried to draw the attention of uninformed caregivers. She persistently made vocalizations, moved repeatedly in the direction of her outdoor enclosure, formed the sign "hide" (by covering her eyes with her hand), pointed in the direction of the hidden objects, and tried to communicate by selecting the appropriate lexigrams on her keyboard (such as the symbols for "stick," "hide," and "blueberries"). She thus successfully initiated symbolic communication with uninformed humans about objects removed in time (she had to recall the object that had been hidden the day before) and space (these were beyond her sight and reach).

Some of the reports by ape language researchers suggest that nonhuman great apes may be remarkably creative in producing new signing combinations. A famous example is the combination "water bird," which was formed by Washoe upon seeing a swan. Critics have remarked that these were simply independent signs for separate objects, not a novel signing combination to describe the swan; however, in support of Washoe, it has been asserted that she *consistently* signed "water bird" for swans, whether they were in or out of the water. Also, such criticism may be less easily applied to combinations such as "white tiger" (by the gorilla Koko, to indicate a zebra), "rock berry" (by Washoe, for a Brazil nut), "cry hurt food" (by the chimpanzee Lucy, for radishes), and "eye drink" (by the orangutan Chantek, for contact lens solution).

What about the presence of syntax or grammar? Most language-trained apes seem to produce combinations of around three signs, though these may also consist of up to six or seven symbols. To meet the requirement of syntax, there must be indications of linguistic rules; in other words, the combinations of signs or lexigrams must reveal some order. Some indications indeed point in the direction of a rudimentary syntax. In Washoe's signing, for example, the subject precedes the action in almost 90 percent of her combinations. Washoe thus typically signs "you me go" or "you me out," but "out you me Dennis" is the exception. Roger Fouts writes that Washoe understands differences of meaning according to the position of the subject and object (1997). He illustrates this with the examples "me tickle you" and "you tickle me." The chimpanzee Ai has learned to indicate on a computer console, through keys, the quantity (!), color, and kind of object shown by Tetsuro Matsuzawa. Ai is familiar with lexigrams, Arabic numbers, and Japanese *kanji* characters. Although she was free to choose the order of the keys, she nearly always selected color/object/number and object/color/number among six possible alternatives. Matsuzawa has commented that she may be utilizing "a rudimentary form of her own 'grammar' to describe the perceptual world" (1989).

Sue Savage-Rumbaugh emphasizes that we should not only look at the combinations one can produce, but also to the comprehension of such combinations. In a test with 660 different sentences, the bonobo Kanzi reacted properly to 72 percent of the requests (a two-and-a-half-year-old human child responded correctly to 66 percent of these sentences). He turned out to understand quite complex sentences, such as "You can have some cereal if you give Austin your monster mask to play with." When asked "Can you throw a potato to the turtle?" he made no mistakes, such as throwing both items or throwing the turtle toward the potato. Some of his reactions were quite surprising, though; for example, when asked to put water on the carrots, he threw them outdoors in the rain.

Whether we can say that nonhuman great apes can learn language depends, ultimately, upon how "language" is defined. Nonhuman great apes appear to be capable of using several hundred symbols in a meaningful way. There are also indications of rudimentary syntax. This suggests that what makes humans unique in connection with language may rather be a difference in degree of complexity. Marc Hauser, Noam

Chomsky, and Tecumseh Fitch hypothesize that "recursion" is the only uniquely human component of the faculty of language. This capacity allows us to produce an—in principle—infinite number of combinations with a limited set of elements. For example, any possible "longest" sentence can still be made longer by adding "Mary thinks that . . ." Some commentators have suggested that the linguistic capacities of nonhuman great apes have resulted in redefining language in terms of what distinguishes humans from nonhuman apes. We may only wonder how central "recursion" will become in language definitions during the coming years.

See also

Ethics and Animal Protection—*Great Ape Project*

Further Resources

Candland, D. K. (1993). *Feral children & clever animals: Reflections on human nature*. New York and Oxford: Oxford University Press.

Cavalieri, P., & Singer, P. (Eds.). (1993). *The great ape project: Equality beyond humanity*. London: Fourth Estate.

Fouts, R., & Mills, S. T. (1997). *Next of kin: My conversations with chimpanzees*. New York: Avon Books.

Gardner, R. A., & Gardner, B. T. (1969, August 15). Teaching sign language to a chimpanzee. *Science, 165*, 664–72.

Hauser, M. D., Chomsky, N., & Fitch, W. T. (2002, November 22). *The faculty of language: What is it, who has it, and how did it evolve?* Retrieved from http://www.sciencemag.org/cgi/content/full/298/5598/1569.

Matsuzawa, T. (1989). Spontaneous pattern construction in a chimpanzee. In P. Heltne & L. Marquardt (Eds.), *Understanding chimpanzees* (pp. 252–65). Cambridge, MA and London: Harvard University Press and The Chicago Academy of Sciences.

Menzel, C. (1999). Unprompted recall and reporting of hidden objects by a chimpanzee (Pan troglodytes) after extended delays. *Journal of Comparative Psychology, 113*(4), 426–34.

Neisser, A. (1983). *The other side of silence: Sign language and the deaf community in America*. New York: Alfred A. Knopf.

Patterson, F., & Linden, E. (1981). *The education of Koko*. New York: Holt, Rinehart and Winston.

Peterson, D., & Goodall, J. (1993) *Visions of Caliban: On chimpanzees and people*. Boston and New York: Houghton Mifflin Company.

Pinker, S. (1994). *The language instinct*. New York: Harper Perennial.

Savage-Rumbaugh, S., & Lewin, R. (1994). *Kanzi: The ape at the brink of the human mind*. London: Doubleday.

Sebeok, T. A., & Umiker-Sebeok, J. (1979, November). Performing animals: Secrets of the trade. *Psychology Today*, 78–91.

Stokoe, W. C. (1983). Apes who sign and critics who don't. In J. de Luce & H. Wilder (Eds.), *Language in primates: Perspectives and implications* (pp. 147–58). New York: Springer-Verlag.

Temerlin, M. (1975). *Lucy: Growing up human*. Palo Alto, CA: Science and Behavior Books.

Terrace, H. (1979, 1987). *Nim: A chimpanzee who learned sign language*. New York: Columbia University Press.

Terrace, H., Petitto, L. A., Sanders, R. J., & Bever, T. G. (1979, November 23). Can an ape create a sentence? *Science, 206*, 891–902.

Wallman, J. (1992). *Aping language*. Cambridge: Cambridge University Press.

Koen Margodt

Nim Chimpsky

Koen Margodt

The chimpanzee Nim Chimpsky (1973–2000) participated for the first four years of his life in a sign-language research project organized by psychologist Herbert Terrace (Columbia University). After the termination of this project, Terrace turned into a major critic of Nim's linguistic capacities—and ape language research in general. Nim thus arrived at the center of one of the most remarkable and controversial debates of primatology.

Nim, named after the linguist Noam Chomsky, was born in 1973 at the Institute for Primate Studies (Norman, Oklahoma). He was the seventh infant that was taken away from his mother for language research. "Project Nim" began when Nim was two weeks old and lasted until 1977, when the project ran out of funding. In *Nim: A Chimpanzee Who Learned Sign Language* (1979), Terrace vividly describes how Nim acquired a vocabulary of 125 signs. However, toward the end of the book, Terrace expresses skepticism based upon his analysis of three and one-half hours of videotapes of the signing of Nim and his teachers. These were based upon nine sessions recorded between the ages of 26 and 44 months. Also in 1979, an article by Terrace and his colleagues was published in *Science* that was a major force of criticism—it stated that Nim's and other great apes' signing was not spontaneous and was based merely upon imitating humans (see accompanying entry, Great Apes and Language Research).

Careful reading of the *Science* article reveals that more than half of Nim's signing, nevertheless, was spontaneously initiated by Nim or composed of other signs than the ones used by the humans participating in the conversation. Terrace's book gives many examples that would withstand video analysis; for example, Nim spontaneously drew the attention of his teachers to objects in his environment by naming the objects. He would also roll over on the ground and request "You tickle me." On another occasion, while waiting for a traffic light with Terrace, Nim drew Terrace's attention by signing repeatedly "drink" and pointing to a bus driver sipping his coffee. And Nim invented a sign for "hand cream" by rubbing his hands together.

Nim's physical and social living conditions were far from ideal and need to be taken into account to understand his signing behavior. Chimpanzees are highly curious beings, but, during weekdays, Nim was taught sign language for five to six hours a day in a concrete classroom of barely six square meters. After two hours of learning, Nim was allowed a break in a small playroom. Terrace later "wondered how I and the other teachers could have spent so much time in these oppressive rooms." Chimpanzees develop strong social bonds that may last a lifetime (sometimes more than fifty years), but Nim had some sixty teachers within only four years. Even his principal eight caregivers were present for only parts of these four years, and Terrace was too busy with many other occupations to be present enough for Nim's developmental well-being. All four of his main caregivers at the Delafield house left around August and September 1976; in particular, when Laura Petitto left, Nim became "depressed and inconsolable." Terrace recognized that "undoubtedly the loss of Nim's immediate family at Delafield at a critical stage of his growth had a permanent adverse effect on his social, linguistic, and emotional development." Nevertheless, at least four of the ten videotapes used for the *Science* article were recorded somewhere between September 1976 and September 1977.

At the end of September 1977, Nim was returned to the Primate Institute in Oklahoma. After a quarantine period in total isolation, he was put in a cage with three chimpanzees

(continues)

Nim Chimpsky (continued)

and later moved to a small island with nine chimpanzees. In 1982, Nim and the other Oklahoma chimpanzees were sold, for biomedical research, to the Laboratory for Experimental Medicine and Surgery in Primates (LEMSIP) at the New York University Medical School. Nim and his brother Ally were returned to Oklahoma after protests by animal advocates, joined by Terrace. In the end, Ally was apparently sent, for biomedical research, to the White Sands Research Center in Alamogordo, New Mexico. Nim was retired to the Black Beauty Ranch of The Fund for Animals in Texas, which was founded by the animal advocate Cleveland Amory (1917–1998). He lived there, together with three other chimpanzees, until he died of heart failure on March 10, 2000.

See also

Ethics and Animal Protection—*Great Ape Project*

Further Resources

Fouts, R., & Mills, S. T. (1997). *Next of kin: My conversations with chimpanzees.* New York: Avon Books.

Peterson, D., & Goodall, J. (1993). *Visions of Caliban: On chimpanzees and people.* Boston and New York: Houghton Mifflin Company.

Petitto, L. A. (2000, March 23). *Impish chimp changed science.* Retrieved February 18, 2007, from http://www.mcgill.ca/reporter/32/13/chimpsky.

Probst, M. (2000, March 10). *Nim, world's most famous sign language chimp, dies at age 26.* Retrieved from http://www.fund.org/library/documentViewer.asp?ID=96&table=documents.

———. (2000, March).*Nim Chimpsky.* Retrieved from http://www.fund.org/ranch/documentViewer.asp?ID=42.

Terrace, H. S. (1979, 1987). *Nim: A chimpanzee who learned sign language.* New York: Columbia University Press.

Terrace, H. S., Petitto, L. A., Sanders, R. J., & Bever, T. G. (1979, November 23). Can an ape create a sentence? *Science, 206,* 891–902.

Communication and Language
Human-Horse Communication

Humans cannot "speak" horse, and horses do not use verbal language as a means of communication. This means that, together, humans and horses must co-create a system of communication using a medium that they both are able to understand. For both species, the body is a tool through which they can communicate a wide range of emotions and desires, and this makes possible the creation of shared meaning. Though different from the elements of spoken language, a complex embodied language system can be learned by both horses and humans, thus enabling each to express a subjective presence to the other and work together in a goal-oriented fashion.

In order for humans to be effective communicators with horses, we need to develop a focused awareness of our bodies, knowing that our physical movements and expressions are always translating an idea or feeling to the horse. Horses, in general, have highly sensitive bodies, because that *is* their vehicle for communication. Because horses rely on their bodies to transmit and receive information, they are highly skilled at reading (and using) body language. Given their tremendous bodily sensitivity, horses are always keenly aware of others' body language—human and nonhuman alike. People, whether we are aware of it or not, are always communicating ideas and feelings by way of our bodies. Therefore, with the understanding that horses send and collect ideas through *their* bodies, humans must develop a hyperawareness of our own, in order to become effective communicators with horses. Without this developed bodily awareness, it would be very difficult—if not impossible—for humans to understand why the horse responds to the person in the way that she or he does.

Embodied communication between humans and horses is akin to when humans are engaged in verbal conversation with one another. For example, when conversing, we are often mindful about the words we choose to convey various ideas. Similarly, humans who work with horses develop a similar heightened awareness of their body language (rather than spoken words) and are careful to think about the messages they are conveying, or intending to convey, to the horse by way of their bodies. As humans develop a more acute tactile sense, not only are we more effective with our bodies, but we are also better able to "tune in" to the horse's body to understand what is being communicated to us. Doing both simultaneously enables the horse and the human to engage in a two-way conversation.

Learning this new language and understanding this form of communication takes both time and experience. It involves subtle (sometimes micro-) movements that both human and horse use to communicate intentions or emotions. Many people who ride horses use the assistance of a professional horse trainer to help them refine their riding and horsemanship skills. Professional trainers are interpreters, in a sense—they teach the rider proper bodily form and how to use her "aids" (her legs, seat, and hands) to communicate her intentions to the horse. Conversely, the trainer also helps the rider understand what the horse is communicating, so that, ultimately, the horse and rider can be completely united when working together. In this way, communication between horse and rider is truly a body-to-body process.

It is important to emphasize that human-horse communication is not just a one-way relationship of humans merely imitating "horse language." Horses, too, are thinking, emotional, decision-making beings who, like humans, develop ways to communicate their subjective experiences to their human partners. This manner of communication between the two is a cyclical and dynamic process, and both species are full participants in it. Thus, it is not only necessary for humans to learn and refine ways of communicating with horses, but it is also important to think about the horse as an *active member* in the communication process.

Horses use various parts of their bodies, along with a wide range of movements, to communicate a feeling or desire. For example, the ears of a horse are very expressive, and different positions can tell a human whether the horse is relaxed, curious, scared, angry, or listening. Tenseness in a horse's body can signal fear and anxiety, or a constant swishing or wringing of the tail can indicate emotional agitation or physical discomfort. These are all signals, among many others, that humans who work with horses need to be acutely aware of in order to assess a horse's emotional state.

For humans, learning how horses communicate with other horses is an important part of this language-building process. By learning how horses communicate with one another, humans can use that knowledge as a basis for the development of a communication style that horses will understand. This comes from watching horses interact with each other, as well as by learning from more advanced horse people about the meanings of certain body gestures and signals given off by horses.

This cocreative process that results in a shared language system between humans and horses is important for several reasons. First, effective communication between horses and humans helps to ensure safe and humane interactions for both species. The average horse weighs about 1,200 pounds, and the average human's weight is a mere 10 to 20 percent of that. Humans need to communicate effectively to the horse, as well as understand what the horse communicates to *them*, so that no misunderstandings cause one or the other to react in a manner that may be harmful to both. Second, when the horse and human are effectively communicating with each other, they can work together in a goal-oriented fashion. Without the establishment of a shared language system, humans and horses would experience constant conflict—which, more often than not, is at the expense of the horse's well-being. Finally, and more importantly, creating a system of communication helps horses and humans develop a deeper understanding of each other.

The Grammar of Human-Horse Communication

Through this multidimensional system of a shared body language, horses and humans are able to develop an understanding of one another. Undoubtedly, the elements and rules of a body language are different from those of a verbal language, and although body language is traditionally not thought of as a complicated form of communication, for horses and humans it is clearly a style of communication that enables complex working and emotional relationships between the two. Initially, both humans and horses must learn a basic system of communication. This system is taught to almost all young (or "green") horses and beginning riders. When horses first begin working with humans, they are taught a basic vocabulary of bodily cues. In general, the cues work within a system of pressure and release. For example, horses are taught that pressure on the right side of their body from a rider's leg (or from a person's hand when the person is standing on the ground) is a signal to move left; once the horse moves, the pressure is released, to communicate to the horse that that movement was the desired outcome. The same basic cues are taught to a person learning how to work with a horse. Putting pressure on the left side of the horse's body will signal to her or him to move right, and, once the horse moves right, the pressure should be released.

Beginning riders are usually taught through instruction from an experienced or professional horse person, who pairs up the rider with a well-schooled, often older, horse. Similarly, young horses generally learn from knowledgeable and more experienced horse persons. Because of a horse's large size, it can be a dangerous endeavor for even the most experienced horse person to work with a horse that has not learned the basics of human-horse grammar. People who "start" young horses generally have a wealth of knowledge about the human-horse communication process. Likewise, novice horse people who do not understand the elements and rules of human-horse communication run the risk of putting their well-being in jeopardy, and that is why they are often paired with an older, more schooled horse.

The basic cues of pressure and release become the alphabet of body language—the foundation from which a more sophisticated use of the language system can grow. The language becomes more complex and nuanced as the vocabularies of both horse and human expand. Gradually, horse and rider can synchronize various cues at once. The more humans and horses engage with one another, the more refined, clear, and subtle their ability to communicate becomes.

For the human partner, it takes a great deal of effort and time to learn to use his body in new and different ways. Similar to learning a verbal language, at first using/speaking the "words" is awkward and crude, but the more humans refine their ability to speak words and string them together in a meaningful way, the more subtle and smooth their speaking becomes. To be sure, part of the process of learning to verbalize words is training the mouth, the lips, and the tongue to move in particular ways in order to produce certain sounds. Likewise, it takes time and dedication to develop the tremendous bodily discipline necessary to be an intentional and effective communicator with horses. As a result, for both horses and humans, as their understanding of body language in the human-horse communication process becomes less crude, the more subtle and refined their expressions become.

In addition to a well-developed embodied-language system of touch, pressure and release, movement, and body posture, humans also draw on the experience of bodily feeling sensations as a resource to guide their interactions with horses. This requires the human to be keenly aware of bodily sensations—the experience of "feeling" states in bodies—as a source of information. "Feel" is another element of the human-horse communication process, whether riding a horse or working with it from the ground. Horse people often speak of having "a good feel of the horse" or how much the horse "feels off" of the person. This concept of "feel" is based, in part, on the idea that horses pick up on feelings and thoughts through the body-to-body connection with humans. Feel, therefore, is partly tactile in the sense of body-to-body contact, and it is also the experience of internal embodied sensations.

The bodily experience of "feel" becomes another way in which shared meaning is created between horse and rider. Empathy—the embodied sensation of feeling of another's emotions—is used in this human-horse context, but it is employed in the service of communication. Empathy is not used only for the sake of feeling *for* the horse, but rather, it is used as a resource to feel *of* the horse. This difference between feeling *for* the horse and feeling *of* the horse is an important distinction in the human-horse context. To feel *for* the horse is to use the experience of embodied empathy for the purposes of compassion and understanding. To feel *of* the horse is to use empathy at the level of embodied sensation for the purpose of communication. Emotional empathy, in the traditional sense, is a loss of—or forgetting of—*the self* in order to experience *the other*. In the human-horse interaction, however, the concept of "feel" is more utilitarian, because it is a mode through which communication can happen. When working with horses, humans treat nonverbal embodied sensations—in the same way as words—as a source of information to help structure any given interaction. As discussed above, given the tactile sensitivity of horses, they have a unique ability to pick up emotions and "feeling sensations," particularly when connected body to body with humans.

Ultimately, the desired outcome is for a horse and human combination to have a "good feel of each other." Horse people talk about having a good feel of the horse, or a horse having a good feel of the human. When a human has a good feel of the horse, it means that she or he is keenly aware of the exchange of feeling sensations between herself or himself and the horse and is able to effectively communicate meanings to the animal through this mode of sensation. When a horse has a good feel of its rider, it means that, through the exchange of internal sensations, the horse has a clear understanding of the rider's intent and literally *feels good* about working with that specific rider. However, having a good feel is not always a given—or even automatic. "Good feel" between humans and horses, like all other elements of human-horse communication, takes time and conscious, skillful effort to develop. In developing her communication skills, the ultimate goal for the human is to have a unison of emotional and physical control while engaged in the communication process with a horse.

Horse people often say that the best riders and horses are the ones who can go around the arena and make it look effortless, without any visible signs of communication taking place. Well-developed riders and horses learn how to communicate and understand each other on such subtle levels that, to someone watching, it can look like no discussion is taking place between the two—just two united bodies moving together seemingly effortlessly and silently. To do this takes body discipline and a well-developed understanding of one's body as being a vehicle for receiving and communicating different signals. The embodied experience of human-horse communication is similar to Kenneth Shapiro's concept, "kinesthetic empathy." The "empathic experience," he says, "involves appropriating a second body that then becomes my auxiliary focus. Through my lived body, I accompany yours as it intends an object" (1990). When horse and rider are moving together, the rider must make the horse's body her focal point by way of her *own* body, as both literally accompany each other in a shared embodied experience. Without this empathic basis, horse and rider would be disjointed, in conflict, and unable to have a shared experience of "other." This learning is part of the cocreative process of the human-horse language system that makes communication and emotional relationships between the two species possible.

It can be tempting to take a human-centered approach to studying this language, but it must be acknowledged that horses are active participants in this communication process. To suggest that humans are entirely responsible would misconstrue the dynamic nature of this form of communication. The development of successful human-horse partnerships involves a complex system of negotiations, within which both the rider and the horse engage in a process of give and take. Acknowledging the complexity of human-horse communication requires an understanding of horses as sentient beings that live valuable lives of their own. In addition, understanding the process of this embodied-language system illustrates, in part, the dynamic relationships humans share with their horse companions and highlights its unique qualities, benefits, and intricacies.

Further Resources

Brandt, K. (2004). A language of their own: An interactionist approach to human-horse communication. *Society & Animals, 12,* 299–316.

McLean, A. N. (2005). The positive aspects of correct negative reinforcement. *Anthrozoös, 18,* 245–54.

Shapiro, K. J. (1990). Understanding dogs through kinesthetic empathy, social construction, and history. *Anthrozoös, 3,* 184–95.

Wipper, A. (2000). The partnership: The horse-rider relationship in eventing. *Symbolic Interaction, 23,* 47–72.

Keri Brandt

Communication and Language
Interspecies Communication—N'Kisi The Parrot: A Personal Essay

If animals could really communicate with us, what would they want to say? To find out, I have been working to teach a young parrot to communicate in spoken English. "N'Kisi" (in-KEE-see) is a hand-raised, domestically bred Congo African Grey parrot that is now one of the most skilled language-using animals. He does not simply mimic, as most parrots are thought to do, but he can actually understand and use language appropriately. N'Kisi was not *trained* to perform behaviors by rote or on demand, like the animals used in entertainment or in laboratory research studies. Instead, he has been *taught* the meaning and purpose of language, and encouraged to express himself, in a nurturing environment of human-animal social relationships. N'Kisi speaks in original sentences—showing an understanding of grammatical rules—and can conjugate verbs to talk about the past, present, and future. He generally initiates comments about whatever is happening in his environment, and he has expressed love, compassion, logic, and a sense of humor. N'Kisi's remarkable abilities have provided us with a window into the animal mind and an opportunity to explore meaningful communication with another intelligent species.

At age 9, N'Kisi has a vocabulary of over 1,200 words, which he can use in context. For comparison, Koko, the famous sign-language-using gorilla, learned hand signs for 1,000 words by age 29, and Alex, the parrot, uses around 250 words after 23 years of teaching. However, the counting methods used by each researcher are different, which is an element that must be taken into account. Since N'Kisi may be the first animal known to conjugate verbs and use plurals and other inflections, in order to document this finding, we count all forms of words that are used correctly. Parrots can have a lifespan comparable to humans, and N'Kisi's language skills continue to develop. He is learning around 100 new words per year (depending on the amount of teaching). Eventually, animals in language studies tend to reach a plateau after a certain point, and their language and cognitive skills may never progress beyond a level comparable to that of a human child. Language-using animals are extremely rare; with only a handful of individuals of various species documented in intensive studies around the world, each one is a unique case study, and each can shed light on another aspect of the animal mind.

What is most notable about N'Kisi, compared to other language-using animals, is that he speaks in sentences, showing a grasp of grammar and syntax in formulating his own novel constructions. N'Kisi has created complex original sentences up to sixteen words long. For example, referring to my treatment for a serious illness, N'Kisi said: "Remember, we had the sick, but then we had to go to the doctor, my body." (He often spoke of this as though it affected him, too.) While the grammar isn't perfect, and there are some interesting errors, N'Kisi's linguistic skills are clear when we compare this with the longest recorded primate "sentence," from Nim Chimpsky: "Give orange me give eat orange me eat orange give me eat orange give me you." Although Nim's ability to communicate his intended meaning is amazing (animals in language studies often use their skills to request favorite foods), this string of repeated words lacks grammatical structure. Due to results like this, many scientists have thought that animals are not capable of using syntax, and this has been considered a definitive aspect of human uniqueness. Indeed, after Jane Goodall's discovery that chimpanzees use tools, language was thought to be the fundamental dividing line between humans and animals. And after animals in interspecies communication studies demonstrated some rudimentary language abilities, syntax became seen as the new boundary. However, new research is emerging that suggests that some species, such as starlings and parrots, may have some capabilities in this area, too.

Regarding animal language, syntax has been described as the ability to combine known vocabulary words in a potentially infinite number of new phrases, following the laws of grammar. N'Kisi's ability to form novel sentences with appropriate grammatical structure is thus a significant advance. One of our most remarkable conversations occurred when N'Kisi was only about three years old, after we had rented a car and taken him for a ride around a beautiful park:

N'Kisi: "Remember, we went in a car."
Aimee: "Yeah, I remember!"
N'Kisi: "We went for a ride in a car."
Aimee: "Yeah! Did you like it?"
N'Kisi: "I like that—wanna go out in the car."
Aimee: "We can't; we don't have a car now."
N'Kisi: "Wanna go in a car right now."
Aimee: "I'm sorry, we can't right now—maybe we can go again later."
N'Kisi: "Why can't I go in a car now?"
Aimee: "Because we don't have one."
N'Kisi: "Let's get a car."
Aimee: "No, 'Kisi, we can't get a car now."
N'Kisi: "I want a car!"
Aimee: "I'm sorry, baby, not today."
N'Kisi: "Hurry up, wanna go in a car! Remember? We were in a car!"

What is most striking about this conversation, besides his accurate conjugation of verb tenses, is that N'Kisi seems to have an understanding of time and can speak about things in the past, present, and future. Animals were not thought to be capable of communicating about anything that was not present or occurring at the time, but N'Kisi often talks about experiences that have made an impression on him, sometimes for months afterward, as well as things that are going to happen or that he'd like to do.

No other animal has been known to conjugate verbs. However, this may also be due to the type of language used in prior studies. The American Sign Language (hand) and lexigram (abstract symbol) systems used in studies with primates do not have all of the syntactical elements of spoken language, and other animals, including parrots, have been taught using a simplified laboratory "pidgin" that excludes most grammatical details.

When composing sentences, N'Kisi sometimes makes the kind of mistakes small children do, such as saying, "Jane flied in a airplane." Although there is nothing about this in the existing literature on animal communication, linguists who study children's language consider such "generalization errors" for irregular verb tenses to be an important milestone in language acquisition, as they show an understanding and independent application of linguistic rules. N'Kisi has also invented original expressions for new things he does not know the name of, such as calling the decongestant aromatherapy oils I was using to help me get over a cold "pretty smell medicine." Now, when I am getting sick, he will tell me, "Gotta put smell medicine."

These breakthroughs may have been possible because N'Kisi's use of language was not constrained to fit any pre-determined, task-based criteria, as in most laboratory studies. Instead, our project is based on the model of field research, observing and documenting his spontaneous behavior in a more naturalistic way over the course of a long-term study. Our work together is communication-based, rather than task-based, and no behaviorist training techniques are used—such as treat rewards for the production of desired responses. N'Kisi has never been taught the concept of performing for a reward, and he would not understand it. He is allowed to use the tools of language freely to communicate for his own ends. One drawback of this method, however, compared to the traditional use of rewards, is that N'Kisi does not say anything "on cue." All of his comments are entirely self-generated, and he says what he wants, when he wants to. But the decision to relinquish a framework of control has opened the door for N'Kisi's creativity.

N'Kisi is capable of associating words and pictures with things they represent. When the renowned primatologist Jane Goodall first met N'Kisi, she said "You must be N'Kisi—I've heard a lot about you." He replied immediately, "Got a chimp?" N'Kisi had never said the word "chimp" before; but he had been shown a book with pictures of Jane Goodall with chimpanzees the day before, and was told she was coming to visit him. His sense of humor was also evident when he met another African Grey, Endora. The first time N'Kisi saw Endora hanging upside down by her claws from the top of her cage and swinging back and forth to get attention, he remarked, "You gotta put this bird on the camera!"—suggesting a sense of the purpose of filming, to show extraordinary things.

Another time, a necklace I was wearing broke, and as I bent to pick up the beads, he said "Oh, no, you broke your new necklace." N'Kisi often expresses concern for others; for example, when our parrot Pooka is bothering another bird, N'Kisi will tell him to "be more gentle." While N'Kisi can be altruistic, he may also be deceptive: after he saw his flock mate Endora getting punished for making an ear splitting whistle, he would make the whistle himself and then yell, "Endora!" Because he often chided other birds for misbehaving, this tricked us into covering her—until we learned to see whose beak the sound was really coming from. Deception is thought to require a "theory of mind," the ability to form a concept of what someone else is thinking.

Sometimes, N'Kisi likes to engage in pretend play and act out situations with his toys. Once, as I was videotaping him, he pretended to put batteries in a toy that needed new ones. He reached around and pecked the place in the back where the batteries go and then announced, "There—I put a battery." Pretending is another activity that requires an understanding of abstract concepts. I have been documenting N'Kisi's language for over six years and have recorded thousands of his comments, many of which reveal aspects of intelligence and awareness that animals were not thought to be capable of—especially a species so different from us.

N'Kisi was not taught using the "model-rival" method, used with Alex and other parrots, which requires two researchers to act out exercises in object labeling. He was systematically taught in much the same way as a human child. Human mothers instinctively talk to babies in a special way that seems to facilitate learning. Linguists call this simplified, singsong mode of speaking to young children "Motherese." I adapted some of these techniques, using insights gained from over fifteen years of working with parrots. From the moment N'Kisi's teaching began, when he was five months old, every word he ever heard was used in context. Activities were explained to him in a natural fashion, and he observed and participated in our social interactions. This gave him the opportunity to discern the contextual meanings of words by observing recurrent patterns.

Two basic teaching methods were used: *sentence frames* and *cognitive mapping*. The first describes objects and activities in a series of sentences that repeat a keyword. For example, if I was getting a glass of water, I'd show him and say, "Look, that's my water. Oh, boy, yummy water! You want some water?"—and so on. N'Kisi not only had an opportunity to learn the target word, but he was also getting lessons in forming sentences.

Cognitive mapping reinforces meanings that may not yet be fully understood. If N'Kisi said the word "water," I would get some water and give him some, or show and describe things that are not appropriate to give him. A key aspect of these methods is repetition. The target word is repeated a few times, in varied sentences, but *only* in reference to what is occurring *at the time*, to enable a conceptual link to be made between word and thing. One unintentional side effect of this method is that N'Kisi's own speech often reflects this same "Motherese" style of repetition.

How is N'Kisi able to learn aspects of a human language? His ability to do so suggests that he has the cognitive hardware required to understand the rational scaffolding of language. But why would a nonhuman animal have evolved this capability? There are several theories that may account for this. One is that animals that have to remember a large number of seasonal food sources across a large territory range—such as trees that fruit at particular times—must create complex mental maps, and this cognitive function may be similar to the one necessary for language. Another theory is that social animals must keep track of all of the other animals in the group—and the shifting politics of their social hierarchy—and this may require similar cognitive skills. There is also the possibility that animals may have some form of rudimentary language of their own. Researchers have discovered that several species, including vervet monkeys, prairie dogs, and dolphins, have various types of calls that have specific meanings—such as alarm calls for different types of predators, contact calls, and identifying calls unique to individual animals. Parrots have different types of natural calls that seem to serve specific functions, and they also imitate the calls of other species in the wild. African Grey parrots live in large flocks with complex social interactions that have yet to be studied. They are considered one of the most intelligent species of birds, with a brain-to-body size ratio comparable to that of higher primates.

N'Kisi has a vocabulary of over 1,200 words, making him one of the most skilled language-using animals. Courtesy of Aimee Morgana/The N'Kisi Partnership.

Success in teaching a parrot to use language will depend on several factors. The most important is the depth and intensity of the parrot's relationship with the human teacher. Many species of birds form closely bonded relationships with another individual that may be lifelong, and this devoted camaraderie can be transferred to a human companion, just as a pet dog will transfer its instinctive pack loyalty to its human family. Teaching a parrot to speak takes a lot of time and patience. Some birds are also more vocally inclined than others. This may have more to do with control of the syrinx, the organ analogous to the human larynx that parrots use to make sounds, than it does with innate intelligence. Just as some people can sing better than others, control of the syrinx can vary as an individual talent, even within the same species, and some parrots have more difficulty than others in reproducing human-speech sounds.

It was important that N'Kisi have relationships with other parrots, in order to learn to communicate with others of his own kind, as well. He lives in a mixed-species "flock" of humans, other parrots, and reptiles. This has also allowed me to observe the calls and gestures parrots use to communicate with each other, and their interrelational dynamics. N'Kisi's abilities have flourished in this intimate family group, which provides him with social engagement and a sense of security and self-confidence. Due to my background as an artist, I knew that self-confidence is an important prerequisite for creativity. Hoping to encourage his creativity and help N'Kisi feel more empowered to use language for his own ends, I made the unusual decision to give him "dominance" in our relationship. He is housed with a high perch, so that his eye level is higher than anyone else's. In wild flocks, dominant parrots compete for the highest position in roosting trees, where they keep watch as sentinels and give warnings to protect the flock. N'Kisi's emotional level seems comparable to a human toddler's, and he is extremely sensitive and shy. Taking into account his social needs, I have worked to establish mutual trust and engage him in our dialogue as a willing partner. This dialogue is entirely dependent upon our close relationship, which has facilitated and motivated the desire to communicate.

As our closeness grew, one of my most startling discoveries was that sometimes N'Kisi seemed to vocalize my unspoken thoughts, and he even described images I was looking at in other rooms. To learn more about this curious phenomenon, biologist Rupert Sheldrake and I designed a series of carefully controlled double-blind tests. The results were highly significant when statistically compared to chance, and they seemed to support the hypothesis that N'Kisi was indeed able to describe what I was looking at in a distant enclosed room, even though there was no possibility for any cues through normal means. These findings are controversial, and challenge many dominant beliefs and scientific theories. However, experiments in quantum physics have confirmed the existence of "non-local" phenomena, although the mechanism of how widely separated particles can "communicate" instantly regardless of distance has yet to be explained. This may be related to telepathy, or even the "group mind" of synchronously wheeling flocks of birds. Advances in quantum research may open this new frontier for further scientific study.

N'Kisi and Aimee Morgana in a teaching session. Courtesy of Aimee Morgana/The N'Kisi Partnership.

N'Kisi's abilities continue to advance. He is now learning phonics and has spelled some simple words, such as "cat"; perhaps, someday, he may learn to read a little. He is fascinated with the idea of writing, and has even expressed interest in writing a book. He talks about this often; during one videotaping session he said, "I write here, and made book. Let me write book. I wish I could be on a book. I wanna be on a book." I am currently writing a book to share his story.

N'Kisi's accomplishments may help further our understanding of the other minds that share this planet—and our own. We are only beginning to learn about the intelligence of other animals, and yet, tragically, many of the most intelligent species, including African Grey parrots, are now endangered in the wild. Greys and other threatened species of parrots are still being captured and exported for the pet trade; tens of thousands may die in the process every year. And many people who purchase parrots as pets are unable to deal with a parrot's demanding social needs, causing a growing problem of abandoned parrots, with limited sanctuaries for their long-term care. Hopefully, learning about N'Kisi's abilities will help inspire people to protect these intelligent species and their habitats, so that their wild societies can survive for future generations. Language-using animals are like animal ambassadors, bridging the worlds of other species with our own. At the heart of every successful study in interspecies communication is an extraordinary human-animal relationship.

See also

Communication and Language—*Telepathic Communication Systems between People and Animals*

Further Resources

Gentner, T., Fenn, K., Margoliash, D., & Nusbaum, H. (2006). Recursive syntactic pattern learning by songbirds. *Nature, 440,* 1204–07.

Goodall, J., & Bekoff, M. (2002). *The ten trusts* (pp. 47–49). San Francisco: Harper SF.

Hauser, M. D., Chomsky, N., & Fitch, W. T. (2002). The faculty of language: What is it, who has it, and how did it evolve? *Science, 298,* 1569–79.

O'Hanlon, E. (2004, February). Polly wants a dictionary. *BBC Wildlife Magazine,* 58–60.

Patterson, F., & Linden, E. (1981). *The education of Koko.* New York: Holt Rinehart Winston.

Pepperberg, I. M. (1999). *The Alex studies.* Cambridge, MA: Harvard University Press.

Sheldrake, R. (2003). *The sense of being stared at* (pp. 23–27, 300–305). New York: Crown.

Sheldrake, R., & Morgana, A. (2003). Testing a language-using parrot for telepathy. *The Journal of Scientific Exploration, 17*(4), 601–16.

Terrace, H. (1979). *Nim.* New York: Knopf.

To see N'Kisi on video, see *Jane Goodall's When Animals Talk.* (2005). Discovery/Animal Planet.

To hear N'Kisi talk in a teaching session, go to http://www.sheldrake.org/nkisi/nkisi1_text.html

Aimee Morgana

Communication and Language
Similarities in Vocal Learning between Animals and Humans

Human parents are ceaselessly fascinated by their children's acquisition of language. In two to three years—just a fraction of the human juvenile period—most individuals progress from simple, nonverbal distress communication (crying) to the expression of complex thoughts in long series of words. These words are assembled and inflected according to specific rules and are selected from a learned vocabulary that quickly numbers into the thousands. Learning a language—and speaking it—comprises one of the most outstanding indications of the power and versatility of human intelligence. Thus, for over a thousand years, "dumb" has meant not only "mute," but also "stupid."

For nearly as long, humans have referred to animals as "the dumb beasts." We have not thought that animals are quiet, of course, but we have often considered their sounds to be devoid of real communication. (Otherwise, as Descartes wrote with a hint of sour grapes, the animals would have found a way to let us know what they were saying!) Starting there, we would naturally conclude that animal sounds are automatic and unlearned, and require no cognition.

Behavioral scientists know better. For one thing, even unlearned vocalizations can be effective in communication. Second, we should take animal cognition seriously, whether or not animals learn their sounds. Third, many animals do learn their sounds, as well as when to make them and what they mean. This does not make human language any less phenomenal, and there are many aspects of human language that still appear to be unparalleled. However, we must be very careful when we talk about the uniqueness of human language, because the study of vocal learning has established several links between humans and other animals.

Vocal plasticity is the ability of an organism to change something about its vocalizations over time. All vocalizing animals have some plasticity. Crickets, for instance, provide the best-studied example of unlearned, automatic animal communication (though it is not vocal). But even crickets will stop singing when a large animal approaches them, which demonstrates plasticity in the timing of their song. So we humans share vocal plasticity, at its basic level, even with the insects.

The most expedient and versatile sort of vocal plasticity is known as *vocal learning*. Most of this learning is social, meaning that the animal gets its information from other animals. There are two classes of social learning in animal communication: *contextual learning* and *production learning*. Contextual learning is when an animal learns either how to use its sounds (such as when to make them) or what certain sounds mean. Production learning is when an animal learns how to produce its sounds properly. Humans are highly dependent on both contextual and production learning. When an infant learns that "milk" refers to the drink, or that saying this word to a parent helps milk to appear, this is contextual learning. When an infant actually learns to say the word—to make the sound "milk"—this is production learning. In nature, contextual learning is much more common than production learning. Any animal that can be trained to vocalize on demand or respond to a particular word is using at least a rudimentary form of contextual learning. Any species with individual recognition on the basis of voice, as when king penguin (*Aptenodytes patagonicus*) chicks and parents locate each other, uses a more advanced form of contextual learning. Young vervet monkeys are using contextual learning when they discover that a particular call in their repertoire should be given whenever a predatory bird flies overhead, whereas another is appropriate when a dangerous snake is observed on the ground. Despite its commonality, contextual learning has only recently been an explicit focus of communication research, but some fascinating examples have been discovered so far. It has been demonstrated in monkeys and apes, whales and dolphins, seals, bats, dogs, cats, horses, and many birds.

Most of this entry focuses on production learning, since that is what is more specifically meant by "vocal learning"—literally, learning to vocalize. The discovery of instances of production learning continues: at last count, it has evolved independently at least eight times, but only in birds and mammals. In birds, it is apparently ubiquitous in the Oscine Passerines ("songbirds") and parrots, and it is also found in hummingbirds; future research is needed to confirm suggestions that it exists in a few fowls, toucans, and sub-Oscine Passerines. Among mammals, at least some cetaceans (whales and dolphins), seals, bats, elephants, and humans show production learning. The elephant case is based on a single sample of imitated sounds by adult African elephants. Some animals rival humans in their mastery of vocal learning, mimicking an astonishing variety of sounds—including human voices and machinery. Examples are mynahs (*Gracula religiosa*), lyrebirds

(*Menura superba*), grey parrots (*Psittacus erithacus*), and harbor seals (*Phoca vitulina*). Communication in three groups of animals has been studied extensively enough to explore similarities between human and nonhuman vocal learning.

Primates

Humans are unique among primates in the widespread and clear use of learned vocalizations. The calls of several primates differ between social groups—or change while the animal is developing—but other processes besides learning might explain these situations. For instance, a social group might share similar songs because of the influence of a common physical environment or shared genes. In chimpanzees (*Pan troglodytes*) and Japanese macaques (*Macaca fuscata*), the calls of two individuals can be more similar when they are communicating with each other than when they are communicating with other individuals, a phenomenon known as vocal convergence; whether this involves production learning is unclear. Surprisingly, our closest relative, the chimpanzee, is unable to imitate new sounds like humans do after nine months, even though the chimp and the other great apes can imitate various other kinds of actions. These differences are reflected in the different ways the brain controls vocalization in humans versus other primates. Lesions to certain parts of the brain, such as the primary motor cortex, lead to speech disorders in humans, but do not produce similar effects in monkeys. This suggests that they use different parts of their brains in vocalizing than we do, despite our close evolutionary relationship. Moreover, certain genes—such as *FOXP2*—that are involved in vocalization have evolved since our divergence from the chimpanzee.

Dolphins

Bottlenose dolphins (*Tursiops truncatus*) are the only nonhuman mammal that is known to interact by matching each other with learned sounds. Dolphins use several kinds of whistles to communicate. An individual's most common whistle type, its "signature whistle," is used in matching, usually when the other dolphin is out of sight. Signature whistles are thus thought to function in group cohesion and individual recognition. Dolphins appear to converge with other members of their social group in certain aspects of their whistles, so the whistle may also signal group membership. A few other whistle types also have known functions, such as mother-calf communication. Like humans, dolphins have a period early in their life (usually within the first two months in dolphins) when vocal learning is very important and basic communication skills develop in the presence of other individuals. Also like humans, dolphins remain open to learning and producing new whistles throughout their lives.

Songbirds

Vocal learning is better understood in songbirds than in any other animal; in some ways, we know more about how birds learn song than about how humans learn language. A fuller account of the relationship between birdsong and human speech can be found elsewhere. In general, although the functions and significance of songbird and human vocalization can be very different, the process of learning is very similar and uses much of the same neural equipment. Some of the more striking of these similarities are as follows:

- Two pathways in the brain are important to vocal learning that are apparently absent from birds or mammals that do not learn their vocalizations. A posterior pathway appears to be directly involved in vocal production, whereas an anterior pathway is a loop that involves thinking centers and is probably more responsible for flexibility or plasticity.

- Young individuals imitate the sounds produced by older individuals, often parents.
- Two developmental periods can be distinguished: an early *perceptual* phase where the young are listening and remembering, but not yet making sounds, which guides a later *sensorimotor* phase where the learner is practicing out loud, as well as listening.
- Early experience is crucial, and the tendency to learn declines as an individual gets older.
- Early vocalizations are diverse ("babbling" in humans, "subsong" in birds) compared to the fewer and more refined sounds that are ultimately preserved in the repertoire for life.
- Auditory feedback, where a young individual hears itself vocalize, is necessary. Its importance decreases as the individual gets older.
- Social stimulation—interaction with other members of the species beyond hearing them sing—aids learning.
- An instinctive tendency to listen for and imitate the "right" sounds. Development is a process of selecting, or shaping, the repertoire according to what the individual hears in its models.

Considering the diversity of vocal learners both avian and mammalian, some similarities appear to be widespread. First, within a species, the fastest learners are the individuals that experience the most social interaction and information such as what a vocalization is about or when to do it. Second, individuals often develop distinctiveness in their vocalizations, resulting in the ability to tell each other apart. Third, vocal convergence within groups is common, resulting in "dialects," or different vocal styles between groups. Many birds, and a few mammals, have been found to exhibit this phenomenon, including humpback whales (*Megaptera movaeangliae*), bottlenose dolphins, greater spear-nosed bats (*Phyllostomus hastatus*), and greater horseshoe bats (*Rhinolophus ferrumequinum*). Fourth, dialects can change through time by cultural evolution, as has been studied in depth in indigo buntings (*Passerina cyanea*) and humpback whales. Fifth, vocal learning is usually an indicator of social status or membership. When they open their mouths, individuals proclaim what group is theirs, as well as how socially proficient they are.

Why did vocal learning evolve, in humans as well as in other animals? One hypothesis is that social species that are also highly mobile need to learn local dialects in order to communicate well in local conditions. These conditions might include the acoustic properties of the environment, as well as the nature of social interactions. Individual recognition is often important, for instance, when the social system is fluid and the group members need to constantly update each other as to who is around. In these cases, the small genetic differences between individuals result in differences in vocalization without learning, and this is the extent of the variation for most animal species. But learning can diversify vocalizations far beyond what genetic differences alone could accomplish. Perhaps the extent to which vocal qualities can change via genetic differences is particularly constrained in animals with certain traits or lifestyles. For instance, the structure of organs in small animals might not be able to change very much without compromising function; and the rigors of powered flight might impose limitations, as well. These factors might explain why vocal learning has evolved as an alternative means of vocal diversity in hummingbirds, songbirds, and bats. One hypothesis for why there are so many vocal-learning marine mammals as compared to terrestrial mammals is that water pressure changes the nature of the vocal organs at different depths and speeds of movement, obscuring small genetic differences and rendering vocal learning more important if individuals need to be distinctive. As for us humans, we have a fluid social system, where groups are very important to our survival. In this situation, we find it important to know and to convey our group membership and status, and to communicate a multitude of other social facts to one another. These social facts, and many other less central aspects of human culture, change during our lifetimes and over the generations. Perhaps only by learning our speech can we be flexible enough to adjust our communication to such changes.

See also

Communication and Language—*Birdsong and Human Speech*

Further Resources

Boughman, J. W. (1998). Vocal learning by greater spear-nosed bats. *Proceedings of the Royal Society of London, Series B, 265,* 227–33.

Brainard, M. S., & Doupe, A. J. (2002). What songbirds tell us about learning. *Nature, 417,* 351–58.

Janik, V. M. (2000). Whistle matching in wild bottlenose dolphins (*Tursiops truncatus*). *Science, 289,* 1355–57.

Janik, V. M., & Slater, P. J. B. (2000). The different roles of social learning in vocal communication. *Animal Behaviour, 60,* 1–11.

Kuhl, P. K. (2000). A new view of language acquisition. *Proceedings of the National Academy of Sciences of the United States of America, 97,* 11850–57.

Moore, B. R. (2004). The evolution of learning. *Biological Reviews (Cambridge), 79,* 301–35.

Pinker, S. (2000). *The language instinct.* New York: HarperPerennial.

Snowdon, C. T., & Hausberger, M. (Eds.). (1997). *Social influences on vocal development.* Cambridge: Cambridge University Press.

David C. Lahti

Communication and Language
Telepathic Communication Systems between People and Animals

Many people have noticed that their companion animals seem to "read their minds." This perceptiveness may well depend on a combination of influences, such as the observation of body language, hearing particular words, and learning the owners' routines. In addition, the animals may be able to pick up intentions directly by a kind of telepathy. But surprisingly little research has been done on this subject—biologists have been inhibited by the taboo against "the paranormal," and psychical researchers and parapsychologists have, with few exceptions, confined their attention to human beings (exceptions include Duval & Montredon in 1968, and Schmidt in 1970). On the other hand, committed skeptics believe that any mysterious connections currently unknown to science are impossible, or too unlikely to merit serious attention (Marks, 2000).

Recent random household surveys in England and the United States have shown that many pet owners believe that their animals are sometimes telepathic with them. An average of 48 percent of dog owners and 33 percent of cat owners said that their pets responded to their thoughts or silent commands (Sheldrake, Lawlor, & Turney, 1998).

Many people experienced with animals take telepathy for granted, and there is a wealth of anecdotal experience that points to the reality of telepathic influences. Much of the evidence for animal telepathy is anecdotal, in the sense that it consists of unpublished stories; I have built up a large database of such accounts, with over 1,500 case histories, and they provide the outlines of a natural history of what people believe about the telepathic abilities of their companion animals.

The most common kinds of seemingly telepathic response are: the anticipation of owners coming home; the anticipation of owners going away; the anticipation of being fed; cats knowing when their owners intend to take them to the vet and disappearing; dogs knowing when their owners are planning to take them for a walk; and animals getting excited when their owner is on the telephone, even before the telephone has been answered, while the dogs ignore incoming calls from other people. Many horse trainers and riders believe that their horses can pick up their intentions telepathically. Some companion animals seem to know when a particular person is on the telephone. Some react when their owner is in distress in a distant place—or dying (Sheldrake, 1999a).

Cats That Disappear before Visits to the Vet

Some cats strongly dislike going to the vet. Dozens of cat owners have told me that their cats simply vanish when they are due to be taken for their appointments. Experienced owners of such cats try to avoid giving away any clues, but their efforts are often in vain. This is inconvenient, not only for the owners, but also for the vets. Some advise people to keep their cats shut up indoors before the appointment, especially when injections or operations are involved. But some cats still escape.

How common is this kind of behavior? My assistants and I carried out a survey of the veterinary clinics listed in the North London Yellow Pages telephone directory. We interviewed the vets themselves, or their nurses or receptionists, asking whether they found that some cat owners canceled appointments because the cat had disappeared. Sixty-four out of sixty-five clinics said that they had cancellations of this kind quite frequently; the remaining clinic had given up an appointment system for cats—people simply had to turn up with their cats, and thus the problem of missed appointments had been resolved (Sheldrake, 1999a).

Although there was general agreement that some cats do, indeed, pick up their owners' intentions, there were a variety of opinions as to how they might do it. It is usually hard, if not impossible, to disentangle the ways in which animals pick up their owner's intentions when they are nearby. But what if a person decides to take the cat to the vet in the cat's absence? For example, if a person rings the vet from her place of work to make an appointment for that same evening, will the cat have disappeared when she goes home to collect it? Several reports on my database claim that this is the case, but there have been no experimental studies of this phenomenon to date.

Animals Who Know When Their Owners Are Coming Home

Many pet owners have also observed that their animals seem to anticipate the arrival of a member of the household, sometimes ten minutes or more in advance. The pets typically show their anticipation by going to wait at a door or window. In random household surveys in Britain and the United States, 51 percent of dog owners said they had noticed such anticipatory behavior, and 30 percent of cat owners said the same (Sheldrake 1999a). Some parrot and parakeet owners have also observed that their birds become excited before a member of the family returns.

As skeptics rightly point out, some of these responses could be explained in terms of routine expectations, subtle sensory cues, chance coincidence, and selective memory—or put down to the imaginations of pet owners.

The only way to obtain rigorous evidence is to make detailed and systematic observations in situations in which the owners are miles away from the homes, when they come back at nonroutine times, and when no one at home knows when to expect them. In tests that my colleagues and I have carried out, the owner went at least five miles away from home, so there could be no question that the dog was reacting to direct sensory clues from the person. While the owner was away from home, the place where the dog usually waited for its owner was continuously filmed on time-coded videotape for the duration of the test. The person then returned at a randomly selected time, unknown to the people at home, and traveling in an unfamiliar vehicle. Over 100 trials with a dog called Jaytee, belonging to Pam Smart of Ramsbottom, Lancashire, England, have shown that the dog did indeed anticipate her return by fifteen minutes or more, even when she was returning at randomly selected times in a taxi from over five miles away and no one at home knew when to expect her. The dog usually reacted before she actually began her journey and seemed to be responding to her *intention* to return. However, Jaytee sometimes failed to anticipate Pam's return, especially when he was sick or distracted by a bitch in heat in a nearby flat. Nevertheless, taking all the results together, this effect was highly significant statistically, with odds against it being a chance effect of more than 100,000 to one. Jaytee showed the same anticipatory behavior even when tested by skeptics (Sheldrake and Smart, 2000a), although the skeptics themselves (Wiseman, Smith, & Milton, 1998) claimed that the dog had failed their tests. However, they were able to make this claim only by ignoring most of their own data (Sheldrake, 1999b, 2000; see also the reply of Wiseman, Smith, & Milton, 2000).

Although Jaytee is the dog that has been investigated most, we have obtained very similar results in videotaped tests with several other dogs (Sheldrake and Smart, 2000b).

Other Kinds of Telepathic Communication between People and Pets

Other kinds of animal telepathy can also be investigated experimentally; for example, the apparent ability of dogs to know when they are going to be taken for walks. In these experiments, the dog is kept in a separate room or outbuilding and videotaped continuously while its owner, at a randomly selected time, thinks about taking it for a walk, and then, five minutes later, does so. Preliminary experiments have shown dogs exhibiting obvious excitement when their owner is thinking about taking them out, although they could not have known this by normal sensory means. They did not manifest such excitement at other times (Sheldrake, 1999a).

Of all the seemingly telepathic phenomena described by cat owners, the ability to summon a cat mentally is one of the most common. Some people say that they have only to think about their cat, and, within a minute or two, it appears at the door. It would be possible to investigate these claims experimentally by filming the doorway continuously, and asking the cat owner to silently call the cat at randomly chosen times.

In my database, there are dozens of reports of animals that responded to the telephone when a particular person was ringing, before the receiver was even picked up.

Most of these reports concern cats, but some involve dogs, and there are also a few with parrots. In all cases, the person calling was someone to whom the animal was very attached. Typically, the animal showed obvious signs of excitement when the phone was ringing (Sheldrake, 1999a). This occurred only with calls from that particular person and not with calls from strangers. Most of the cats said to respond to telephone calls from particular people reacted when the telephone began to ring, but some did so even before the ringing began.

As far as I know, there have been no experimental investigations of animal telepathy in connection with telephone calls. However, experimental research on the human ability to detect who is about to ring shows that some people do, indeed, seem to detect the identity of the caller telepathically, but only when the caller is a familiar person (Sheldrake, 2003; Sheldrake and Smart, 2003).

Experiments with a Language-Using Parrot

The fact that some parrots can use language meaningfully has been established beyond reasonable doubt by Pepperberg (1999). Her pioneering work inspired Aimee Morgana to train her African Grey, N'Kisi (pronounced "in-KEE-see"), in the contextual use of language, starting when he was five months old. She used two teaching techniques known as "sentence frames" and "cognitive mapping." In sentence frames, words were taught by repeating them in various sentences such as, "Want some water? Look, I have some water." Cognitive mapping reinforced meanings that might not yet be fully understood. For example, if N'Kisi said "water," Aimee would show him a glass of water. By the time he was eight years old, he had a contextual vocabulary of more than 1,200 words. He apparently understood the meanings of words and used his language skills to make relevant comments. He ordinarily spoke in grammatical sentences, and by January 2002, Morgana had recorded more than 7,000 original sentences. Although Morgana's primary focus was on the meaningful use of language, she noticed that N'Kisi often seemed to say things that referred to her thoughts and intentions.

In 2000, Morgana began keeping a detailed log of seemingly telepathic incidents, and has continued to do so. By January 2002, she had recorded 630 such incidents. Here are two examples:

"I was thinking of calling Rob, and picked up the phone to do so, and N'Kisi said, 'Hi, Rob,' as I had the phone in my hand and was moving toward the Rolodex to look up his number."

"We were watching the end credits of a Jackie Chan movie, edited to a musical soundtrack. There was an image of [Chan] lying on his back on a girder way up on a tall skyscraper. It was scary due to the height, and N'Kisi said, 'Don't fall down.' Then the movie cut to a commercial with a musical soundtrack, and as an image of a car appeared, N'Kisi said, "There's my car." (N'Kisi's cage was at the other end of the room, behind the TV. He could not see the screen, and there were no sources of reflection.)

Of all the various incidents, perhaps the most remarkable occurred when N'Kisi interrupted Aimée's dreams. (He usually slept by her bed.) For example: "I was dreaming that I was working with the audiotape deck. N'Kisi, sleeping by my head, said out loud, 'You gotta push the button,' as I was doing exactly that in my dream. His speech woke me up." On another occasion, "I was on the couch napping, and I dreamed I was in the bathroom holding a brown-dropper medicine bottle. N'Kisi woke me up by saying, 'See, that's a bottle.'"

Clearly it was important to try to test this apparent telepathic communication in controlled experiments. We developed a procedure that could work fairly naturally in

N'Kisi's familiar environment. Morgana had noticed that N'Kisi seemed to respond to moments of discovery, as if he "surfed the leading edge" of her consciousness. Therefore, methods of testing for telepathy that used repetitive images, such as playing cards or Zener cards, were not likely to work. In order to preserve an element of surprise, we designed an experiment in which Morgana was filmed as she opened sealed envelopes one at a time, each containing a different photograph. Meanwhile, N'Kisi was alone in a different room, unable to see or hear Morgana, and was filmed continuously to record his behavior and speech. A third person, not otherwise involved in the test, selected these photographs on the basis of a list of keywords that were part of N'Kisi's vocabulary, sealed them in opaque envelopes and randomized them.

The tapes of N'Kisi's comments were transcribed blind by three independent people. The transcripts, which were in good agreement with each other, were then compared with the images Aimée was looking at in the synchronized videotapes. In many cases, N'Kisi's comments corresponded to the images Morgana was seeing. For example, when she was looking at a picture of flowers, he said, "That's a pic of flowers." When she was looking at picture of someone talking on a mobile phone, he said, "Whatcha doin' on the phone?" and made a series of noises like a phone being dialed.

We considered it a "hit" when N'Kisi said a predefined keyword that corresponded to an image representing that keyword. There were twenty-three hits out of seventy-one trials. The analysis of the results by randomized permutation analysis showed that hits were far more frequent than would have been expected by chance ($p<0.0002$) (Sheldrake & Morgana, 2003)

Telepathy from Animals to People

If telepathy can take place from people to animals, it would be surprising if it did not occur in the opposite direction. In my database, there are more than 1,500 accounts of seemingly telepathic or psychic influences of owners on their pets, and seventy-three cases where the influence seems to flow the other way. People seem much less sensitive to these influences than their animals—or pay little attention to them. But seventy-three is still a large number of cases, and, presumably, many people who have not written to me have had similar experiences.

Out of these seventy-three cases, ten concern deaths or accidents in distant places. Most of the sixty-three other cases involve calls for help, with the majority from cats. The most common situation in which this occurs is when the cat is outdoors and wants to be let in. Some cat owners say they not only know when a cat wants to come in, but also which of several cats is silently calling them (Sheldrake, 1999).

Cats that roam freely have a tendency to get lost, sometimes because neighbors unintentionally shut them into sheds or garages. Some cat owners have found that they can somehow know where the lost cat is. Some stories about the rescue of lost cats are quite dramatic and seem to show that the cat in some way draws the owner toward it (Sheldrake, 1999).

The picking up of distress signals from animals is not confined to dogs and cats, but also seems to happen with cows, sheep, and horses (Sheldrake, 1999).

Animal Communicators

In addition to these communications between domesticated animals and their owners, there is a long tradition of communication with animals by shamans in tribal societies. In fiction, stories such as those of Dr. Doolittle have a deep appeal to our imaginations.

Some people make a living as "animal communicators" and claim to pick up telepathically what people's pets are thinking and feeling. Some give counseling and advice for a fee, either in person or over the telephone. However, some of these so-called animal communications may well be a projection of the communicator's own thoughts, rather than genuine cases of telepathy. Animal communicators themselves are well aware of the problems. Penelope Smith of Point Reyes, California, who has trained hundreds of people in "inter-species telepathic communication" in her workshops, has seen people "mixing their communication abilities with their own agendas or emotional short-comings" (Smith, 1989).

Professional animal communicators are often ready to venture information about animals' feelings, and some may play a valuable role in counseling the animals' owners, but they are often reluctant to provide information that can be more immediately verified. For me, the most interesting of these apparent communications are those that can be tested empirically, and the most impressive ability is the finding of lost pets. Arthur Myers (1997) interviewed many communicators to see how successful they had been at finding lost animals telepathically. Most told him that they try to avoid such jobs, but he did find a few cases where communicators had apparently been able to locate lost animals.

There is much potential for further research on animal telepathy, and if domestic animals are telepathic with their human owners, it seems very likely that animals are telepathic with each other in the wild. Field observations by the naturalist William Long (1919) suggest that the coordination of flocks of birds and herds of animals may involve telepathy, as may communication between members of a pack of wolves, but here again there is much scope for further research.

Research into animal telepathy should enable human telepathy to be seen in an evolutionary light. The investigation of animal telepathy would also build bridges between parapsychology and biology, to their mutual benefit.

Telepathy from people to animals usually occurs only when there are close emotional bonds; this may well be an important factor in human telepathy, too. My own hypothesis is that these bonds depend on fields that link together members of a social group, called "social fields" (Sheldrake, 1999b, 2003). These are one type of a more general class of fields, called "morphic fields" (Sheldrake, 1988). These bonds continue to link members of the social group together, even when they are far apart, beyond the range of sensory communication, and can serve as a medium through which telepathic communications can pass.

See also

Communication and Language—*Interspecies Communications*
Communication and Language—*N'Kisi The Parrot: A Personal Essay*

Further Resources

Duvall, P., & Montredon, E. (1968). ESP experiments with mice. *Journal of Parapsychology, 32,* 153–66.

Long, W. J. (1919/2005) *How animals talk.* Rochester, VT: Bear & Co.

Marks, D. (2000). *The psychology of the psychic.* Amherst, NY: Prometheus Books.

Myers, A. (1997). *Communicating with animals.* Chicago: Contemporary Books.

Pepperberg, I. M. (1999). *The Alex studies: Cognitive and communicative abilities of grey parrots.* Cambridge, MA: Harvard University Press.

Schmidt, H. (1970). PK experiments with animals. *Journal of Parapsychology, 36,* 577–88.

Sheldrake, R. (1988). *The presence of the past.* New York: Times Books.
———. (1999a). *Dogs that know when their owners are coming home, and other unexplained powers of animals.* New York: Crown.
———. (1999b). Commentary on a paper by Wiseman, Smith, & Milton on the 'psychic pet' phenomenon. *Journal of the Society for Psychical Research, 63,* 306–11.
———. (2000). The 'psychic pet' phenomenon. *Journal of the Society for Psychical Research, 64,* 126–28.
———. (2003). *The sense of being stared at, and other aspects of the extended mind.* New York: Crown.
Sheldrake, R., Lawlor, C., & Turney, J. (1998). Perceptive pets: A survey in London. *Biology Forum, 91,* 57–74.
Sheldrake, R., & Morgana, A. (2003). Testing a language-using animal for telepathy. *Journal of Scientific Exploration, 17,* 601–15.
Sheldrake, R., & Smart, P. (1998). A dog that seems to know when his owner is returning: Preliminary investigations. *Journal of the Society for Psychical Research, 62,* 220–32.
———. (2000a). A dog that seems to know when his owner is coming home: Videotaped experiments and observations. *Journal of Scientific Exploration, 14,* 233–55.
———. (2000b). Testing a return-anticipating dog, Kane. *Anthrozoös, 13,* 203–12.
———. (2003). Videotaped experiments on telephone telepathy. *Journal of Parapsychology, 67,* 187–206.
Smith, P. (1989). *Animal talk: Interspecies telepathic communication.* Point Reyes, CA: Pegasus.
Wiseman, R., Smith, M., & Milton, J. (1998). Can animals detect when their owners are returning home? An experimental test of the 'psychic pet' phenomenon. *British Journal of Psychology, 89,* 453–62.
———. (2000). The 'psychic pet' phenomenon: A reply to Rupert Sheldrake. *Journal of the Society for Psychical Research, 64,* 46–49.

Rupert Sheldrake

Conservation and Environment

Advertising and the Endangered Saimaa Ringed Seal in Finland

Nature conservationists and the business world are eager to exploit the good image of the protected Saimaa ringed seal (*Phoca hispida saimensis*). For example, business enterprises and environmental organizations often use images of the seal in logos and commercials.

The Saimaa ringed seal is the only endemic mammal in Finland. At the moment, nearly 280 seals live in the Saimaa lake-complex (which has an area of approximately 4,500 km^2) in the southeast part of the country. This seal is also one of the most endangered ringed seals in the world. It is a close relative to the ringed seals living in the Baltic Sea. After the last Ice Age, about 10,000 years ago, some of the seals living in the Baltic Sea were landlocked when the Saimaa lake-complex was formed. This was the initial stage in the development of the Saimaa ringed seal as a subspecies. At present, it is one of the very few ringed seals that have physiologically adapted to living in freshwater.

For thousands of years, the Saimaa ringed seal has been an important animal for local people. Some 4,000 years ago, human settlement began to extend to new areas that emerged as a result of land-uplift. At that time, the ringed seal was an important source of meat, oil, and leather. During recent centuries, the human population around the Saimaa lake-complex considered the local ringed seal as a competitor for fish resources. Bounties were paid for killing seals until the second half of the twentieth century.

Hunting this seal has been banned since 1955. At the moment, fishery is the greatest threat to the Saimaa ringed seal, and around fifteen pups and one to three adults are annually drowned in gill nets. The Habitat Directive of the European Union categorized the Saimaa ringed seal as needing strict protection (Council Directive 92/43/EEC). This seal is also listed in the Federal Register of the U.S. Department of Commerce as an endangered and threatened animal and in the Red Data Book of IUCN (the World Conservation Union) as endangered. Almost all the breeding areas of this seal have been included in the European ecological network—Natura 2000. Thus, the change of attitude from regarding this seal as a prey animal to regarding it as a protected one happened in a few decades. This change has been widely exploited in various parts of Finnish society.

The Saimaa ringed seal provides an example of how general support for an endangered and cute-looking animal is exploited in environmental policy and by the business world. With the emergence of environmental awareness during the last decades of the twentieth century, local problems concerning pollution and eutrophication were at the center of discussions concerning the condition and future of our environment. Pollution of the lakes was one of the most evident problems.

At that time, nature conservationists evaluated the success of environmental policies and the condition of lakes largely through the viability and size of the seal stock. The number of pups born annually and, for example, the methylmercury and nickel content of the tissues of the Saimaa ringed seal were important issues in this context. This seal has been a central motif of the logo of the Finnish Association for Nature Conservation for decades. The logo is used on the official papers of the association and on products sold by it, such as t-shirts. Critics have pointed out that the strong link between the ringed seal and nature conservation has sometimes led to an underestimation of more extensive problems in environmental policy.

The business world has named supermarkets, coffee bars, restaurants, and wind electricity after the Finnish name of this seal or its pup. Similarly, conservation biologists have named their research boats after the Finnish name of this ringed seal. At present, the seal is considered to be important also for the development of regional tourism, and it is often mentioned in tourism commercials. Environmentalists would like to restrict the use of the name of the Saimaa ringed seal in commercials and in society in general to cases where the emphasis is on environment sustainability. However, the problem is that this is to a large extent a question of attitudes, and thus, everyone defines sustainability according to his or her own criteria.

See also

Media and Film—*Advertising and Animals*

Further Resources

Auvinen, H., Jurvelius, J., Koskela, J., & Sipilä, T. (2005). Comparative use of vendace by humans and Saimaa ringed seal in Lake Pihlajavesi, Finland. *Biological Conservation, 125,* 381–89.

Hyvärinen, H., Sipilä, T., Kunnasranta, M., & Koskela, J. (1998). Mercury pollution and the Saimaa ringed seal (*Phoca hispida saimensis*). *Marine Pollution Bulletin, 36,* 76–81.

IUCN. (1996). *1996 IUCN red list of threatened animals.* Gland, Switzerland, and Cambridge, UK: World Conservation Union, IUCN.

Kokko, H., Lindström, J., Ranta, E., Sipilä, T., & Koskela, J. (1998). Estimating the demographic effective population size of the Saimaa ringed seal (*Phoca hispida saimensis* Nordq.). *Animal Conservation, 1,* 47–54.

Kostamo, A., Medvedev, N., Pellinen, J., Hyvärinen, H., & Kukkonen, J. V. (2000). Analysis of organochlorine compounds and extractable organic halogen in three subspecies of ringed seal from Northeast Europe. *Environmental Toxicology and Chemistry, 19,* 848–54.

Sipilä, T., & Hyvärinen, H. (1997). Status and biology of Saimaa (*Phoca hispida saimensis*) and Ladoga (*Phoca hispida ladogensis*) ringed seals. In C. Lydersen & M.-P. Heide-Jörgensen (Eds.), *The seals of northern hemisphere* (pp. 83–99). North Atlantic Marine Mammal Commission, Scientific Publication 1.

Tonder, M., & and Jurvelius, J. (2004). Attitudes towards fishery and conservation of the Saimaa ringed seal in Lake Pihlajavesi, Finland. *Environmental Conservation, 31*(2), 122–29.

Juha Jurvelius

Conservation and Environment
African Wild Dogs and Humans

Within the past 200 years, since the arrival of white hunters and settlers in Africa, *Lycaon pictus*, the African wild dog, also know as the Cape hunting dog or painted hunting dog (Lycaon) endemic to that continent, has experienced remarkable changes in some people's attitude and behavior toward the species.

White hunters considered Lycaon a cruel, ruthless, profligate killer that should be exterminated, an attitude shared by some early conservationists, even in national parks, up to the mid-1970s. This view changed soon after when field research revealed more about the behavior and importance of Lycaon in different ecosystems and when it was recognized that Lycaon had become a highly endangered species, with possibly fewer than 6,000 free living individuals that needed both protection and possibly active conservation. Now considered an important and charismatic species, the Lycaon attracts the attention of writers, filmmakers, behavioral biologists, veterinarians, epidemiologists, geneticists, and mathematical model makers. The species also attracts an increasing number of tourists who have added Lycaon to their "must see" list of safari species.

Although Lycaon has coexisted with evolving humans for at least one million years in Africa, there is no evidence from Stone Age or later cave art of interaction between Lycaon and early hunter-gatherers. However, humans have probably opportunistically exploited Lycaon by stealing their kills. For Lycaon, such prey loss was of little consequence, but for humans, this provided an important protein-rich, high-energy food source long before development of stone tools and bows and arrows enabled humans to effectively hunt fast-moving and wary prey themselves. So for over a million years, evolving Lycaon and humans coexisted in Africa, with the latter benefiting by scavenging wild dog kills as a source of "bush meat," a tradition that continues today.

Mainly because of its unique social behavior and consequent inability to accept a surrogate non-Lycaon as leader of its pack, unlike the gray wolf, with which it shared a common ancestor some three million years ago probably in North America, the Lycaon has not been domesticated. There is still a common misconception that Lycaon is just a feral domestic dog, but on the basis of anatomical and genetic criteria, it is a different genus.

Three major events changed relationships between people and Lycaon over the last 10,000 years. The first was the introduction of domestic animals, the second was the arrival in the nineteenth century of white hunters and farmers, and the third was the largely twentieth-century development of scientific wildlife research, the conservation movement, and mass tourism.

The first major change in some indigenous African people's relationship with Lycaon occurred within the last 10,000 years, when domesticated herbivores bred from wild ancestors and domesticated wolves (i.e., domestic dogs) from Eurasia were first brought to North Africa. The arrival of such "foreign" domestic cattle, sheep, goats, and dogs in Africa led to the emergence of new semi-nomadic pastoralist and settled farming communities in North Africa dependent on domestic stock.

These rapid changes in human behavior led to changes in the relationship between people and wildlife, particularly with the fresh meat–dependent hypercarnivore Lycaon. Unlike indigenous wildlife, domestic herbivorous stock are ill-adapted to being hunted and, when kept in large concentrations, are vulnerable to predators such as Lycaon. Pastoralists and farmers, unlike hunter-gatherer communities, were now on the losing side against Lycaon, a new experience for most indigenous peoples.

Pastoralists, however, developed a very effective system of husbandry. They constructed permanent or temporary protective enclosures (bomas) in which a number of families lived and to which, before nightfall, herdsmen and their domestic guard dogs brought the stock from their daylight grazing areas.

These new human societies flourished, and their cultures spread from North to East Africa despite the indigenous carnivores and some inevitable stock losses to them. The present-day Maasai of East Africa usually have few problems with Lycaon predation, but others with less-well-developed husbandry techniques sometimes suffer the consequences. Maasai consider Lycaon beneficial because they help to keep wildebeest herds moving, thus reducing the possibility of a virus that causes bovine malignant catarrhal fever contaminating the grasslands and so infecting their cattle. Thus pastoralists, like the hunter-gatherers, learned to coexist with Lycaon and other predators that coevolved with various antelope species on the open African savannahs.

The second change in relationships between people and Lycaon in Africa took place in the nineteenth century with the arrival of white hunters and farmers. White hunters considered Lycaon as vermin and shot whole packs on sight, not because Lycaon posed a danger to the people—there are no confirmed reports of human deaths caused by Lycaon—but because Lycaon disturbed the game and therefore their sport and because they disapproved of Lycaon's method of killing prey by disemboweling, a technique that enables Lycaon to kill large prey.

The expansion of the white settlers' arable and stock farming considerably reduced the areas of land available to wildlife herbivores and their predators, leading to rapid decline in the numbers of both.

In East Africa, Maasai pastoralists lost grazing lands to white farmers and were prevented from grazing their stock in game reserves and the newly established national parks from which they were evicted. In southern Africa, destruction of migrant herds of herbivores in the nineteenth and early twentieth centuries due to habitat loss led to rapid declines in Lycaon populations in many countries. In South Africa, the Lycaon population was reduced to a single self-sustaining population that still persists in Kruger National Park.

The third change in relationship between people and Lycaon came with the beginning of the study of wild mammal behavior and demography in their natural habitat and the parallel rise of the wildlife conservation movement.

In the 1960s, Western scientists began studying the behavior and demography of free-living Lycaon and other mammals in Africa. The attitude of early conservationists toward predators was, however, mainly negative, following the precedent set by the white hunters. Some of those hunters even became park wardens who encouraged park rangers to shoot Lycaon as vermin, as was the case in Serengeti National Park up to 1973. Lycaon packs were not difficult to shoot because they were easily approached in vehicles; pups and yearlings were naturally curious; and once one dog was shot, the others would not immediately run away.

By the mid-1970s, the attitudes of conservationists and wardens in national parks toward Lycaon, if not the attitudes of farmers adjacent to these protected areas, had changed, particularly when people realized that Lycaon had become a highly endangered species and as such had commercial potential both as a tourist attraction and as a generator of funding for continuing scientific research and wildlife conservation.

Licenses to hunt Lycaon are no longer issued, and the species is now officially legally protected over much of its greatly reduced range, although such protection is rarely effective or enforced. Lycaon are probably extinct in twelve African countries and are "rare" or "vagrant" in another twelve African countries. Outside protected areas, legal protection probably inhibits local people from taking part in conservation activities rather than encouraging them to do so.

For the field scientists, the development of radio telemetry and 4 × 4 vehicles and the availability of small aircraft made it possible to aerially track and monitor the wide-ranging Lycaon packs in their often huge home ranges (up to 1,500 km^2) in East Africa.

Though the Lycaon was once an obscure, little-known, and rarely seen species, interest in the study of Lycaon became a growth industry and the subject of many academic studies and subsequent scientific papers, popular books, articles in wildlife magazines, and wildlife films. Conferences on Lycaon behavior and conservation began, and the International Union for the Conservation of Nature Species Survival Commission (IUCN SSC) set up a Lycaon group that produces specific conservation action plans for this and other species.

In the mid-1980s, the ability of diseases such as rabies and canine distemper to wipe out a whole pack or most of a pack of Lycaon, both in the wild and in captivity, led to the experimental rabies vaccination of Lycaon in captivity. To counter the threat of rabies, claimed to originate from domestic dogs living around the protected area, most of the individuals in the free-living radio-collared study packs in two sectors of the Serengeti ecosystem (the Serengeti in Tanzania and the Maasai Mara area of Kenya) were vaccinated against rabies between 1989 and 1990. All the study packs died by mid-1991, with rabies the only confirmed cause of death in the three packs from which tissue samples were obtained. Similar experimental rabies vaccinations of Lycaon in southern African have also unfortunately proved ineffective in preventing rabies mortality.

In the early 1990s, it was noted that the sudden losses of whole study packs of wild dogs in protected areas in the Serengeti and Maasai Mara region post-1985 were coincident with the introduction and increased use of routine invasive research techniques (collectively termed "handling"), including immobilization, radio collaring, blood sampling, and rabies vaccination. A statistically significant reduction in longevity of both handled individuals and handled packs, as compared with those unhandled, suggested that a causal, possibly stress-related relationship existed between handling and reduced longevity in the Serengeti ecosystem.

The suggested adverse effects of handling, including vaccination of whole packs against a perceived disease threat from canine rabies possibly transmitted by domestic dogs, led to a vigorous debate that began in the early 1990s but unfortunately remains unresolved.

Despite this rapid change in fortune for Lycaon in protected areas, some farmers with land adjacent to such areas still snare, poison, or shoot Lycaon, so packs that include either farm or ranch land in their extensive home ranges and emigrant single-sex groups that move out of reserves to find partners to form new packs are at risk. Despite persecution, however, Lycaon packs still persist and den in non-protected areas such as farms and ranch lands from which packs and groups have been captured and translocated to protected areas.

Humans' attitudes toward predators have changed over time with continuing research. The much-maligned spotted hyena, once commonly characterized as a skulking, cowardly scavenger, is now known to be a highly efficient hunter from which, together with other smaller predators including Lycaon, the so-called noble hunter, the lion, frequently scavenges. Many problems with people's perceptions of wildlife persist, particularly when those perceptions are based on human values that lead to value judgments regarding the "morality" of the behavior and hunting techniques of predators.

So successful has breeding of Lycaon in captivity and private game reserves been that the lack of sufficiently large reserves for this highly mobile species, with yearlings of both sexes emigrating each year from successfully breeding packs, that there is now a problem of where to put the surplus dogs. Euthanasia of "surplus" Lycaon has recently been considered an option in South Africa.

With climate changes, rising sea levels, a rapidly increasing human population, and demand for even more of the existing land mass to be made available for housing, agricultural, and industrial expansion, we are running out of space for wildlife even in the vast continent of Africa. Species such as Lycaon that require large tracts of land, a plentiful supply of wild prey, and denning areas undisturbed by man may be doomed eventually to a "safari park" existence despite humans' recently acquired desire and attempts in various ways to conserve them in their natural habitat.

Further Resources

Adams, J. S., & McShane, T. O. (1992). *The myth of wild Africa*. New York and London: Norton.

Bonner, R. (1993). *At the hand of man*. New York: Simon & Schuster.

Burrows, R. (1995). Demographic changes and social consequences in wild dogs. In A. R. E. Sinclair & P. Arcese (Eds.), *Dynamics, management, and conservation of an ecosystem* (pp. 400–20). Chicago and London: University of Chicago Press.

Burrows, R., Hofer, H., & East, M. L. (1995). Population dynamics, intervention and survival in African wild dogs *(Lycaon pictus)*. Proceedings of the Royal Society of London B, *262*, 235–45.

McNutt, J., & Boggs, L. (1966). *Running wild*. Washington, DC: Smithsonian Institution Press.

Schaller, G. B. (1972). *The Serengeti lion*. Chicago and London: University of Chicago Press.

Scott, J. (1991). *Painted wolves*. London: Hamish Hamilton.

Van Lawick, H. (1973). *Solo: The story of an African wild dog*. London: Collins.

Van Lawick-Goodall, H., & Van Lawick-Goodall, J. (1970). *Innocent killers*. London: Collins.

Woodroffe, R., Ginsberg, J. R., & Macdonald, D. W. (1997). *The African wild dog—Status survey and conservation action plan*. Gland, Switzerland: IUCN.

Woodroffe, R., McNutt, J. W., & Mills, M. G. L. (2004). African wild dog. In C. Sillero-Zubiri, M. Hoffman, & D. W. Macdonald (Eds.), *Canids: Foxes, wolves, jackals and dogs* (pp. 174–83). Gland, Switzerland: IUCN.

Roger Burrows

Conservation and Environment
Animal Conservationists

Animal conservationists and preservationists are those who work to protect animals—primarily those that are threatened or endangered—as well as the habitats and biodiversity upon which they (and we) depend. A distinction is typically made between conservation and preservation. Conservationists attempt to manage and sustain the animals and habitats humans use for the future; preservationists work to maintain animals and habitats in a natural state. This article speaks broadly of both groups.

Animal conservationists face a daunting struggle. In 2002, the United Nations Environment Programme (UNEP) forecasted that about one in eight bird species and one in four mammal species (as well as over 5,000 plant species) will face extinction in the next thirty years, largely from habitat loss and the introduction of invasive species. A very conservative lower limit is that a species goes extinct every ten minutes (Wilson, 1993). Of course, the loss of so much biodiversity and the fragmentation of habitats could trigger additional environmental deterioration and extinctions (Tilman et al., 2002; 2005). Brown bears, red wolves, silver sharks, black rhinoceroses, blue whales, California condors, Hawaiian monk seals, Rhode Island skinks, Alabama sturgeon, Colorado pikeminnows, Mississippi sandhill cranes, Texas ocelots, and Florida cougars are but a few of those endangered animals that animal conservationists are fighting to save.

Margaret "Mardy" Murie, known by many as the "Grandmother of the Conservation Movement," and her husband Olaus were primary driving forces in early conservation efforts in the United States. As the leaders of the Wilderness Society, they impelled the U.S. government to create the Arctic National Wildlife Refuge in 1960. Through their lobbying efforts, the refuge was expanded in 1980. The refuge boasts the greatest diversity of plants and animals preserved in any arctic region. Mardy also conceived of the Wildness Act of 1964, which guaranteed the protection of over nine million acres and set up the National Wilderness Preservation System. The act also established the legal definition of wilderness for the United States: "an area where the earth and its community of life are untrammeled by man, where man himself is a visitor who does not remain."

Animal conservation came into its own near the middle of the twentieth century. Many of the movement's pioneers—such as Sir Dietrich Brandis, Berthold Ribbentrop, Dr. Ferdinand Vandeveer Hayden, Robert Marshall, Gifford Pinchot, John Muir, Mina Hall, and Harriet Hemenway—straddled both environmental and animal conservation; the distinction is often moot even today. For example, Rosalie Edge used her experience fighting for women's rights to begin to protect animals in the 1930s. Working largely within ornithology, Edge increased people's awareness of the importance of predators. She also spearheaded the creation of a wildlife refuge in Pennsylvania, the Hawk Mountain Sanctuary.

Marine biologist Rachel Carson was also among the early animal conservationists who also fought for the environment. Carson published articles and several books about nature and conservation, most notably *Silent Spring* in 1962, which addressed DDT use, but also raised awareness about the interconnectedness of life. Chemical companies and others aggressively challenged *Silent Spring* and Carson herself. While suffering from the breast cancer that would eventually take her life, Carson prevailed against those who disputed her claims; DDT was banned in 1972, and concern over our impact on animals and the environment gained considerable prestige.

The widest-ranging law protecting animals in the United States, the Endangered Species Act of 1973, reflects this union between animal and environmental conservation. The act not only prevents "taking" (i.e., harming or removing) endangered animals, but also protects the land they need to survive.

Several animal conservationists began as scientists; often their research opened their eyes to what was happening to animals. British archaeologist Dr. Louis Leakey, for example, mentored four pioneering animal conservationists. Each of Leakey's protégés took a different approach to animal conservation. The first was Dame Dr. Jane Goodall. First hired to be Leakey's secretary, Leakey saw the potential Goodall held and helped her begin her groundbreaking research on chimpanzees in Tanzania. From the beginning, Goodall laid the groundwork for protecting the chimpanzees by working closely with local communities. In 1977 she also founded the Jane Goodall Institute for Wildlife Research, Education, and Conservation (JGI). JGI predicts that chimpanzees may be extinct in fifteen years.

Leakey next asked Dian Fossey to study the endangered mountain gorillas in the forests of the Virunga volcanoes in Rwanda. Dr. Fossey had a talent for winning the trust of the gorillas she studied; however, she did not establish close ties with the local communities. Instead, she advocated "active conservation"—such as patrols to prevent poaching and sanctions to prevent tourism and the gorilla-infecting diseases and environmental damage tourists could bring. Fossey's activism was brought to a tragic end when she was murdered in her Rwanda cabin in 1985. According to the United Nations' Great Apes Survival Project, all species of gorillas face extinction by the middle of the twenty-first century, victims of habitat loss and the wars that have claimed millions of human lives as well. Among the last words Fossey wrote in her journal were "When you realize the value of all life, you dwell less on what is past and concentrate more on the preservation of the future."

Dr. Biruté Galdikas, Leakey's third protégé, studies the endangered orangutans in the tropical forests of the Republic of Indonesia. Because orangutans' habitat is difficult and remote, Galdikas was also not able to establish strong ties with neighboring communities. However, Galdikas generally takes a broad view to save orangutans and their diminishing habitat, including in her efforts the preservation of the people and cultures that are intertwined with the orangutan habitat. She also helps rehabilitate young orangutans to the wild. She cofounded Orangutan Foundation International in 1986. Unless there is dramatic change, orangutans may be extinct in the wild within this first decade of the twenty-first (van Schaik, Monk, & Robertson, 2001).

Louis Leakey's son, Dr. Richard Leakey, became the head of the Kenyan Wildlife Service (KWS). Richard Leakey took bold steps to help the Kenyans protect the animals—especially the endangered elephants—in their care. He armed park guards and gave them orders to shoot poachers on sight. In 1989 Leakey and Kenyan president Daniel Moi burned a pile of over 2,000 confiscated elephant tusks to alert the world to the horrors of the ivory trade and thus devalue the ivory. Their daring and costly plan worked: the price of ivory dropped 97 percent. Across Africa, Moi and Leakey's gesture galvanized efforts to protect elephants from poaching. Leakey's bold efforts and incorruptible character won him more than great strides against poaching; they also won him enemies. Four years after being appointed to head the KWS, Leakey lost both of his legs when his single-engine plane crashed; sabotage was suspected but never proven. Sometimes from within and sometimes from without the Kenyan government, Leakey continues to fight to protect African animals and lands.

Dr. Cynthia Moss is another individual who has played a prominent role in elephant research and conservation. Moss was employed as a journalist when she went on a long trip to Africa. There, she fell in love with elephants, quit her journalism job, and eventually founded the Amboseli Elephant Research Project in Kenya. Her research brought to light the amazingly complex emotional and family lives of these intelligent, sensitive giants.

Jacques-Yves Cousteau, coinventor of the aqua-lung, was an effective defender of the "blue continent." The award-winning films Cousteau and his colleagues made while traveling in the ex–World War II minesweeper *Calypso* did much to raise awareness and interest in marine animals. Cousteau used his prestige to protect the animals he loved, such as by organizing a successful campaign to prevent the dumping of radioactive waste in the ocean.

Other figures among the growing number of important animal conservationists include U.S. Supreme Court Justice William Douglas; journalist Joseph Glascott; World Wildlife Fund U.S. president Kathryn Fuller; biologist Dr. Marc Bekoff; marine biologists Drs. Sidney Holt, Roger Payne, and Idelisa de Calventi Bonnelly; forest preservationist Seub Nakhasathien; zoologist Dr. Hideo Obara; sea turtle activist Lily Venizelos; and biologist Dr. Edward O. Wilson.

Animal conservationists help animals in more and more ways. In addition to research and information campaigns, animal conservationists preserve "animal banks"—such as the Madras Crocodile Bank Trust—to preserve animals in captivity while they fight ignorance and overdevelopment to reintroduce them to the wild. Conservationists work to create habitats safe for both humans and other animals, for example, advocating ecopassages that allow animals to cross highways safely. Some, like the Center for Reproduction of Endangered Species in San Diego, conserve animal DNA. Others rehabilitate wounded animals or try to help wild animals behaviorally scarred when someone tried to make them into pets. Animal conservationists lobby to protect and preserve the land upon which countless animals depend.

Early animal conservation and preservation was largely driven by a handful of exceptional luminaries. The field has evolved into a complex array of research and programs that rely on the efforts of a broad, dedicated base. Most animal conservationists make a difference through small, local actions that never see the limelight. And, of course, everyone's actions affect wildlife directly or indirectly.

Kenyan president Daniel Arap Moi ignites $3 million of ivory and rhino horns confiscated from poachers by Kenya National Park game wardens, 1989. ©Tom Stoddart/Getty Images.

Further Reading

Bekoff, M. (Ed.). (2000). *The smile of a dolphin: Remarkable accounts of animal emotions*. New York: Discovery Books.

Fouts, R. (1998). *Next of kin: My conversations with chimpanzees*. New York: HarperCollins.

Leakey, R. (2001). *Wildlife wars: My battle to save Kenya's elephants*. London: Macmillan Books.

Leopold, A. (1949). The land ethic. In *A Sand County Almanac*. Oxford: Oxford University Press.

Tilman, D., May, R. M., Lehman, C. L., & Nowak, M. A. (2002). Habitat destruction and the extinction debt. *Nature*, *371*, 65–66.

Tilman, D., Polasky, S., & Lehman, C. (2005). Diversity, productivity and temporal stability in the economies of humans and nature. *Journal of Environmental Economics and Management*, *49*, 405–26.

United Nations Environment Programme, *Global 500 Roll of Honour for Environmental Achievement*.

Van Schaik, C. P., Monk, K. A., & Robertson, J. M. Y. (2001). Dramatic decline in orangutan numbers in the Leuser Ecosystem, Northern Sumatra. *Oryx*, *35*(1), 14.

Wilson, E. O. (1993). Biophilia and the conservation ethic. In S. R. Kellert & and E. O. Wilson (Eds.), *The biophilia hypothesis*. Washington, DC: Island Press.

William Ellery Samuels and Susan Frey

Conservation and Environment
Animal Reintroduction and Ecological Restoration

In the twentieth century, when animal reintroduction became a component of ecological restoration, it was commonly assumed that the goal of environmental management is to maximize resource output by protecting selected species. This type of resource management, then, generally required human control of ecosystems and animals. More recently, there has been an evolution in the methodology and philosophy of restoration, from quantitatively driven restorations of species and ecosystems toward more holistic restoration schemes that allow more autonomy for nonhuman animals and the character of their environments. What began as the maintenance of ideally small museum-piece habitats and populations of charismatic large carnivores has evolved into a broadly interdisciplinary field that now faces the daunting moral challenge of weighing animal interests against human development goals.

One of the first large-scale, systematic efforts at restoring a particular ecosystem to a previous state was the University of Wisconsin Arboretum, founded through the vision and effort of Aldo Leopold. Leopold's now famous "conversion" moment, shooting a she-wolf and arriving by her side just in time to see the "green fire" die in her eyes, was the impetus for his influential "land ethic," in which the unit of moral considerability was the ecosystem. Leopold's famous arboretum and other early restorations, however, sought to revive long-defunct *plant*, not animal, regimes. This brought ecological restoration into the public eye, but did little to quell public ambiguity toward wild animals.

The first policy providing extended protections to animals was the landmark Endangered Species Act (ESA) in 1973. Besides placing prohibitions on the killing of animals listed as threatened or endangered, the legislation allowed for the proactive reintroduction of species such as wolves. Because the ESA focuses on quantitative measurements of success and not on systemic problem solving, however, reintroduction of many species has proved problematic. For example, the red wolf reintroduction program snagged on the genetic purity of the captive breeding pairs, and debate ensued over whether the ESA could support reintroduction of an animal derived from the introgression of two species (*Canis latrans*, coyote; and *Canis lupus*, gray wolf), which were forced by anthropogenic habitat loss to interbreed. The Florida panther introduction met with similar resistance when Texan pumas were imported to broaden the breeders' gene pool. The gray wolves introduced in the Northern Rockies have also proved a challenge, arousing the ire of cattle ranchers and others intent on keeping the wolves confined to human-imposed boundaries. Especially in the case of these predator animals, reintroductions must contend with public belief in the dangers of the "howling wilderness" and its animal occupants, a sentiment that has roots in the early colonial and frontier periods. In some cases, local populations resist reintroduction and introduction of predators, but scientists insist that they are necessary for ecological integrity. A decade after the first broad policy protections, as these crises in animal management grew, some scientists began to talk about natural values in addition to the typical instrumental or utilitarian language used by most resource managers.

In the 1980s, two distinct fields of study emerged, both with political teeth aimed at more effective restorations. Michael Soulé and others founded the Society of Conservation Biology in 1986, and the Society for Ecological Restoration followed closely in 1988. A fault line between these two fields persists on the relative importance of floral versus faunal regimes to the re-creation of desired ecological conditions. Generally speaking, restoration ecologists typically focus on floral regimes and soil composition, with the assumption that these foundations are necessary for a sustainable restoration. Conservation biologists instead argue that the disruption of plant succession is largely a result of the removal of predator species, which often maintain appropriate population distributions throughout other trophic levels.

Because piecemeal policies have proven ineffective, some conservation biologists have proposed centralized, top-down schemes that seek to restore large core wilderness areas with (when possible) the historically appropriate megafauna. This "re-wilding" of regions, or "continental conservation," focuses on restoring entire ecosystems, rather than a single charismatic plant or animal (or even plant regimes). Thus far, the approach seems largely irreconcilable with current federal and state policy structures. No agency is authorized to engineer such broadly conceived restorations. In addition, creating large core wilderness areas off-limits to human interference may do little to endear these troubled animals to the public.

Animal reintroductions are largely governmental undertakings because of the significant resources required to maintain and monitor animal populations. Even many restoration ecologists, typically concerned primarily with plant regimes, are advocating more holistic restorations that, at least in theory, aim at cultivating a new vision of relationship with nonhuman nature in general and animals in particular.

Restoration ecologist William Jordan III has argued that local rituals of restoration can evoke affective reactions that share common emotive referents with religious and spiritual belief and practice. He hopes such emotive and religious perceptions allow humans to view restoration holistically and to grow affinities for some local animal species. Similarly, the continental conservation schemes offered by some conservation biologists also require new perspectives on human and animal relationships, which trade some human control for increased animal autonomy.

New adaptive management methodologies are bringing a systemic perspective to all resource management fields, including both conservation biology and restoration ecology. Adaptive management explicitly rejects reliance on quantitative measure of success and instead attempts to manage for multiple futures. Such actively adaptive management schemes now include traditional ecological knowledge (TEK), which involves epistemologies specific to particular places and cultures. In many cases, these indigenous worldviews challenge Western suppositions of human exceptionalism. Some scholars suggest that the emerging drift toward holistic, adaptive restoration and reintroduction projects in the natural sciences as well as the growing acknowledgment of local knowledges provide parallel trends that are strong correctives for conventional restoration ecology and animal reintroduction. That animals have value outside of their potential as resources is a significant conceptual shift among "resource managers." The new visions offered by both camps are often couched in terms that suggest that animals have value because of their contribution to their ecosystems or that they have value as individual sentient creatures. In these instances, the value of animals is not equated with their use value to humans. It remains to be seen if such scientifically informed views about the ecological, and even moral, value of animals will have a critical impact on public opinion and management policy.

Further Resources

Foreman, D. (2004). *Rewilding North America: A vision for conservation in the twenty-first century*. Washington, DC: Island Press.

Higgs, E. (2003). *Nature by design: People, natural process, and ecological restoration*. Cambridge: MIT Press.

Jordan, W., III. (2003). *The sunflower forest: Ecological restoration and the new communion with nature*. Berkeley: University of California Press.

Maehr, D., Noss, R. F., & Larkin, J. L. (Eds.). (2001). *Large mammal restoration: Ecological and sociological challenges in the twenty-first century*. Washington, DC: Island Press.

Soulé, M. (Ed.). (1986). *Conservation biology: The science of scarcity and diversity*. Sunderland, MA: Sinauer.

Soulé, M., & Terborgh, J. (1995). *Continental conservation: Scientific foundations of regional reserve networks*. Washington, DC: Island Press.

Lucas F. Johnston

Conservation and Environment
Arctic Char, Zander Fish, and Global Warming

Arctic char and zander are fish that live under extreme conditions in Finnish lakes in Northern Europe near the Arctic Circle. Both fish species are thus vulnerable to changes in climatic temperature. The Arctic char is a colored, streamlined sixty- to seventy-centimeter-long predator fish living in the deepest and coolest (under 15°C) waters of lakes. The zander is a predator living in the warmest water layers. Its optimum temperature is 19–24°C.

Because the lakes of Finland are shallow (mean depth 7 m, maximum depth of 100 m), changes in the climatic temperature have a rapid influence on their water temperature and oxygen situation. This changes the living conditions and thus geographic distribution of fish, including the char and zander. At the moment, the forecast for Finland is that in winter the air temperature may rise by as much as 10°C during the following eighty years. This could have a drastic influence on the snow and ice conditions of the lakes. To this day, every Finnish lake has an ice-coat during the winter.

Finland's climate is generally affected by the Gulf Stream. Therefore, we must keep in mind also another scenario; although global climatic warming increases precipitation, the direction and speed of the Gulf Stream may change, and this may have a cooling effect and thus counteract global warming.

After the last Ice Age about 10,000 years ago, char was one of the first fish species that colonized the cool freshwaters created by the receding ice sheet in northern Europe. At that time, zander lived in warm waters far south. At the moment, the Arctic char has the northernmost range of any freshwater fish with a circumpolar distribution. Southern stocks of the species are found in deep lakes in Scandinavia, Russia, and the British Isles as well as in alpine lakes in central Europe. Isolated populations occur also on the eastern coast of North America. In Finland, char live in the southern borders of its continuous circumpolar distribution, and zander occur in the northern borders of their distribution. The present northern distribution of zander is mainly determined by temperature and corresponds to the July 15°C isotherm.

In order to ensure fishable stocks, introductions and stockings of zander were necessary in many lakes in the 1980s and 1990s. During recent years, it has become obvious that zander are spawning naturally in many of these lakes and that the fish species is spreading northward. The withdrawing of char from the warming areas has not yet been very obvious. However, the apparent low rates of reproduction for char in Finnish lakes may be an indirect indication of changes in the species' living conditions.

At the moment, pollution and eutrophication are no longer major problems in Finnish inland waters; thus, climatic warming is one of the most evident factors changing the living conditions of Arctic char and zander. When the temperature of lake waters increases, the deepest, coolest waters form the last refuges for Artic char, and zander penetrates to the north with the increasing water temperature of lakes. The distribution of zander may cover the whole of Finland by 2060. We can extrapolate that climatic warming changes the distribution of freshwater fish species in global scale, and it may locally lead to extinction of some species—including the Arctic char in Finland.

The zander, a winner in global warming. Courtesy of Juha Jurvelius.

Further Resources

Crawshaw, L. I. (1977). Physiological and behavioral reactions of fishes to temperature change. *Journal of Fisheries Research Board of Canada, 34*(5), 730–34.

Donner, J. (1995). *The Quaternary history of Scandinavia.* New York: Cambridge University Press.

Keskinen, T., & Marjomäki, T. J. (2004). Diet and prey size spectrum of pike-perch in lakes in central Finland. *Journal of Fish Biology, 65*(4), 1147–53.

Lappalainen, J., & Lehtonen, H. (1997). Temperature habitats for freshwater fishes in a warming climate. *Boreal Environmental Research, 2*(1), 69–84.

Lehtonen, H. (1996). Potential effects of global warming on northern European freshwater fish and fisheries. *Fisheries Management and Ecology, 3*(1), 59–71.

Lyytikäinen, T., Koskela, J., & Rissanen, I. (1997). Thermal resistance and upper lethal temperatures of underyearling Lake Inari Arctic char. *Journal of Fish Biology, 51*(3), 515–25.

Magnuson, J. J., Crowder, L. B., & Medwick, P. A. (1979). Temperature as an ecological resource. *American Zoologist, 19*, 331–43.

Primmer, C. R., Aho, T., Piironen, J., Estoup, A., Cornuet, J.-M., & Ranta, E. (1999). Microsatellite analysis of hatchery stocks and natural populations of Artic char, *Salvelinus alpinus*, from the Nordic region: Implications for conservation. *Hereditas, 130*(3), 277–89.

Juha Jurvelius

Conservation and Environment
Bears and Humans

The relationship between bears and people is complex and ancient. The eight species of bear found worldwide today inhabit diverse habitats in Europe, Asia, and North and South America, from the high Arctic to the tropical rain forest, and in all of those places, bears interact with people. Almost universally, bears are strong symbols, something particularly apparent in popular culture and among circumpolar indigenous peoples. Much of this symbolic affiliation may be a result of similarities with our own species; bears can stand on their hind legs, they care for their young, and they are omnivorous, adaptable, and able to learn from experience and observation. The bear's annual hibernation is mirrored in aboriginal ceremonies that celebrate a cycle of death and rebirth (Rockwell, 1991).

Interactions between bears and humans reflect much of the range of human-wildlife interactions (Bath & Enck, 2003). Bears compete with people for natural resources, prey on livestock and crops, and under certain circumstances can cause injury or death to humans. Besides being killed or otherwise managed to resolve such conflicts, most bear species are hunted, whether because hunters want their hides for trophies, for subsistence uses such as the traditional polar bear (*Ursus maritimus*) skin pants of the Greenland Inuit, or for meat. In southeast Asia, there is an active trade in live bears as pets and in bear parts, particularly gall bladders, which are used in traditional Chinese medicine (Servheen et al., 1999). In China bears are also "farmed" to produce bile, though the bears involved are not domesticated (Servheen et al., 1999). Bears are also still captured and trained to perform as "dancing bears" in parts of southwest Asia (Servheen, 1990). Conversely, wild bears are also popular tourist attractions in national parks and other reserves in North America, and in some locations, bear viewing supports a lucrative, though small-scale, economy (Lemelin, 2005). Where bear populations are threatened—usually because of habitat loss or excessive killing—high-profile conservation programs involving extensive scientific research are often put in place. Because of this mixture of positive and negative identifications and societal perceptions about what bears are, these varied forms of human-bear interactions have at times been the subject of intense controversy.

Prehistory

For at least 80,000 years, Neanderthals and modern humans interacted with brown bears (*Ursus arctos*) and the now-extinct cave bears (*Ursus spelaeus* and *Ursus deningeri*) in Eurasia. (In accordance with common usage, I generally refer to most North American members of the species *Ursus arctos* as grizzly bears and to Eurasian and coastal Alaskan members as brown bears.) In the early twentieth century, finds of cave bear bones and skulls apparently buried in an organized manner in caves in the Swiss Alps led to much speculation about the existence of a Neanderthal cave bear cult. Although this idea has since been demonstrated to be false (an artifact of bones being better preserved in some parts of caves than others), bears were nevertheless important to prehistoric man because they were hunted and clearly depicted in cave paintings as early as 32,000 years ago (Kurtén, 1995).

Bears in Aboriginal Cultures

In 1926, A. Irving Hallowell surveyed the different customs that indigenous people throughout North America and northern Eurasia displayed surrounding bears. His thorough examination revealed several surprisingly common behaviors across that vast geographic and cultural space. Rather than naming bears directly, people referred to bears in honorific terms that usually signified kinship, such as "grandfather." Talking directly to the bear, however—usually offering a conciliatory address to a bear that was, or was about to be, killed—was common. Similarly, specific ceremonial activities were performed when a bear was killed. These rituals ranged in complexity up to the elaborate festivals of the Ainu (Japan) and Amur-region peoples (eastern Russia), in which a captive-raised bear was sacrificed and eaten. Slain bears were often offered gifts such as food, tobacco, or even hunting tools. Throughout the entire study region, Hallowell documented numerous specific taboos that went along with cooking and eating bears, often prohibiting women from eating specific parts or even any of the bear's meat.

Taken together, these observations suggest a clear pattern of what indigenous peoples in widely different circumstances came to feel was an appropriate way to interact with bears, developing rules of interaction over time. These rules clearly recognized the danger to humans that bears could pose, especially when hunting them without firearms. This probably helped to reinforce the need to be careful and humble around bears—avoiding potentially dangerous mistakes from overconfidence and also establishing principles of reciprocity and respect. Hunters often remark that skinned bears resemble people, but ideas of kinship likely go deeper than just physical resemblance. In Catherine McClellan's (1975) documentation of the Athapaskan (Yukon/Alaska) story of *The Girl Who Married the Bear,* individual people and bears become one another and intermarry, yet must observe rules of behavior between their species or meet with disastrous fates. In many aboriginal cultures, such rules about how to demonstrate respect for bears still remain strong and have been widely documented in northern Canada and Alaska (Georgette, 2001; Jans, 2005; Nelson, 1973; Tanner, 1979; Rockwell, 1991). Communication between aboriginal and Euro-American cultures can be difficult, and it appears possible that some controversies over bear management and conservation efforts may have their origins at least partially in different cultural conceptions of what "respect for bears" actually means. For example, many conventional practices in bear research, such as fitting radio-collars on bears to track their movements, may be seen as deeply disrespectful by aboriginal people yet absolutely necessary by biologists (Loon & Georgette, 2001).

A History of Decline

In historic times, most bear species have suffered significant declines in their numbers and ranges, mainly because of expanding human populations and conflicts with agricultural activities (Mattson, 1990). Bears were popular animals in Roman circuses, where they fought both men and other animals and were even used to execute criminals. (Such "bear-baiting" continued even into late nineteenth-century California, with bear-and-bull fights, Storer & Tevis, 1955.) Bears were apparently captured in large numbers from around the Mediterranean for such performances, and throughout Roman Europe, they were hunted both recreationally and to limit their numbers (Toynbee, 1973). In Europe and the United States, this decline accelerated over time and was particularly acute for brown bears: by the mid-twentieth century in the continental United States, grizzlies survived only in national parks and wilderness areas (Mattson & Merrill, 2002).

Bear-Human Conflicts and Their Management

Evidently, people are more dangerous to bears than the other way around, yet injuries to people do occur, and our innate fear of being eaten by a larger animal leads us to pay disproportionate attention to the risks that bears pose (Kruuk, 2002). On the night of August 13, 1967, two women were killed by grizzly bears in different regions of Glacier National Park, Montana (Herrero, 2002). Subsequently, a public debate erupted over whether grizzlies should be extirpated from national parks for the sake of public safety, and conflicts between humans and bears began to receive more focused attention. Another controversy followed when biologists John and Frank Craighead challenged the U.S. National Park Service's decision to rapidly close garbage dumps in Yellowstone National Park to prevent grizzlies from feeding there, without allowing them time to adapt. The Craigheads' grizzly research project was terminated, and the short-term dispute was settled by a Committee of the National Academy of Sciences, but acrimony persisted long afterward (Craighead, 1979; Craighead et al., 1995).

Current scientific understanding of bear-human conflicts is largely based on the work of Stephen Herrero, whose 1970 analysis of grizzly bear–inflicted injuries and its later expansion into his book *Bear Attacks: Their Causes and Avoidance* (2002) essentially defined the field. His findings are reassuring for people who fear injury by bears when traveling in wilderness areas: attacks on humans by any species of bear are statistically rare events, and there are many things that people and management agencies can do to reduce the risks of unwanted bear-human interactions. Perhaps most importantly, Herrero and others have clearly established that access to human food and garbage leads to food-conditioning of individual bears (when a bear forms a simple association between people and food), which dramatically increases the likelihood of bears coming into conflict with people (Gilbert, 1989; Herrero, 2002; Herrero & Fleck, 1990).

Another factor that can influence bear-human interactions is habituation, which occurs when a bear loses its avoidance and escape responses after repeated (and nonnegative) interactions with people—for example, at certain bear-viewing locations in Alaska. However, behavior of habituated bears can be indistinguishable from bears that are naturally tolerant of people (Herrero et al., 2005), and habituation may or may not be desirable in specific circumstances, but a combination of food conditioning and habituation is particularly dangerous. A recent development that builds on these ideas and should provide useful guidance to people in the field is overt reaction distance: the distance at which a bear overtly reacts to a person or another bear (Herrero et al., 2005). Overt reaction distance reflects the outward signs of how a bear is responding in a given situation because it is difficult to observe any internal responses that a bear may be having, and the reaction distance appears to be shorter in areas with abundant food and many other bears than in places where bear population densities are lower. This likely explains why most bear-viewing operations in the rich environments of coastal Alaska and British Columbia are able to create safe environments for people to watch bears at relatively close range.

This behavioral paradigm for understanding bear–human interactions dominates bear-management philosophy and techniques throughout North America and has permitted significant advances to be made in reducing both human injury and bear mortality (Herrero & Fleck, 1990). Bear-human conflicts create significant institutional and even personal liabilities, and immense investment has been made in preventing bears from becoming food-conditioned and habituated in parks and other jurisdictions throughout North America. Controversially, Craighead et al. (1995) advanced the idea that grizzly bears have evolved to take advantage of ecocenters: highly localized, nutritious food sources that bring bears from a wide area into close proximity at seasonal intervals and that may (e.g., garbage dumps) or may not (e.g., salmon runs) have human origins. They argue that anthropogenic ecocenters are now so important to some bear populations, such as in Yellowstone, that creating and manipulating ecocenters must now be considered valid conservation tactics, though they stress using them strategically to avoid food conditioning and habituation.

In comparison with the well-studied situations involving grizzly and black bears, conflicts between humans and polar bears are more poorly understood and may differ in several important ways. Polar bears are almost entirely carnivorous, subsisting mainly on seals hunted from the sea ice. Although food conditioning and habituation do occur in polar bears, at times they will also prey on humans, a behavior much more rarely observed in other bear species. Though polar bear–human conflicts are very rare, the rates of such conflicts in national parks in the Canadian Arctic may be influenced more by the absence of sea ice (forcing bears onto land and into proximity with people) than rates of park visitation (which is the case with grizzly and black bears in more southern parks) (Clark, 2003). As the already-warming Arctic climate leads to earlier breakup of sea ice, polar bear–human conflicts are expected to increase in the future (Stirling et al., 1999).

Conserving Bears

The goal of bear conservation is generally to limit mortality to a level that the bear population can sustain without declining (Servheen et al., 1999). Because bears reproduce slowly in comparison with many other hunted mammals, the number of bears that can be hunted is usually low, especially given that other sources of bear mortality, such as killing in defense of life or property, must also be considered. By the mid-1990s, computerized geographic information systems had led to a spatially oriented approach to conservation of bears and other large carnivores, based on designating large core areas of protected habitat—usually parks and legislated wilderness areas—and adjoining multiple-use buffer zones (e.g., Paquet & Hackman, 1995). In principle, those core areas should be large enough to contain minimum viable populations of bears and should be linked to one another by corridors that bears can travel, maintaining genetic diversity and providing a source for recolonizing areas from which bears may become extirpated (Boyce et al.,

An adult male polar bear reacts with a threat display to the approach of a photographer's vehicle, near Churchill, Manitoba, Canada. Courtesy of Douglas Clark.

2001). This spatial orientation is also an integral component of initiatives to link protected core areas with corridors across entire bioregions, such as the Yellowstone-to-Yukon Conservation Initiative (Y2Y) in the United States and Canada.

Experience now suggests that under some circumstances, such large ecosystem-scale conservation efforts are vulnerable to failure, largely because of insufficient involvement and support by local people who share those same landscapes with bears. Current thinking about conservation programs for bears (and other large carnivores) is now beginning to move away from large-scale wilderness designation (while recognizing this is still important) and toward locally based, community-scale participatory efforts, with particular emphasis on reducing bear-human conflicts (Clark et al., 2005). Although the nature of bear-human interactions has changed relatively little over time, the ways that they are perceived and being managed are evolving. Bears are becoming increasingly valued more for their aesthetic qualities and ecological importance, and generally less for their material values. Worldwide, conservation of bear populations is unlikely to be achieved everywhere in the same way; it will require a diversity of ideas and knowledge from people whose experiences, knowledge, and values reflect the full range of bear-human interactions.

Further Resources

Bath, A. J., & Enck, J. W. (2003). Wildlife-human interactions in national parks in Canada and the USA. *Social Science Research Review, 4*(1).

Boyce, M. S., Blanchard, B. M., Knight, R. R., & Servheen, C. (2001). *Population viability for grizzly bears: A critical review*. International Association for Bear Research and Management, Monograph Series No. 4.

Clark, D. (2003). Polar bear–human interactions in Canadian national parks, 1986–2000. *Ursus, 14*(1), 65–71.

Clark, T. W., Rutherford, M. B., & Casey, D. (Eds.). 2005. *Coexisting with large carnivores: Lessons from Greater Yellowstone*. Washington, DC: Island Press.

Craighead, F. (1979). *Track of the grizzly*. San Francisco: Sierra Club Books.

Craighead, F., Sumner, J. S., & Mitchell, J. A. (1995). *The grizzly bears of Yellowstone: Their ecology in the Yellowstone ecosystem, 1959–1992*. Washington, DC: Island Press.

Georgette, S. (2001). *Brown bears on the Northern Seward Peninsula, Alaska: Traditional knowledge and subsistence uses in Deering and Shishmaref*. Technical Paper No. 248. Juneau: Alaska Department of Fish and Game, Division of Subsistence.

Gilbert, B. K. (1989). Behavioural plasticity and bear-human conflicts. In M. Bromley (Ed.), *Bear-people conflicts: Proceedings of a symposium on management strategies* (pp. 1–8). Yellowknife: Northwest Territories Dept. of Renewable Resources.

Hallowell, I. A. (1926). Bear ceremonialism in the northern hemisphere. *American Anthropologist, 28*, 1–175.

Herrero, S. (1970). Human injury inflicted by grizzly bears. *Science, 170*, 593–98.

———. (2002). *Bear attacks: Their causes and avoidance* (2nd ed.). Piscataway, NJ: Winchester Press.

Herrero, S., & Fleck, S. (1990). Injury to people inflicted by black, grizzly or polar bears: Recent trends and new insights. *International Conference on Bear Research and Management, 8*, 25–32.

Herrero, S., Smith, T., DeBruyn, T., Gunther, K., & Matt, C. A. (2005). Brown bear habituation to people—Safety, risks, and benefits. *Wildlife Society Bulletin, 33*(1), 362–73.

Jans, N. (2005). *The grizzly maze: Timothy Treadwell's fatal obsession with Alaskan bears*. New York: Penguin.

Kruuk, H. (2002). *Hunter and hunted: Relationships between carnivores and people*. Cambridge: Cambridge University Press.

Kurtén, B. (1995). *The cave bear story: Life and death of a vanished animal*. New York: Columbia University Press.

Lemelin, R. H. (2005). Wildlife tourism at the edge of chaos: Complex interactions between humans and polar bears in Churchill, Manitoba. In F. Berkes, H. Fast, M. Manseau, & A. Diduck (Eds.), *Breaking ice: Renewable resource and ocean management in the Canadian north* (pp. 183–202). Calgary: University of Calgary/Arctic Institute of North America.

Loon, H., & Georgette, S. (1989). *Contemporary brown bear use in northwest Alaska*. Technical Paper No. 163. Kotzebue: Alaska Department of Fish and Game, Division of Subsistence.

Mattson, D. J. (1990). Human impacts on bear habitat use. *International Conference on Bear Research and Management, 8*, 33–56.

Mattson, D. J., & Merrill, T. (2002). Extirpation of grizzly bears in the contiguous United States, 1850–2000. *Conservation Biology, 16*, 1123–36.

McClellan, C. (1975). *The girl who married the bear: A masterpiece of Indian oral tradition*. Publications in Ethnology, No. 2. Ottawa: National Museum of Man.

Nelson, R. K. (1973). *Hunters of the Northern Forest: Designs for survival among the Alaskan Kutchin*. Chicago: University of Chicago Press.

Paquet, P., & Hackman, A. (1995). *Large carnivore conservation in the Rocky Mountains*. Toronto: World Wildlife Fund Canada.

Rockwell, D. (1991). *Giving voice to bear: North American Indian rituals, myths and images of the bear*. Niwot: Roberts Reinhardt.

Servheen, C. (1990). *The status and conservation of the bears of the world*. International Association for Bear Research and Management, Monograph Series No. 2.

Servheen, C., Herrero, S., & Peyton, B. (Eds.) (1999). *Bears: Status, survey and conservation action plan*. Gland: IUCN.

Stirling, I., Lunn, N. J., & Iacozza, J. (1999). Long-term trends in the population ecology of polar bears in western Hudson Bay in relation to climatic change. *Arctic, 52,* 294–306.
Storer, T. I., & Tevis, L. P., Jr. (1955). *California grizzly.* Berkeley: University of California Press.
Tanner, A. (1979). *Bringing home animals: Religious ideology and mode of production of the Mistassini Cree hunters.* Social and Economic Studies No. 23. St. Johns: Institute of Social and Economic Research, Memorial University of Newfoundland.
Toynbee, J. M. C. (1973). *Animals in Roman life and art.* London: Camelot Press.

Douglas Clark

Conservation and Environment
Behavioral Ecology and Conservation

When we are told about an endangered animal or habitat, most of the information we receive is about its ecology: how animals' interactions with their environments affect their abundance and distribution. Hunters have studied ecology for thousands of years to help predict how the quantity and location of animals change in response to changes in seasons, food supply, competition from other animals, and many other factors.

However, the most sensitive environmental response we can observe is an animal's *behavior*. An animal must constantly decide how to make the best use of its energy. If it detects a meaningful change in its immediate surroundings, such as a signal indicating a predator or a prey item, it may immediately change its behavior. Therefore, studying an animal's behavior can give you a rapid indication of the impact of ecological changes. By combining those observations with precise monitoring of multiple ecological changes, you can learn which impacts are causing the behavioral change and gain a better understanding of the animal's worldview.

The skills to understand animal behavior are useful in environmental conservation. For example, a Nisga'a tribal fisherman in British Columbia, Canada, observed that the Dungeness crabs he normally picked were behaving strangely; they were "marching past the dock at the mouth of the Nass River, rather than staying in the deep water of Alice Arm" (Snively & Corsiglia, 1998, p. 22). He hypothesized that the crabs' behavior was a result of water pollution by a nearby molybdenum mine, and later analysis supported his hypothesis.

Behavioral ecology is important because it integrates signals from many levels of biological organization. A single behavior is affected by the chemical signals between cells in the organs, by that organism's interactions with others of the same species and other species, and by interactions with its changing environment. Studying behavioral ecology can provide important information about the condition of species from the biochemical to the ecosystem levels. Behavioral ecology is also important because it enables understanding of how people and other animals are linked through evolution and how conservation can improve the condition of all species in an environment.

Four questions that can help explain any species' behavior are as follows:

1. Function—how does the behavior affect the individual's survival and reproduction?
2. Causation—what stimuli tend to cause the behavior, and how does the behavior change in response to new events?

3. Development—how is the behavior different in individuals of different ages, and what early life experiences are necessary to develop the behavior?
4. Evolution—is the behavior similar in other related species, and how might it have evolved through the process of phylogeny? (Tinbergen, 1963)

Linking People and Fish

Over ten thousand years, members of indigenous tribes living near the "bloodline"—called the Columbia River by later settlers—learned to predict the time of year when each of seven salmon species migrated up the river to spawn in the headwaters where they were born. Each spring, when the Chinook salmon began their run, tribes from throughout the region came together to catch, prepare, eat, and preserve salmon (Yakama Tribal Fisheries, 2003). Today, that river is polluted by an estimated 20 million pounds of toxins a year from agriculture, mining, paper mills, and other industries. Tribal scientists took the lead to discover ninety-two chemicals in the Columbia River (EPA, 2002). Now, many other scientists are showing how that pollution is affecting the appearance and behavior of those fish. For example, polycyclic aromatic hydrocarbons emitted through car exhaust can be washed by rain into fish spawning grounds. These chemicals reduce the transmission of nerve impulses in the heart of a fish embryo, causing spinal curvature, facial deformation, and reduced swimming performance in adult fish (Incardona, 2004).

Many chemical processes have been conserved throughout the process of evolution and as a result are found in the cells of all vertebrates, including fish and people. Regulatory chemicals include important hormones such as testosterone, progesterone, insulin, acetylcholine, norepinephrine, dopamine, histamine, serotonin, melatonin, vasopressin, oxytocin, cortisol, thyroxine, and prostaglandins. These chemicals are responsible for many ecologically relevant behaviors in fish and people, including learning, polymorphism, sexual dimorphism, dispersal within habitats, schooling, migration, foraging, predator/antipredator behavior, communication, territoriality, dominance, mating systems, mate choice, and parental care (Adkins-Regan, 2002). Toxins interfere with those chemicals in similar ways, creating learning disabilities, attention problems, deformities, and changed neurotransmission in fish and people (Klaasen & Watkins, 2003). Evolution is therefore a useful framework for understanding how the lives of people and other animals have become linked from the chemical to the ecosystem levels.

Using Behavioral Ecology to Study Human Impacts

Each person needs to study the lives of the animals in our "footprint," the area of land we impact through the activities that generate the resources we use. Those activities include producing food, energy, consumer goods, and recreation and transportation services. That footprint may be local, but it likely includes other regions and countries (Wackernagel, 1996). We compete with other animals for resources, but the average North American uses 100 times more resources per pound of body weight than any animal. By studying a species' behavior, we can better understand how our actions change their lives—over the course of a day, a season, a year, a lifetime, or generations—and how we can reduce that impact.

We can begin by studying an animal's natural range of behavioral responses to environmental changes. For example, if the habitat degrades and the food supply decreases, animals may compete to determine who can stay and who must travel farther to find new food. As global warming changes the onset of seasons, the food plants of some birds mature earlier, which requires that they migrate earlier to their nesting grounds.

Because most people and other animals live in constantly changing environments, evolution has favored traits and abilities that enable us to choose among many survival strategies:

- Optimality—the need to choose the greatest benefit in return for an investment of time and energy, such as feeding in the patch with the highest-quality food
- Life history—abilities that vary based on developmental changes, such as selecting a mate when reaching reproductive maturity
- Conditionality—strategies that change in different environmental contexts, such as hibernation in winter
- Contingency—the need to weigh many factors in making a decision, such as choosing between feeding and avoiding a predator
- Natural selection—the process by which traits are inherited that optimize survival in a particular ecological context, such as thick fur
- Sexual selection—the process by which males compete for access to females who choose traits that result in improved offspring survival (Krebs & Davies, 1993, 1997)

By studying the behaviors of different animals, we can gain insights into their complex decision-making processes. By comparing our own decision-making processes, we can better understand what we have in common. Because people and animals share a common environment, the key conservation decision we must make is how we can live together. To help make that decision, we can engage some of our most important evolved abilities.

Two Evolved Conservation Strategies

When people (and many animals) make decisions that benefit others and ourselves, we use two decision-making strategies: reciprocal altruism and kin-selected altruism (Hamilton, 1996; Trivers, 2002). Reciprocal altruism is an expectation that "if you help me now, I will help you later." People use that strategy with friends and in business. We calculate how much material help we will give another person now and how much material help we will receive later. If we get back at least as much as we give, we consider it a fair exchange. If we do not, we feel cheated and want to punish the cheater. Animals use that strategy when sharing food, grooming, giving alarm calls, caring for others' offspring, and showing their fitness to a mate.

People tend to use the reciprocity strategy to address environmental problems. For example, if we spend money cleaning up pollution in the river now, the suggestion is that the river will provide *us* with clean water in the future. However, people often "cheat" or "free-ride," getting benefits from the environment without reciprocating, such as cutting trees without planting seedlings or polluting without cleaning up. They do this to maximize their environmental benefits while minimizing their costs. This is often successful because the environment cannot immediately detect and punish cheaters to enforce the promise of reciprocity. Therefore, other people must take this role—to monitor the environment to detect impacts and enforce reciprocity through sanctions, to sustain healthy local habitats.

Kin-selected altruism results in a very different set of decision-making strategies that we use with our families and with those who feel like relatives, such as close friends and pets. We calculate how much material help we will give in return for the nonmaterial benefit of increasing our relatives' well-being in ways that will increase chances of genetic survival. For example, a parent may spend time teaching a child to cook or repair a car, to increase that child's ability to survive and to raise children of his or her own. Parents do not expect repayment for the time and money they spend. The increased well-being and survival of that child and future offspring is considered its own reward, and parents receive those nonmaterial rewards immediately by seeing the child practice his or her new survival skills. If parents do not invest that kind of effort with their children, we say they are practicing child neglect, which is a crime.

Can We Perceive Kin-Like Relationships with Other Species?

Can the powerful decision-making strategy of kin-selected altruism be engaged for conservation? Will we invest material help in return for increased well-being and survival of other animals, even though we are not genetically related? This is possible, if the cues of relatedness that evolved in human societies can be perceived through interactions with other species. Some family members perceive those cues with their pets, some hunters perceive them with their prey species, and some biologists perceive them with the species they are studying. All of these perceptions of kin-like relationships require sustained contact, which can engage other processes.

The central process for perceiving relationship with others is empathy. Empathy is a sympathetic response to another's emotional state because of that person's distressful situation. Empathy is a neurological, physiological, and behavioral response in social animals that primes an individual who pays attention to another's state to change its state to become more like that of the other (Preston & DeWaal, 2002). For example, if you see someone emotionally close to you in a state of distress, your state would tend to change toward increased distress. Empathy evolved partly because a parent who empathized with her child could tell if that child was in danger and would immediately bring him close. Empathy is also co-opted by prey to tell if a predator is hunting them and by predators to tell if a prey is weak and easier to hunt. Empathy therefore gives us information useful in many contexts, and we can use that information to help, to hurt, or to do nothing at all.

Empathic helping depends on five factors:

1. How much familiarity we have with another—if we have spent enough time together to feel like kin
2. How much similarity we have with another—if we share a similar identity and social role such as age, gender, and class that helps us feel more alike
3. How much shared experience we have with the situation of distress that the other is experiencing now
4. How accurately we can detect the amount of distress the other is feeling—if they are just upset or if their life is in danger
5. How much ability we have to reduce the other's distress—and the distress we feel

Can we feel empathy with an individual of *another species*? Will we use the information we gain through empathy to help care for those other species without asking for anything in return except for their increased survival? Can other species engage our mechanisms of kin-selected altruism? The conditions for empathy can be met through sustained observation of animal behavior—the same processes we use to develop relationships with other people.

Most of us feel empathy toward our pets. They feel like family because we spend time together, they have a social role in our lives, we share many of the same stressful situations, we can often tell how they are feeling, and we can usually help them if they are in distress. But can we also feel empathy toward wild animals, especially those who look very different, live in very different environments, and communicate very differently—such as wild fish in a stream?

Because of evolutionary conservation and convergence, people and fish can meet many conditions for empathy. We have degrees of similarity in our

- organs for sight, smell, hearing, touch, and taste
- structures for propulsion, respiration, digestion, and reproduction
- methods for navigation and social communication
- life history stages from embryo to juvenile to adult
- hierarchical social structures
- motivations of hunger, fear, reproduction, and aggression
- behaviors of foraging, avoiding threats, and competing for resources and reproductive opportunities
- habitats that change with climate, altitude, season, and time of day and that contain different microhabitat types and different prey species (Pitcher, 1992; Wootton, 1999)

Therefore, we may be able to use these areas of similarity to better understand the lives of fish, and with increased understanding of their worldview, we may empathize with fish in distress and decide to help them.

Using Behavioral Ecology for Fish Conservation

We may meet the conditions for similarity, but can we put that theory into practice? I conducted research in northwest Montana with 1,200 people ages eight to adult, studying the impacts of people and introduced species on the behavior of native fish. The purpose was to investigate whether kin-selected altruistic mental mechanisms (or behavioral strategies) can be utilized to motivate conservation. I found that the challenge was to increase *familiarity* with other species through sustained observation of the natural lives of animals by *groups* of people. Group observation—in the field, through video, or in microcosms—is essential for several reasons:

- Groups of people are needed for effective conservation practices, and their observations will help inform and motivate their conservation actions.
- Groups with many different ages, experiences, and points of view can observe and interpret more of an animal's complex interactions with its environment than can any one individual.
- Groups can share their observations, interpretations, and conservation plans with each other, increasing social cohesion by working together to improve the quality of their shared environment.

To increase group familiarity with animal responses to human impacts, we used the following sequence of field and laboratory methods.

1. **Observing the Range of Animal-Habitat Interactions**
   If you want to empathize with an animal, you must first try to understand its worldview. To do that, first identify the important environmental characteristics that an animal must detect and interpret to make decisions that will improve the survival of itself and its offspring. For a fish, these characteristics include shelter, water flow

rate, temperature, turbidity, prey availability, predation risk, and predator and prey signals such as sounds, scents, or images.

Montana participants observed individual fish in the field and in large aquariums designed to look like natural habitats, and they hypothesized about the fish's behavioral responses to changes in those characteristics. Sustained observation helped us understand the problems fish can solve, using the abilities that are adapted to the environment within which they evolved. Repeated observation also helped us understand that fish could not solve some new problems created by people.

2. **Identifying Human Impacts on Animals and Habitats**
   Human impacts often create environmental changes that are beyond an animal's evolved ability to respond and that evoke a high-stress response. For fish, those changes include increased turbidity so that it is impossible to see prey or detect predators; increased temperature that increases metabolic rate and the need for more food; and habitat simplification that reduces the types and amounts of insect prey. Montana participants compared the important environmental characteristics of an ideal or "reference" habitat with those of a human-impacted habitat to help understand how those impacts affected fish. They also created small and non-harmful versions of those impacts in aquaria—such as increasing silt in the water—to observe the fish's response to increased turbidity that was common in local streams impacted by logging or ranching. They found that fish stopped feeding until the water was clear enough to detect predators.
3. **Measuring Animal Responses to Human Impacts**
   Animals and people have many similar responses to stressful impacts. These include increases in heart rate, respiration rate, fin movement, stress, and aggression; reductions in activity level, feeding, territory defense, swimming speed, weight, number of offspring, survival of offspring, and population size; and changes in color, type of movement, territory size, location in the habitat, prey choice, and nutritional quality. Montana participants found that an animal population will quickly recover from a brief "pulse" disturbance such as a change in the locations of rocks and logs in a stream. They inferred that a population may not recover from a continual "press" disturbance, such as permanently removing rocks and logs that create microhabitats or covering spawning gravel with sediment. By understanding an animal's worldview and measuring its responses to human impacts, we can better understand how our actions cause declines in animal health and population size.
4. **Implementing a Conservation Plan**
   Just as a parent wants a child to be able to survive independently, conservation must improve an animal's ability to survive. By observing an animal's range of behaviors, we can recognize that each animal needs a variety of environmental characteristics. It needs different places to complete its daily, seasonal, and life history tasks: to be safe, to find high-quality food, to select and keep a mate, and to raise offspring that can grow to reproductive age. By becoming familiar with the needs of local animals, we can identify conservation actions that will help them. Empathy is just a step toward understanding how improved natural habitats provide ecosystem services that are healthy for all organisms in an area, including people.

Montana participants recognized that in our urbanized world, few people could see the impact of their ecological footprint on other species. By the time these participants had observed enough fish behavior to develop a local conservation plan, they had learned much more than the average person about their local species. They then faced the challenge of involving many others in observing and learning about behavioral ecology of native fish in their local socio-ecological systems.

The science of behavioral ecology enables us to observe how animals and people are linked, by observing animals' behavioral responses to our actions. We know that many human societies and habitats are more complex than those of many animals and that human brains, minds, languages, and cultures have coevolved to respond to that complexity. Yet many behavioral responses have been time-tested through the process of natural selection and found to be reliable for both animals and people. Therefore, animals' behaviors can provide honest signals about our impact on an animal's level of stress, its ability to complete daily tasks such as feeding, and its ability to survive and reproduce. That knowledge can help us design adequate conservation plans for the other species in our local environment.

As we become more familiar with the behavior of other local species, we gain an understanding of our similar needs and how animals express their distress at not being able to meet those needs. The challenge then is to learn many conservation strategies that are effective in helping improve the well-being of other local species. One of the rewards is gaining small insights into the worldviews of the other species that share our environment and better appreciating the grandeur of life. Empathy has been designed by evolution to help us better care for our children and other kin. As you become more familiar with the animals in your locale, you can see how the benefits they provide also improve the well-being of your family in ways that deserve care.

Further Resources

Adkins-Regan, E., & Weber, D. N. (2002). Mechanisms of behavior. In G. Dell'Omo (Ed.), *Behavioural ecotoxicology*. New York: Wiley.

Hamilton, W. D. (1996). *Narrow roads of gene land: The collected papers of W. D. Hamilton*. New York: Freeman.

Incardona, J. P., Collier, T. K., & Scholz, N. L. Defects in cardiac function precede morphological abnormalities in fish embryos exposed to polycyclic aromatic hydrocarbons. *Toxicology and Applied Pharmacology, 196*, 191–205.

Kirschner, M., & Gerhart, J. C. (2005). *The plausibility of life: Resolving Darwin's dilemma*. New Haven: Yale University Press.

Klaasen, C. D., & Watkins, J. B., III. (2003). *Casaret & Doull's essentials of toxicology*. New York: McGraw-Hill.

Krebs, J. R., & Davies, N. B. (1993). *An introduction to behavioural ecology* (3rd ed.). Malden, MA: Blackwell.

———. (Eds.). (1997). *Behavioural ecology: An evolutionary approach* (4th ed.). Malden, MA: Blackwell.

Pitcher, T. (1992). *Behaviour of teleost fishes*. New York: Springer.

Preston, S., & deWaal, F. B. M. (2002). Empathy: Its ultimate and proximate bases. *Behavioral and Brain Sciences, 25*(1).

Snively, G., & Corsiglia, J. (1998). *Rediscovering indigenous science: Implications for science education*. Paper presented at the National Association for Research in Science Teaching, San Diego, CA.

Tinbergen, N. (1963). On aims and methods in ethology. *Zeitschrift für Tierpsychologie, 20*, 410–33.

Trivers, R. (2002). *Natural selection and social theory: Selected papers of Robert Trivers*. New York: Oxford University Press.

U.S. Environmental Protection Agency (EPA). (2002). *Columbia River Basin Fish Contaminant Survey*. Seattle: EPA Region 10.

Wackernagel, M., & Rees, W. (1996). *Our ecological footprint: Reducing human impact on the earth.* Gabriola Island, BC: New Society.
Wootton, R. J. (1999). *Ecology of teleost fishes.* New York: Springer.
Yakama Tribal Fisheries. (2003). *Sacred salmon: A gift to sustain life.* Pablo, MT: KSKC-TV.

Michael LaFlamme

Conservation and Environment
Birds and Recreationists

People are spending increased amounts of leisure time and money traveling to areas throughout the United States and the world in pursuit of outdoor recreational opportunities. In their travels, they often visit remote areas, not to hunt or kill the animals they see, but rather to photograph and experience wildlife. This type of recreation, which involves wildlife without removing or destroying it, is termed "nonconsumptive wildlife-oriented recreation," or "wildlife watching." According to the U.S. Department of the Interior, over 66 million people spend about $40 billion each year watching wildlife in the United States. Almost 46 million bird watchers in the United States spend their free time visiting national parks and forests, as well as state-owned public areas, in search of rare, unique, or beautiful birds. Although wildlife-oriented recreation is theoretically not supposed to affect wildlife, it does impact certain animals. As more people use public areas for recreation, scientists and wildlife managers have become increasingly concerned about adverse effects on wildlife in these areas.

Since the early 1900s, researchers have been aware of the problems caused by human disturbance. Stephen Boyle and Fred Samson compiled a list of more than 500 studies that examined the effects of nonconsumptive wildlife-oriented recreation on wildlife; most of these effects were negative and came from activities such as hiking, camping, and wildlife observation. Most studies involved birds and mammals, although many other species are probably affected, but need more research.

Birds are not all equally vulnerable to the presence of recreationists. Some species seem to tolerate human presence well, despite high levels of activities. Those species that do not react negatively to the presence of humans are said to have become habituated to humans. Through habituation, birds minimize their responses to humans, thereby saving energy and devoting time to other activities, such as feeding or caring for their young. Whether a particular individual can habituate to human disturbance depends on the species, its social organization, its environment, the season, and the experiences the bird has had with people. Some species such as great crested grebes (*Podiceps cristatus*) readily breed in suburban parks. Other birds, including a variety of sea gulls, may even become aggressive toward people. In winter ski resorts, birds can habituate to skiers, ignoring people unless skiers stop too close to them. In protected areas, with regular exposure to people who pose no threat, individual birds may become habituated because they have learned that humans are not dangerous. For example, some species such as the black-crested titmouse (*Baeolophus bicolor*), Carolina chickadee (*Poecile carolinensis*), and blue jay (*Cyanocitta cristata*) approach people in parks and accept food from their hands.

Habituation, however, does have its limits. Excessive closeness or persistent disturbance can still impact birds. If humans interrupt their feeding or nesting activities or chase them, birds will become skittish and avoid those areas. Even if people do not harass them, birds must constantly move out of the path of strollers, joggers, and vehicles and will fly away when frightened, wasting valuable time and energy.

Although some birds are capable of habituation, other species are negatively affected by humans. Because many species are present in the vast majority of places visited by people, recreationist-caused disturbances greatly affect sensitive individuals. Park visitors do not even need to be overly active; the mere presence of people, such as wildlife observers, photographers, and researchers, in an area can affect birds, a phenomenon termed "human intrusion." Intrusion can change bird behavior and affect how the birds are distributed throughout an area. Disturbance causes short- and long-term effects, depending on the time of year, the species in question, how often people are present, and the type of recreational activity. Birds may be more wary during breeding season, in comparison with winter. Solitary birds may respond differently than birds in flocks, and large birds (such as eagles or herons) may be more leery of humans than small species (such as sparrows). Also, birds that live in dense brush may react differently toward humans than birds that live high in trees or in open areas. Bird watchers and photographers unintentionally harass rare birds by surrounding and approaching them, which makes it hard for the birds to forage for food or care for their young. Birds may be exposed to disturbance year round because people use public areas throughout the year, but for different activities (for example, hiking in the summer, skiing during wintertime).

Human disturbance can affect a bird's habitat by changing the soil and vegetation characteristics, which affects the bird's use of that area. Recreationists damage vegetation by straying off trails or collecting firewood in unauthorized areas. Excessive hiking tramples vegetation, leading to patches of bare soil that become eroded. Cross-country bicyclists cause major damage when they leave designated trails, particularly on steep hillsides that are easily eroded. Fewer prey species, such as earthworms and spiders that birds depend on for food, are found in heavily used areas.

The effects of disturbance on birds depend on a variety of factors, including the type, frequency, and degree of disturbance, the species involved, and the history of persecution of birds in an area. Birds with young are particularly stressed because they must spend time avoiding humans instead of caring for their chicks. If it is disturbed enough, a bird may abandon its nest and nestlings, leaving the offspring to die. The ecologist Jonathan Bart found that nesting American robins (*Turdus migratorius*), Eastern bluebirds (*Sialia sialis*), and mourning doves (*Zenaida macroura*) are sensitive to human disturbance, and chick mortality for these species is higher when people visit their nests. Research conducted by Wayne C. Weber shows that some bird species, such as black-billed magpies (*Pica pica*) and Northwestern crows (*Corvus caurinus*), cope with human disturbance by building their nests higher in trees in disturbed areas. If tall trees are unavailable, birds may build their nests in areas with thicker cover that are more difficult to locate, or they may completely leave an area.

Depending on the type of disturbance, a bird's responses can be temporary or permanent. A bird might temporarily avoid the area when humans are present, but use it when people are absent. If the disturbance is severe enough, or the species is particularly sensitive to disturbance, an individual may leave the area permanently. In this case, the species suffers from less available habitat, and the population may decrease to dangerously low levels.

Despite the fact that human intrusion can have severe, negative consequences for some species of birds, people can still use public areas while minimizing damage to birds by implementing a few simple activities. One major way to protect species is to monitor birds to determine whether humans are affecting them. Restricting visitor access from important roosting, feeding, and nesting areas and posting signs telling people why they are not allowed in these areas protects birds and their habitats. Restoring degraded land, replanting vegetation, and maintaining trails prevents further environmental damage. New construction should only be done in already-disturbed areas, and "buffer strips" of vegetation (brush, trees, and so on) should be allowed to grow between recreation areas and bird habitat to provide protection from disturbance. And finally, public education programs should focus on providing information about birds (and other species) within the park and about how people can protect these animals. With a few simple precautionary measures, public places can provide areas that protect wildlife while providing important recreational opportunities for people.

See also

Ecotourism

Further Resources

Bell, S. (1997). *Design for outdoor recreation*. London: Spon Press.

Hammitt, W. E., & Cole, D. (1998). *Wildland recreation: Ecology and management* (2nd ed.). New York: Wiley.

Hampton, B., & Cole, D. (2003). *NOLS Soft Paths: How to enjoy the wilderness without harming it*. Mechanicsburg, PA: Stackpole Books.

Knight, R. L., & Gutzwiller, K. J. (Eds.). (1995). *Wildlife and recreationists: Coexistence through management research*. Washington, DC: Island Press.

Noss, R. F., & Cooperrider, A. (1994). *Saving nature's legacy: Protecting and restoring biodiversity*. Washington, DC: Island Press.

Wilcove, D. S. (2000). *The condor's shadow: The loss and recovery of wildlife in America*. New York: Anchor.

Heidi Marcum

Conservation and Environment
Carson, Rachel

Rachel Carson (1907–64) was concerned about the humane treatment of animals and their welfare both as a biologist and as an ecologist interested in the whole stream of life. In 1960 as she was writing her famous protest against the misuse of pesticides, which became *Silent Spring*, she agreed to contribute the introduction to an educational booklet published by the Animal Welfare Institute in Washington, D.C. The Institute opposed animal experimentation, and the booklet, "Humane Biology Projects," addressed the need for reform in the biology curricula of the nation's high schools. Classroom biology experiments at the time were often accompanied by procedures of systematic cruelty that Carson found both unnatural and abhorrent.

In her introduction, "To Understand Biology," Carson argued that all of life was linked to the earth and that "the essence of life is lived in freedom." She believed that any experiments that required placing animals, insects, reptiles, or amphibians—commonly used in high school biology laboratories—in unnatural restraints, subjecting them to unnatural conditions, or inducing changes in bodily structure produced distorted results. Carson urged teachers of biology to introduce their beginning students first to the true meaning of biology by observing the lives of creatures in relation to each other and their environment. She believed that any other approach compromised a student's emotional response to the mysteries of life.

Echoing sentiments that Carson had first addressed to parents of young children in her article "Help Your Child to Wonder," which would be published posthumously as "The Sense of Wonder" (1965), she was concerned about developing in young people an awareness and reverence for life as a whole. Carson thought deeply about how to inculcate an ethical and moral awareness in young people toward each other and toward the whole of life on earth. It was her belief that no one could understand how all life is linked or experience their full humanity as humans without an understanding of their place in the stream of life.

The Animal Welfare Institute's booklet with Carson's introduction was widely used by high school biology teachers. It contributed to a new awareness of the need for humane treatment of all creatures and extended that ethic to the nation's biology and zoology classrooms.

Rachel Carson's ideas about the humane treatment of animals place her fully in the tradition of Albert Schweitzer and his reverence for all life. Christine Stevens, founder of the Animal Welfare Institute, also introduced Carson to the work of the British animal activist and writer Ruth Harrison. Harrison's book *Animal Machines* exposed the inhumane methods of raising livestock and the deplorable conditions in which they were kept before slaughter. Carson wrote the preface for Harrison's book in 1963.

Harrison's work had been prompted by the new factory farming methods used in rearing animals destined to become human food. It exposed the horrors wrought by the contemporary passion for "intensivism." Harrison described the conditions of those factories where animals lived out their wretched existence without ever feeling the earth beneath their feet—without knowing sunlight, scratching in the dirt, or grazing for natural food. She alerted the public to the cruelty of animals existing in intolerably crowded conditions and raised the question of how animals produced under such circumstances could be safe or acceptable as human food. Clearly they were not.

Rachel Carson approached the problem of agricultural intensivism as a biologist whose special interests were in ecology. She charged that the economic success of such factory farming was possible only with the use of drugs, hormones, and pesticides to keep the whole sad operation going. But in her introduction to Harrison's book, Carson raised the moral question of how far humans have the right to dominate other life forms. Carson believed that humankind could never be at peace unless there was an embrace of Schweitzer's ethic and the adoption of decent consideration for all living creatures.

Carson pointed out that although Harrison's book specifically addressed conditions in Great Britain, its critique applied both to European countries and to the United States. She hoped that it would spark a consumers' revolt against the agricultural industry's practices and, at minimum, force governments to legislate against such practices.

Carson's introduction is a precursor to her own arguments in *Silent Spring* (1962), wherein she challenged the public's complacent belief that "someone" was looking after things. Like Harrison, Carson courageously presented facts to prove that the misuse of chemical pesticides not only polluted the soil, air, and water, but also potentially damaged

Biologist and author Rachel Carson at home with her cat Moppet. ©Alfred Eisenstaedt/Time Life Pictures/Getty Images.

all life. In all her writing, Rachel Carson was concerned with linking humans to the stream of life and with pleading for an end to the modern worship of speed, quantity, and quick and easy profits. The nonhuman and the human world, she believed, were inextricably linked, and the health of one determined the well-being of the other. She argued that the public's "obligation to endure" gave them the right to know what was being put into the environment and to be certain of its safety. In challenging big business and the government science establishment, Carson's brave critique began the modern environmental movement.

See also

Education

Further Resources

Carson, R. (1962/2002). *Silent spring*. Boston: Houghton Mifflin.

———. (1965/2000). *The sense of wonder.* New York: HarperCollins.

———. (1998). *Lost woods. The discovered writing of Rachel Carson* (pp. 193, 194). L. Lear (Ed.). Boston: Beacon Press.

Lear, L. (1997). *Rachel Carson: Witness for nature*. New York: Henry Holt and Company.

Lytle, M. (2007). *Gentle subversive: Rachel Carson, "Silent Spring" and the rise of the environmental movement*. New York: Oxford University Press.

Linda Lear

Conservation and Environment
Conflicts between Humans and Wildlife

Somewhere in the world at this moment, a wolf, tiger, or other large carnivore is stalking sheep or cattle; somewhere an elephant, parrot, or other animal is feasting on crops; and a pelican or shark has captured fish that you or I could have eaten. In that same place or nearby, people are laying traps, clearing forest, or aiming a gun to destroy the wild animals that threaten their lives or livelihoods. It is an ancient problem, but lately it has become more than just competition between people and wildlife.

As people have spread around the world, we have brought our domestic animals and plants with us and have claimed wild forests, game, and fish as our own property. Not surprisingly, these actions have brought us into competition with wildlife that feed on the same foods and use the same space. All manner of wild animals enjoy the same crops we do, eat the young trees eyed by foresters, seek the fish and game that human hunters want, and prey on the livestock we raise; and in the rarest but most frightening of conflicts, animals such as bears, tigers, elephants, and sharks occasionally threaten our own safety. Conflicts between humans and wildlife vary in form and actors around the world, but the basic functional elements recur and have done so for millennia.

Common Patterns in Conflict Worldwide

We are competing for space and resources with other forms of life. The costs of this competition are felt mainly by people and wildlife living on the margins of our heavily populated areas from which we long ago purged most of our competitors. But the people who do live near wildlife have many reasons to fear and resent animals that damage property. Wildlife can be a potent symbol of danger in remote areas and of the lack of control rural people feel over their own destiny. Although economically insignificant on a regional scale, a few farmers, fishers, and foresters can suffer devastating economic losses from wildlife. More worrisome, these conflicts can turn people against wildlife and against conservation itself. The fear and resentment felt by affected communities can undermine wildlife recovering outside of protected areas, as well as the safety of wildlife in national parks.

At the same time, wildlife suffer tremendously from their competition with people—more so than the converse. For example, every year in the United States, almost 2.5 million wild animals are killed by the government to control agricultural damages. Throughout the world, affected human communities use traps, poison, guns, fire, and other means to kill wild animals or clear wild habitat. As our species has expanded and wild places have become small, few, and far between, the total number of wild animals that can cause problems has generally declined. This concentrates conflicts in a few areas and makes the problem a desperate one for wildlife. The traditional response to human-wildlife conflicts—retaliation and habitat transformation—may no longer be wise. Many large, threatening animals are endangered by hunting, fishing, and dwindling wild habitats. As a result, political conflict between people who value wild nature and people who work and live near wildlife is intensifying. Political conflict between people has begun to dominate the discussion of human-wildlife conflict rather than the age-old competition between wildlife and people. The challenge for our generation and the next is to balance wildlife conservation and human lives and livelihoods.

Human-Wildlife Conflict Is Now Political Conflict

When new ideas about wildlife conservation and protection began to clash with traditional ideas about protecting livestock, crops, fisheries, and forestry industries, people argued more than they made progress to solve their local problems. People who loved wolves, elephants, eagles, and other conflict-causing animals would often minimize the losses faced by others and demand more space for wild animals. People who favored farming, hunting, fishing, and other economic activities would often exaggerate the risks they faced and propose radical solutions such as eradication of wildlife. This divisive approach to the problem has been giving way in recent years to more sophisticated and inclusive understanding of the problems. Most people and governments around the world now acknowledge the need for protection of wild animals, especially those in danger of extinction, while also acknowledging the right of people to live without fear for their own safety or economic insecurity. The trick of course is to balance the two. A host of approaches is being tested, and some early successes are emerging.

What Can Be Done about Human-Wildlife Conflicts?

For more than a century, in many areas of the world, biologists and wildlife managers have been studying conflicts between wildlife and people. They have learned a lot about the biology and ecology of conflict-causing wildlife. Over the past twenty years, social scientists and environmental practitioners—people working to solve problems of our environment—have joined the struggle, as it has become obvious that dealing with the many faces of human-wildlife conflict depends as much on understanding wildlife as on understanding humans.

As far as practical solutions to resolving human-wildlife conflicts, the most successful approaches worldwide have combined some form of defense of human property with some means to deter wild animals from humans and their property. From research done in many countries, we have learned that some wild animals develop problem behavior, and others do not, despite having the opportunity. That suggests we should try to shape the behavior of a few individual animals rather than treat the whole species as troublesome. The same goes for some economic activities in wild areas, such as dumping livestock carcasses in the woods, raising crops or livestock without supervision, or competing with wildlife for the last remaining wild foods—these are problem behaviors on the part of humans. Because some human activities exacerbate problems, we need to understand people's values, perceptions, and economics, as well as the behavior and ecology of the wildlife, which demands a new cadre of scientists trained in several disciplines.

We have also found that prevention is less expensive and often more effective than reacting after conflicts arise. Some of our age-old preventive measures are still the best, such as guarding our crops and livestock or separating the places used by wildlife from those we use. When wild food is abundant and human property is well defended, conflicts are rare because wild animals rarely choose the risk path of confronting humans.

Another expression of the new compromises emerging in the management of human-wildlife conflict concerns debate over lethal control of wildlife. Many wildlife conservationists acknowledge the need to kill wildlife if they repeatedly cause problems, just as a growing number of individual producers and industries acknowledge that wild animals play an important role in healthy ecosystems. The ideal may be careful, selective removal of the wildlife that cause problems coupled with strict protection for those wild animals that do not. Similar scrutiny of people's activities near wildlife may be needed.

Many producers and industries would be content to work alongside wildlife as long as they do not lose in the process. For this vision of coexistence to flourish, conservation must pay for itself. In part, that means we, as consumers, must do our part. If we want wildlife to share the land and water with us, we must be willing to pay extra for business practices that are tolerant of wildlife. In the last decade, ecologically sustainable economic practices have begun to spread. One can now buy coffee, fruit, wool, meat, and timber that have been produced in a wildlife-tolerant manner.

In the coming years, we humans will need to improve our technologies and skills so that we can understand human-wildlife conflicts, predict where they will be the worst, work with the stakeholders who face threats from wildlife to reduce conflicts, and measure the success of our efforts, without resorting to "us or them" solutions.

Further Resources

Fascione, N., Delach, A., & Smith, M. E. (Eds.). (2004). *People and predators: From conflict to coexistence*. Washington, DC: Island Press.

Knight, J. (2003). *Waiting for wolves in Japan*. Oxford: Oxford University Press.

Linnell, J. D. C., Aanes, R., Swenson, J. E., Odden, J., & Smith, M. E. "Translocation of carnivores as a method for managing problem animals: A review." *Biodiversity and Conservation, 6,* 1245–57.
Linnell, J. D. C., Odden, J., Smith, M. E., Aanes, R., & Swenson, J. E. (1999). "Large carnivores that kill livestock: Do problem individuals really exist?" *Wildlife Society Bulletin, 27,* 698–705.
Smith, M. E., Linnell, J. D. C., Odden, J., & Swenson, J. E. (2000a). "Review of methods to reduce livestock depredation: I. Guardian animals." *Acta Agriculturae Scandinavica, Section A Animal Science, 50,* 279–90.
———. (2000b). "Review of methods to reduce livestock depredation: II. Aversive conditioning, deterrents and repellents." *Acta Agriculturae Scandinavica, Section A Animal Science, 50,* 304–15.
Treves, A., & Naughton-Treves, L. (1999). "Risk and opportunity for humans coexisting with large carnivores." *Journal of Human Evolution, 36,* 275–82.
Treves, A., Wallace, R. B., Naughton-Treves, L., & Morales, A. (in press). "Co-managing human-wildlife conflicts: A review." *Human Dimensions of Wildlife.*
USDA-WS. (2005). "Wildlife services annual tables." United States Department of Agriculture, Animal Plant Health Inspection Service, Wildlife Services.
Woodroffe, R., Thirgood, S., & Rabinowitz, A. (Eds.) (2005). *People and wildlife, conflict or coexistence?* Cambridge: Cambridge University Press.

Adrian Treves

Conservation and Environment
Conservation and Wildlife Trade in India

People from all walks of life are often thrilled to have a glimpse of wild animals. This fascination may be a result of our own past as hunters and gatherers of the savanna and woodlands. Endangered animals are becoming a commodity in the wildlife pet industry, and developed countries have a higher record of exotic wild pet trade.

Despite legal protection, large numbers of wild creatures are still being smuggled out of tropical countries each year to supply the global market. A man traveling from India to Singapore was caught smuggling more than 1,000 endangered star tortoises in 2003 (BBC News, 2003). Multiple tons of smuggled African elephant tusks and ivory were also seized the same year in Singapore, which is a major port in Asia for the transshipment of illegal timber and wildlife products. The international trade in wildlife is estimated to be over US$20 billion annually, of which an estimated 40 percent is illegal. This global trade in endangered species includes 40,000 primates, ivory from 90,000 African elephants, 4 million live birds, 10 million reptile skins, 15 million furs, and 350 million tropical fish (WWF-Traffic report). The business appears to have profit margins comparable with the illegal narcotic drug trade and arms smuggling. It is slowly wiping out several species of highly endangered animals so that future generations might not have a chance to see them in their natural environments.

Glamorous species such as tigers and giant pandas have already landed on the ten most-wanted species list identified by the World Wildlife Fund in 2000, their survival endangered by illegal trade. Other animals on this global hit list are the hawksbill sea turtle, Sumatran rhinoceros, Tibetan antelope, Asian box turtle, Javon pangolin anteater, and horned parakeet.

In countries such as Hong Kong, Taiwan, and Indonesia, exotic pets are viewed as status symbols among the upper class, even though it is against the law to keep them.

Orangutans, the only great apes from Asia, suffer from this chronic trade, although keeping apes as pets is slowly dying off in Southeast Asia because of strict law enforcement. On the other hand, the exotic pet trade is shifting toward less endangered reptiles from Madagascar and beyond. Thus, active public education is crucial to create and heighten awareness about the deleterious effect of exotic pet trade in species survival.

Nine types of Asian box turtles have been brought to the brink of extinction as a result of high demand for turtles as food. These reptiles have become increasingly popular as pets, sought after by collectors as they become more and more rare. In fact, about ten million turtles are traded in Asia every year, nearly 90 percent of which are shipped or transported by road or even by air to China, where turtle meat is considered a delicacy and the shells are used in traditional medicine (BBC News, 1999). Although several countries have banned the trade in turtles or in specific species, law enforcement is lethargic, and officials are poorly trained. It is estimated that there are nearly ninety species of freshwater turtles and tortoises in Asia—one of the most diverse regions in the world for these reptiles. But about 75 percent of them are now threatened with extinction because of the massive illicit trade. Many of the turtles, which are shipped live, are packed tightly into crates and die on the gruesome journey.

There is no doubt that wildlife is under threat in Asia because of illegal trade. Snakes, monkeys, apes, turtles, birds, and all sorts of endangered animals are being taken out of their habitat to meet the demand in unlawful trade. Rainforest habitat is being lost to logging with poorly planned development, and borders are porous, making smuggling a lot easier. And although supply is dropping, demand remains strong, pushed by poverty and hunger. It appears that countries such as China, Vietnam, Indonesia, Cambodia, and Laos are the vacuum cleaners for wildlife in Asia, and Hong Kong, Taiwan, Singapore, Thailand, and Malaysia play major roles in transit or destination for the illicit trade. It is therefore essential to strengthen law enforcement, promote public education, and eventually eradicate poverty in developing countries to put an end to this trade before more species are added to the extinction list.

India harbors several mega animals such as the tiger, elephant, rhino, snow leopard, and musk deer, which are highly valued in the illegal trade. In order to deal with this problem, the Indian government has taken several initiatives aimed at conserving the biodiversity of the country over the last few decades. It has banned hunting of wild animals and the trade in animal parts, but still the population of tigers is dwindling. The Indian Wildlife Protection Act of 1972 prohibits the hunting of all animals. Poaching of tigers is punishable with a maximum of seven years imprisonment. India has also been a party to the Conventional of International Trade in Endangered Species (CITES) since 1976 and is therefore bound by all its efforts to eliminate trade in tigers and tiger parts.

Endangered snakes are preserved in alcohol and sold in markets across Asia. Courtesy of G. Agoramoorthy.

But in recent years, India's own Bengal tigers have been poisoned, shot, or snared by poachers and sold to traders who smuggle them into China via Nepal. The world's wild tiger population has, because of the growing illegal market, fallen from 100,000 a century ago to fewer than

5,000 now, half of which are in India. In October 2003, one of the biggest seizures in the history of conservation—31 tiger skins, 581 leopard skins, and 778 otter skins—occurred in Angren County, China. All the skins originated in and were smuggled directly from India. The seized tiger skins represent nearly 1 percent of India's entire population of wild tigers (Wildlife Protection Society of India, 2004).

Many believe that the illegal trade is tied to poverty in underdeveloped tropical countries. When more people in these countries have better prospects for their future, probably only then will the future of endangered species be more secured in these places that harbor diverse fauna and flora in Asia.

Further Resources

Agoramoorthy, G. (2003). Wildlife illegal trade. *The Hemispheres Kid,* 4, 26–27.

BBC News Online. (1999, December 4). Asian turtle crisis. Available online at http://news.bbc.co.uk/1/low/world/far_east/550036.stm.

———. (2003, August 27). Rare tortoises escape hand-luggage hell. Available online at www.thewe.cc/contents/more/archive/august2003/rare_tortoises_escape_hand-luggage_hell.htm.

Oldfield, S. (2003). *The trade in wildlife: Regulation for conservation.* London: Earthscan.

Planet Ark. (2000, March 24). Illegal trade threatens 10 "most wanted" species. Available online at http://www.planetark.org/avantgo/dailynewsstory.cfm?newsid=6111.

Wildlife Protection Society of India. (2004, February 18). Tibet wildlife seizure: India confirmed as source of skins. Available online at www.wpsi-india.org/news/18022004.php.

Govindasamy Agoramoorthy

Conservation and Environment

Conservation Conflicts with Hunting Wild Animals in Asia

People in Asia have been hunting and eating wildlife for centuries. In fact, hunting wildlife in tropical forests has been going on for over 100,000 years, but consumption has greatly increased over the past few decades. According to recent estimates, the annual wildlife harvests are 25,000 tons in Borneo; 160,000 tons in the Amazon forest in Brazil and 1–3 million tons in Central Africa. Because of overhunting, many species in Asia are facing extinction threat. Once large animals such as the Malayan tapirs and monkeys disappear, hunters will target smaller animals. In Asia, wildlife meat is a source of protein and cash for several local communities that live around the forest. Most of the people there are poor and marginalized with few options for alternate livelihood. The rapid loss of tropical forest and unsustainable hunting in Vietnam over the last four decades has resulted in the extermination of twelve species of vertebrates.

The demand for wildlife as food is creating serious concerns for many species. Turtles are suffering the most because of high demand. For example, twenty-five tons of turtles are exported each week from Sumatra, Indonesia. If this situation continues, it may literally wipe out all the turtle species on the island of Sumatra. The demand for turtles for food has decimated wild turtle populations across China. It is not the only reptile that suffers because of overharvesting for food. About 10,000 tons of snakes are eaten in China every year. Poisonous snakes are a delicacy in China. The overharvest of snakes in

Asia has resulted in rodent population increases in several countries, which has also ultimately contributed to severe crop damage and economic losses for farmers.

When the viral outbreak known as severe acute respiratory syndrome (SARS) was first reported in 2003 in Asia, scientists tracked the origin of the disease to the wildlife food market in China's Guangdong Province, where it apparently passed from animals. The disease spread rapidly across several countries in Asia, North America, South America, and Europe, killing 813 people. Makeshift food markets in countries such as China, Hong Kong, Taiwan, Vietnam, Cambodia, Laos, Indonesia, Malaysia, Philippines, and Thailand often hold animals in appalling, overcrowded conditions. Reptiles, birds, and mammals are crammed in cages where they can hardly move or breathe. Live animals are often kept in containers in which the animals on the bottom are crushed by the weight of the animals on top of them. Most often animals are not fed, watered, or sheltered, which provides ample opportunities for communicable diseases (parasitic, viral, and bacterial) to propagate. The methods of slaughtering do not follow humane standards, and some countries in Asia lack appropriate animal welfare regulations. Animals often suffer in markets; for example, frogs are frequently skinned alive, and shells of turtles are usually ripped from their backs while they are conscious. Most live birds are placed in plastic bags for sale.

How do we reduce the threat of extinction caused by overharvesting wildlife in Asia for food? Many believe that a few steps can be taken slowly to counter the drastic effect of this overharvesting. A total ban can be implemented to protect highly endangered species. Public awareness regarding hunting and eating wildlife in Asia can be raised where wildlife is sold and eaten, to educate hunters, traders, and consumers on the consequences of unsustainable harvests and biodiversity loss. Capacity building is also a key issue because concerned countries can learn how to enforce management solutions effectively. The hunting and eating of wildlife in Asian countries must be addressed in conjunction with development efforts to improve socioeconomic conditions of the marginalized people who rely on forest resources. Apart from eating wildlife, domestic pets, mainly dogs, are also eaten in Asian countries such as China, Vietnam, Thailand, Cambodia, Laos, and other places. Some animal-protection advocates believe that education is an important tool to bring awareness among children, for example, to discourage them from eating dogs in Asian countries.

Difficult questions include how to protect the Earth's fragile ecosystems in Asia without denying millions of people a chance for a better life. How do we improve the human living standards without damaging the delicate balance that sustains all life on our planet? As long as humanity is divided into the extremes of rich and poor, these two goals cannot be achieved. These disparities have been deepened because of globalization, increased trade, investment, travel, and other border-transcending changes.

Economic trends indicate that the world economy has grown sevenfold since 1950, but this growth has not relieved poverty. About two billion people worldwide are struggling to survive on income of a few dollars a day or less. Hunger is a widespread phenomenon on our planet: some 815 million people are chronically hungry—not because of lack of food but because they cannot afford to buy. Lack of clean water or sanitation kills 1.7 million people each year—90 percent of them children. Nelson Mandela, the former president of South Africa, has said, "Poverty is the greatest assault on human dignity and unemployment makes it worse" (Shone, 2002). Real wildlife conservation consciousness can come about only through informed, educated, and healthy citizens who are able to place biological conservation into social, political, and economic contexts at local and international levels. Balancing economic development and environmental conservation is a daunting task for future world leaders. Biodiversity conservation should be the bottom line for all countries, and it will depend on measures to use its components sustainably and to manage natural resources in ways that minimize adverse impacts on biodiversity.

Wild pheasants on display in a restaurant in southern China. Courtesy of G. Agoramoorthy.

Dogs, "man's best friend," are widely eaten in China. Courtesy of G. Agoramoorthy.

See also

Animals as Food

Further Resources

Robinson, J. G., & Bennett, E. L. (Eds.). (2000). *Hunting for sustainability in tropical forests.* New York: Columbia University Press.

Shone, H. (2002, December). "Ex-president commends Anglo American and De Beers for their AIDS awareness program." *South Africa Sunday Times.*

Govindasamy Agoramoorthy

Conservation and Environment

*Conservation Medicine Links Human and Animal Health with the Environment**

Last June [2004], Jeff Kaminski was a promising graduate student in Virginia Tech's Department of Fisheries and Wildlife Services, conducting field studies in Appalachia on the effects of logging on small mammal populations. In July of that year, he was dead of acute respiratory distress, the victim of hantavirus pulmonary syndrome (HPS), a rare infection spread by exposure to the saliva, feces, or urine of rodents.

HPS was unknown in the United States until 1993, when it erupted without warning in the Southwest. By late 2003, 353 cases had been reported. Thirty-eight percent of the people infected died.

Why did HPS appear in 1993, and could our evolving climate be a factor? Eric Chivian and Sara Sullivan, both of the Center for Health and the Global Environment at Harvard Medical School, point to an unusual confluence of events in the Four Corners area, where New Mexico, Utah, Arizona, and Colorado come together. A six-year drought, they report, ended that year with heavy snow and rainfall. The drought killed off owls, snakes, coyotes, and foxes, natural predators of the native deer mouse, which then enjoyed a tenfold population increase.

Many more deer mice increased the possibility of human exposure. "In this case," the scientists wrote, "a change in climate triggered the outbreak of a highly lethal infectious disease." They added, "It is not known how many viruses or other infectious agents in the environment, potentially harmful to man, are being held in check by the natural regulation afforded by biodiversity."

In other words, the complex web of interlocking species, treasured by environmentalists but frequently disrupted by human activity, may be valuable for a whole new reason: its delicate balance protects our health.

Vampire Bats: Silent Carriers

In 1998 and 1999, a previously unknown but murderous virus outbreak killed more than 100 people (40 percent of those infected) after showing up on the Leong Seng Nam

*Adapted from "Connecting the Dots: The Emerging Science of Conservation Medicine Links Human and Animal Health with the Environment," by Jim Motavalli, in *E/The Environmental Magazine;* November/December 2004, Vol. XV, no. 6. Norwalk, CT 06851. www.emagazine.com. Used by permission.

pig farm in Malaysia. Horses, cats, dogs, and goats were also infected with the virus, which was named "Nipah," after one of the villages affected. The virus soon spread to Singapore, sickening nine slaughterhouse workers who came into contact with Malaysian pigs.

Why did the specter of Nipah virus first make itself known on a remote pig farm in Malaysia? Scientists now say that the world's largest fruit bat, known locally as a flying fox, was the culprit. The pens that once held thousands of pigs are empty now, but still there are the large, overhanging mango and jackfruit trees that attract the bats. Could it be, scientists speculate, that the wholesale burning of millions of acres of forest in neighboring Borneo and Sumatra, destroying fruit trees, forced the increasingly endangered bats to look elsewhere for food?

It seems certain that the bats found a haven at Leong Seng Nam, that they were harboring Nipah virus, and that they then passed the virus on to the penned pigs (possibly by dropping half-eaten fruit). The pigs subsequently spread the virus to the farm workers who worked in close proximity.

The Consortium for Conservation Medicine (CCM), a new coalition based in Palisades, New York, with a wide-ranging mandate, will test the environmental theory of Nipah virus spread with a four-year, $1.4 million grant from the National Institutes of Health. The grant is just one indicator that the scientific community is beginning to understand that some of our most serious health problems may have environmental roots.

Hantavirus in the United States and Nipah virus in Malaysia are different in many ways, but both bring together human health, animal health, and environmental factors, the three interlocking circles of "conservation medicine." As reported in *Environmental Health Perspectives,* nineteenth-century health care practitioners were expected to have training in the natural sciences (as did Charles Darwin, making his pioneering work possible), but specialization in the twentieth century drove the two fields apart. Today, doctors rarely talk to veterinarians, and neither has much interaction with wildlife biologists. Conservation medicine (some like the phrase "ecological medicine" better) is an attempt to bring them back together. The term "conservation medicine" was first used by M. Koch in a 1996 paper titled "Wildlife, People and Development," and the field has grown dramatically since then.

The emerging field of conservation medicine carries with it a sense of urgency, prompted by a wholesale destruction of ecosystems that were still intact in Darwin's day. Diseases shared by humans and animals are called "zoonoses," and three-quarters of all emerging diseases are zoonotic. "Diseases are moving from animals to humans and from one animal species to another at an alarming rate," says Lee Cera, a veterinarian at the Loyola University Stritch School of Medicine and a principal with the Conservation Center of Chicago. "When I went to school we were told, 'This disease won't go from a dog to a cat.' Then all of a sudden a dog virus decimated the lions of the Serengeti. How did it happen? When did it happen?"

Conservation medicine aims to answer these questions. It is an attempt to integrate complementary fields that had previously worked in isolation: human and veterinary medicine; infectious disease research; public health; and environmental science. Wildlife Trust and its partner, CCM, attempted to bring the parties together at a series of landmark conservation medicine symposia at Columbia University in July 2003.

Parasitologist Peter Daszak, executive director of CCM, cites the West Nile Virus as an example of an emerging human disease transmitted from animal carriers, which is encouraged by greater international travel and commerce. There are many other examples: monkeypox, HIV, hantavirus, avian influenza (which recently emerged as a human killer in Vietnam), Marburg, Pfiesteria, Ebola, and Lyme disease. Severe acute respiratory syndrome (SARS), for instance, was found in three species of wild animals tested in a marketplace in China. As *USA Today* described it, "As encounters between man, beast and the germs they carry increase, more strange new diseases can be expected to emerge."

At the heart of the problem are the same environmental issues that set up a conflict between development and ecosystem protection. Loss of animal habitat and increasing human incursion into wilderness areas (often spurred by human population growth) sets up new points of contact. International trade in exotic species breaks down previously existing barriers. Climate change causes species migration. Global travel, including ecotourism (which emphasizes wilderness visits) can move exotic jungle viruses into the modern world, as dramatically documented in Richard Preston's best-selling book *The Hot Zone*. In 1950 three million people a year flew on commercial jets; by 1990, 300 million did. Two million people cross international borders every day, carrying with them huge amounts of agricultural products, live animals, soil, ballast water—and disease-causing microbes.

In 1972 scientist Kent Campbell came down with a serious illness on a visit to Ireland. A senior scientist at the Centers for Disease Control, Thomas Monath, had discovered that rats carried the frightening and frequently lethal African microbial disease Lassa. The disease could be spread by airborne transmission, and Campbell was himself a victim of it. He survived (after being airlifted inside an airtight Apollo space capsule from London to Washington), and the world gained a new perspective on the ability of formerly exotic diseases to get a foothold in the modern world.

According to Louise Taylor of the UK Centre for Tropical Veterinary Medicine, 60 percent of all the 1,415 known species of infectious organisms that affect human health (causing a quarter of the world's deaths) can be transmitted by animals. Approximately 175 of these infectious organisms are linked to diseases that have only recently emerged, or have increased in severity (and geographic distribution) in recent years. There are sixty-three emerging diseases just among marine life, reports the book *Conservation Medicine,* and these include tuberculosis in fur seals and chlamydiosis in sea turtles.

Dying Frogs

Should we be alarmed by the worldwide disappearance of frogs? Are they an indicator species, a harbinger of global environmental crashes ahead? Jasper Carlton, director of the Biodiversity Legal Foundation, told *High Country News* that he believes that frogs and other amphibians are the proverbial canary in the coal mine. "Leopard frogs, boreal toads, spotted frogs and tiger salamanders are experiencing serious declines," he says. "We often attribute species decline to habitat destruction. What is particularly alarming is that many amphibians occupying undisturbed wilderness habitats are also disappearing at a previously unseen rate. These declines appear to be widespread and have been particularly serious for 20 years."

In 1993 environmental officials in Australia asked Rick Speare of James Cook University to help investigate the mysterious disappearance of upland frogs in Queensland. In this case, there was no shortage of evidence because dead or dying frogs littered the O'Keefe Creek study area near Cooktown. Identifying the problem was of crucial importance, not least because the study area is the last known habitat of the sharp-snouted day frog.

The scientists benefited from the opportunity to make on-site pathological examination. Bacterial septicemia was quickly ruled out, with the evidence pointing to a toxic or preacute viral cause. The only consistent lesions found on the specimens studied were heavy—but unfortunately unidentified—skin infections. Such infections had been observed before, but had been dismissed as related to minor parasites.

But further study, in part through a grant from Australia's Nature Conservation Agency, showed that this "minor" skin parasite was, in fact, the primary pathogen. The culprit was identified as an undescribed variation of the amphibian chytrid fungi, or chytridiomycosis. The chytrid has now been identified in twenty-three species of Australian frogs, seven of them endangered. This same parasite caused 100 percent mortality in a mass die-off in Panama and was also identified as a cause of mortality in frogs at the Washington Zoo. It has also been found in Africa and Europe.

What emerged was an international detective story. The parasite varies little from continent to continent, so evidence suggested it had recently migrated around the world. Enter the Consortium for Conservation Medicine, a collaboration of the Wildlife Trust, Harvard Medical School, the Johns Hopkins Bloomberg School of Public Health, the federal National Wildlife Health Center, and the Center for Conservation Medicine at Tufts that "strives to understand the link between anthropogenic environmental change, the health of all species, and the conservation of biodiversity."

Working with a National Science Foundation grant, the Consortium is investigating the possibility that the carrier is *Rana catesbeiana,* a bullfrog that is also a globally traded food item. Although Rana is itself relatively resistant to chytridiomycosis, it may be an efficient carrier of it. The frogs of Australia (part of a pattern of declining amphibian populations around the world since the late 1890s) may be victims of what the Consortium calls "pathogen pollution," the anthropogenic introduction of nonnative hosts or parasites to new locations.

Medical Hubris

Laurie Garrett, author of *The Coming Plague: Newly Emerging Diseases in a World Out of Balance,* writes that in the post–World War II environment, powerful medical weaponry (antibiotics, vaccines, water treatment, anti-malaria drugs) gave scientists confidence that they could eradicate infectious disease from viral, bacterial, or parasitical sources. In 1900 nearly 800 Americans out of every 100,000 died each year of infectious disease. By 1980, the number was down to 36 per 100,000. The Health for All accord, signed in 1978, set a goal of the year 2000 for eliminating many international scourges. But amid all this optimism, the numbers started rising. In 1995, 63 people per 100,000 died.

"The grandiose optimism rested on two false assumptions," Garrett wrote in the journal *Foreign Affairs,* "that microbes were biologically stationary targets, and that diseases could be geographically sequestered." Scientists, she said, "have witnessed an alarming mechanism of microbial adaptation and change. . . . Anything but stationary, microbes and the insects, rodents and other animals that transmit them are in a constant state of biological flux and evolution."

Conservation medicine is a realization that modern science is fighting a new kind of war, one that we are ill equipped to wage. Conservation medicine is still a very small field, but it is increasingly gaining recognition from mainstream funding sources, such as the National Science Foundation, the World Bank, the National Institutes of Health, and private grant-making foundations. Ongoing studies are both uncovering new disease pathways (from animals to humans and vice versa) and helping devise effective treatment. Here are some examples:

- **Rwanda:** The Volcano Veterinary Center was created in 1986 at the request of renowned mountain gorilla researcher Dian Fossey to provide emergency care to Rwanda's sick or injured gorilla population. One possible explanation for a high death rate among mountain gorillas noted in the late 1980s is an outbreak of measles. Mountain gorillas share 97 percent of their genetic makeup with humans and are very susceptible to human diseases. Contact with them has increased exponentially as their fame has grown. Without question, their lives have been disrupted by human contact. Ecotourism is one avenue of contact, and the standards for tourists visiting the great apes are more relaxed than those for visitors to zoos or primate centers. The increasing human population (with a growth rate of 3.7 percent annually) in the region is also a threat. Gorillas have close encounters with trackers, guides, researchers, and veterinarians, not to mention poachers and farmers. Bacteriological studies have shown the presence of salmonella, *Cryptosporidium parvum*, the parasite giardia, and campylobacter among gorilla populations. Gorillas have become habituated to human presence, and "there is a concern that the habituation is enhancing transmission of pathogens infectious to both people and the gorillas," says parasitologist Thaddeus Graczyk of the Johns Hopkins Bloomberg School of Public Health, who also works with penguins that have been infected with avian malaria from North America.

- **New England:** The alarming declines in common loon populations in New England are being studied by Dr. Mark Pokras of the Tufts Center for Conservation Medicine (CCM). Mercury poisoning is believed to be a cause. "The common loon serves as an important environmental sentinel for mercury because, like humans, it feeds on freshwater fish," Tufts CCM reports. The center has documented weight loss and death in common loons resulting from mercury poisoning, which comes from local sources and arrives via aerial transportation. The Wildlife Conservation Society reports that pending Bush administration proposals to relax standards on mercury emissions from coal-fired power plants could further adversely affect common loon populations (already declining precipitously) in the Adirondacks. "Models indicate that, partly due to mercury contamination, reproductive rates of loons may already be too low to maintain their populations in portions of Maine and eastern Canada," says David Evers of the Adirondack Cooperative Loon Program. Another result of human interaction is lead poisoning resulting from ingestion of fishing sinkers. Dr. Pokras has successfully influenced the Massachusetts Fish and Wildlife Agency to regulate lead sinkers in the Quabbin and Wachusett reservoirs, and they have been banned in New Hampshire and Maine.
- **Peru:** Researchers are making a link between destruction of the Amazonian rain forest and an explosion of malaria-bearing mosquitoes that thrive in sunlit ponds, according to a report in the journal *Nature*. A team from Johns Hopkins University collected 15,000 mosquitoes from a jungle road in northeastern Peru and counted how many were *Anopheles darlingi*, which transmits malaria. They then tabulated their results with statistics on deforestation using satellite images. An even 1 percent increase in deforestation increases the number of malaria-bearing mosquitoes by 8 percent, says researcher Jonathan Patz. The study showed that the insects "ran wild" after 30 to 40 percent of the forest was destroyed. Malaria researcher Phil Lounibos of the University of Florida points out that the problem would not be as acute if the *A. darlingi* mosquitoes had not been imported in the first place—a direct result of the establishment of tropical fish farms in Peru.

The Wildlife Trust: Emerging Leaders

Conservation medicine clearly needs a well-organized champion, able to synthesize the vast amounts of new scientific data from disparate sources. That work has fallen to the Wildlife Trust, which shares a leafy campus along the Hudson River in Palisades,

New York, with Columbia University's Lamont-Doherty Earth Observatory and CCM, whose work it fosters.

Wildlife Trust has a long history. The parent organization was founded in 1963 by British naturalist and author Gerald Durrell (brother of Lawrence Durrell, author of the "Alexandria Quartet" books). Mary Pearl, the executive director of Wildlife Trust, describes Durrell as "the Marlon Perkins of England," with a wide following for his animal-themed books. Durrell became convinced that zoos had a responsibility to carry out conservation work, and to that end, he started breeding colonies of endangered animals at Jersey Zoological Park, which he founded. His work pioneered inter-zoo exchanges of animals and scientific information. Today, the British organization he founded continues as the Durrell Wildlife Conservation Trust, whereas the U.S.-based Wildlife Trust that developed from it (founded in 1971 and originally known as Wildlife Preservation Trust International) has undertaken a different mission.

There are many overlaps between the U.S. and British groups, however. The Durrell Trust has been active in attempting to restore critically endangered black lion tamarins (which live on just 2 percent of their historical forest habitat) to the wilds of Brazil. Three were reintroduced in 1999. The Wildlife Trust also works with black lion tamarins (and uses one on its logo), but its work concentrates on improving and connecting isolated pockets of tamarin habitat in Brazil.

The Wildlife Trust is not just the "go to" organization on conservation medicine; it virtually launched the discipline. The Trust conducts original research, bringing together teams of physicians, vets, ecologists, wildlife epidemiologists, and public health officials to study the many strands of emerging diseases.

"We take a complex, multidisciplinary approach," says Pearl, who came to the Trust in 1994 from the Wildlife Conservation Society. "The stumbling block with many scientists is that they focus on pathogens in wild animals without considering the full environmental picture of how they got there."

The same problem exists with media accounts of disease outbreaks. Pearl points to a U.S. outbreak of monkeypox (a squirrel and rat virus that can also affect humans) in the Midwest in 2003. Monkeypox is rare, and it usually occurs only in rain forests in central and western Africa. How did it get here? It turns out that pet traders in Wisconsin brought in an infected rainforest rat from Gambia. It infected the dealer's prairie dogs, which were then sold at a "pet swap" attended by people from other Midwestern states. Some of the prairie dogs got sick with a disease serious enough to kill 10 percent of those infected. The Centers for Disease Control and Prevention eventually confirmed thirty-seven cases in five states.

It was a sensational story, treated as such by the media. "What the news stories don't tell you is that these disease outbreaks are predictable and therefore somewhat preventable," says Pearl. "But we have to systematically address the pathogen pollution that can occur in live animal markets, the exotic pet trade, unmonitored travel from outbreak areas and intensified livestock operations that reduce animal immunity to wildlife disease."

Alonso Aguirre, director for conservation medicine at Wildlife Trust (and both a veterinarian and wildlife biologist), specializes in the diseases of marine animals. He has seen pathogens that not only have the potential to wipe out critically endangered species (such as the fibropapillomatosis tumors that increasingly infect green, loggerhead, and olive ridley turtles), but also have the potential to move back and forth between unrelated animal species and to infect humans. Aguirre cites seal populations with toxoplasmosis (a disease usually found in cats that can be a danger to pregnant women who change litter boxes) and Dutch seals with influenza B virus (identical to the human form that broke out in the Netherlands in 1995). Canine distemper has also infected both seals and dolphins in Western Europe and Russia. One theoretical method for human-seal transmission is marine mammal rehabilitation, which involves physical contact with possibly sick animals.

Aguirre notes that occurrence of the turtle tumors has been associated with such man-made phenomena as heavily polluted coastal areas, high human populations, agricultural runoff, and biotoxin-producing algae. He has made a close study of endangered and declining Hawaiian monk seals, which have many challenges, including human overfishing and a disease that causes blindness. The sight problem ("an ocular condition of unknown etiology") was first noted in twelve female pups brought into captivity for rehabilitation purposes in 1995. "The blindness could have been caused by something human," Aguirre says. "It's possible our activity has introduced a pathogen."

Aguirre cites many such examples of diseases crossing species boundaries and international borders. "We should be very worried," he says, "because several diseases or pathogens have been linked to the wildlife and bushmeat trade, and to wildlife translocations worldwide." In addition to monkeypox, he cites SARS and new HIV-related viruses acquired through African consumption of wild-caught bushmeat. "In addition," he says, "we have ticks with Rift Valley fever carried in with tortoises imported from Africa, and exotic Newcastle disease introduced from the illegal bird trade, both turning up in California."

The solutions are potentially drastic, says Aguirre, who calls for increased enforcement and surveillance of the illegal wildlife trade, especially at airports. "It's a very difficult issue to tackle," he says, "because the trade is also tied in with drug and gun smuggling." He also calls for reform of international endangered species treaties (like the Convention on International Trade in Endangered Species of Wild Fauna and Flora, or CITES) and organizations (like the World Conservation Union) in areas where they have become "politicized, corrupted or inefficient."

Wildlife Trust casts a wide net, with projects ranging from sea turtle health assessments in Long Island Sound (part of a program on the New York bioscape) and manatee rehabilitation and evaluation in Florida to cattle impacts on tapirs in Argentina and flamingo health studies in Chile. Governed by a volunteer board, it has an annual budget of $4 million and an endowment of $7 million.

Frontline Research

The Wildlife Trust administers CCM, which conducts collaborative scientific research on emerging infectious diseases, pathogen pollution, climate change, the health of marine systems, and problems affecting endangered species. CCM works with veterinary and medical students, promotes conservation medicine at workshops and conferences, and informs policy makers through congressional briefings and other forums. CCM is studying West Nile virus through a seven-year, $557,000 contract from the New York State Department of Health. The Consortium is studying how the loss of bird biodiversity could produce high levels of West Nile infection and also how carrier mosquito breeding is affected by local drought and flood cycles. It is also working to keep West Nile out of Hawaii, where it poses a dramatic threat to the islands' remaining native birds.

Another CCM project is based in the Rocky Mountains and studies interspecies disease transmission. Brucellosis, for instance, can move among elk, bison, and cattle, resulting in controversial bison "roundups" when the animals stray out of Yellowstone Park. Other transmittable diseases include anthrax, affecting bison and livestock; rabies; whirling disease (present in trout in Yellowstone); and chronic wasting disease, which affects western deer and elk, bearing similarities to mad cow disease. Under the direction of Colin Gillin, CCM scientists are studying the possible effect of opening wildlife "corridors" on the health of livestock, wildlife, and humans.

And, of course, CCM studies amphibian declines. Peter Daszak was a co-discoverer of chytridiomycosis, the fungal disease that is now associated with mass amphibian die-offs in Panama and Australia. "The fungus has been around for a long time," Daszak says. "It's found on amphibians in museums, and it blocks the way they breathe through their skin. But why did it suddenly spread to devastating effect? We started to see a pattern in the huge trade of amphibians for food, and for the pet trade. Bullfrogs, for instance, don't die from the infection but they spread it."

Monitoring how diseases spread is an important part of CCM's work. West Nile is a passenger in airplanes, says Daszak, and its spread is facilitated by dramatic increases in air travel. He describes a plausible scenario by which West Nile could reach Hawaii. There is an average of one and a half mosquitoes per flight, he says, and although only a tiny percentage are infected with West Nile, the law of averages suggests that a carrier will eventually make it off a plane and bite a bird. "And once one mosquito slips through, it will be only a matter of days before West Nile is all over the island," Daszak says, noting that avian malaria has already wiped out a third of the endemic bird species in Hawaii. (Native birds are surviving at high altitudes, above the reach of mosquitoes, but global warming is sending the biting bugs higher and higher each year.)

As part of its work, CCM has met with key health and military officials in Hawaii and has found them very supportive. "But it's a difficult agenda to push proactively," Daszak says. "If we're successful, nothing happens."

Environmental Factors

In most of the cases CCM studies, environmental factors play a huge role. Nowhere is this more true than with deadly Nipah virus, for which there is no known cure. Daszak says that the widespread deforestation in Sumatra, fueled by paper industry logging, eventually meant that fruit bats "ran out of resources. It's not enough to say, 'Let's eliminate fruit bats.' It's a much more complex situation than that." And now CCM's work has spread to Bangladesh, India, and Madagascar, all of which harbor fruit bats and different strains of virus.

Sharon Collinge of the University of Colorado–Boulder studies incidence of bubonic plague (the same disease that killed a third of Europe's population in the fourteenth century)—not in humans, but in prairie dogs. When it reaches a prairie dog colony, plague can kill 95 to 99 percent of its residents. Plague no longer kills humans quite as efficiently as it once did because it is curable with antibiotics, but ten to twenty cases still occur each year. Between 1957 and 2000, forty-eight plague cases were reported in the United States. (nine of them fatal). In seven of those cases, prairie dogs were implicated as the carrier.

Collinge works in Boulder, which had a 14 percent human population rise between 1990 and 2000. The city's undeveloped growth boundary—an attempt to stop urban sprawl—hosts 218 prairie dog colonies, most of them in close proximity to human populations. The close quarters provide a possible pathway for plague to spread from "dogs" to humans.

Should people be alarmed? Collinge's team hasn't found plague in Boulder's prairie dogs, though outbreaks have occurred elsewhere. Hunters cite their carrying disease as one reason to pull the trigger, but Collinge says that prairie dogs "are not good plague carriers, and getting rid of them would not stop the spread of disease" (which could be taken up by other small animals). Further, she says that prairie dogs are "both keystone species and ecosystem engineers," providing food for eagles and, through their burrowing, increasing the capacity of the soil to hold water and vegetation.

Collinge's work represents an ideal cross-pollination of medical and veterinary sciences. Dr. Mark Pokras of the Tufts University School of Veterinary Medicine says such cross-boundary collaboration is exactly what is needed. "We need to get the vets out of the barn," he says. "We need to change the mindset of all the groups—vets, physicians, scientists, conservation biologists, environmentalists—to be more broadminded and visionary."

Pokras notes that the veterinary profession has always been driven by economic factors, first working on horses in the 1700s and then moving on to cattle, sheep, pigs, and other economically important species in the 1800s. Pets came later, as people acquired the means to keep nonwork animals. Only very recently have vets worked for zoos and nonprofit groups. "Starting in the early 1980s," Pokras says, "a variety of veterinary schools—including Tufts and Cornell—took big steps to get involved in conservation projects. But vets still mostly talk to their peers, and publish in their own journals."

One hurdle, Pokras says, is recognizing the skills of other professions and learning to speak the same language. As noted in *Conservation Medicine,* the first book on the subject, "The rich terminology of the biomedical and veterinary sciences poses particular difficulties for ecologists and conservation biologists." For instance, even the word "ecosystem" has different meanings for the different interest groups.

"We see the world in a different way," Pokras says. "It's like the five blind men and the elephant. But we need to overcome this problem because conservation on a global scale is so complex that no one group has the knowledge, skills and perspective to grasp it all and develop appropriate solutions."

Weapons of Mouse Destruction

An 1860 painting by Gustave Corbet titled *Fox in the Snow* provides a dramatic example of the age-old predator–prey relationship. The fox catches the squirrel, just as it has for centuries. But what happens when humans inadvertently disturb the balance of nature by removing the predators? The role of predators in helping suppress disease is highlighted in work by Richard Ostfeld of the nonprofit Institute of Ecosystem Studies (which combines research and education work). Mammals are the most common reservoirs of zoonotic disease, and rodents play the leading role, implicated in the spread of plague, Lyme disease, hantavirus, and Rocky Mountain spotted fever.

"We know that predators affect prey numbers," Ostfeld says. "If mice are a zoonotic disease reservoir, and the human infection escalates with reservoir abundance, habitats that include foxes would have a lower incidence of disease."

In the case of Lyme disease, for example, a bacterial pathogen causes the disease to occur in white-footed mice, which then pass it on through blacklegged ticks. Without the predators that prey on mice, their population explodes, which increases the chance that infected ticks will cause human Lyme infection. Ostfeld says the process of suburbanization, reducing forests to small fragments, increases risk because these parcels support fewer predators. "The risk of human exposure is four or five times higher in smaller forest fragments less than five acres than it is in larger parcels," he says. "That's where the weapons of mouse destruction come in."

Jeff Kaminski, the Virginia graduate student, was working with white-footed mice. "We're going to see more cases of that kind," Ostfeld says. "Rodents are resilient to human disturbances and they're reservoirs for pathogens that can attack people. As we encroach on and modify natural habitats, allowing rodent populations to explode, these outbreaks will increase. The evidence is very convincing that we're engaging in risky behavior. We need the political will to change how we modify the environment. I'm hopeful that we can stop habitat destruction, because if we reduce habitat fragmentation there's an immediate positive effect. Disease risk can be reduced in decades."

Few people would connect the loss of foxes and other predators to outbreaks of Lyme or West Nile, but that is exactly why conservation medicine is such an important new field. As Ostfeld notes, there are now tens of thousands of Lyme disease cases each year, but West Nile is catching up, spreading "at a phenomenal rate, several hundred miles per year. In five to 10 years it might surpass Lyme." And according to the latest research, fragmentation and loss of biodiversity play a part in both diseases.

Addressing the Problem

A major report on changing ecosystems and their impact on human health was published from the Millennium Ecosystem Assessment, convened by the United Nations, in 2005. Also shedding light on conservation medicine is a new journal titled *EcoHealth*.

The Tufts Center for Conservation Medicine is helping to create the Atlantic Coast–based Seabird Ecological Assessment. It is also part of an ambitious, multiyear research project called Yellowstone to Yukon (Y2Y) that is studying wildlife issues in the last tracts of wilderness in continental North America.

Columbia University's Earth Institute is using science and technology to assist public health efforts, through (among other tools) natural resource management and biodiversity preservation. Its Goddard Institute for Space Studies has examined the impact of global warming on urban environments, using New York City as a model. Cynthia Rosenzweig, a senior Goddard research scientist, says the alarming loss of wetlands in Jamaica Bay is in part due to global warming. "Our researchers realized that something was happening out there that went beyond the usual stresses on this highly manipulated ecosystem," Rosenzweig says.

The Global Fund is a multibillion dollar international financing mechanism designed to help developing countries fight the infectious diseases AIDS, tuberculosis, and malaria with practical initiatives. The Earth Institute's Center for Global Health and Economic Development is supporting the Global Fund through its Harlem–based Access Project, which helps developing countries apply for fund money and then launches on-the-ground programs.

Josh Ruxin, an assistant professor of public health at Columbia, runs the Access Project, which has worked in such African countries as Ethiopia ($11 million in first-round funding), Nigeria ($28 million), Malawi ($42 million), and South Africa ($26 million). Devastated by AIDS and other plagues, many African countries have experienced a steep decline in life expectancy. In Zimbabwe, torn apart by both a political and public health crisis, life expectancy has plummeted more than 40 percent since 1990. In 2000, according to a UN report, the average Zimbabwean could expect to live to be 33.9 years old.

Ruxin, who started the Access Project during an earlier stint at Harvard, says that "the state of public health pedagogy is antiquated and not suitable for the global health disasters we face today. I observed that there were no health systems in place, no money, and no good management expertise to confront these modern scourges."

Ruxin points out that while the Global Fund has received $4 billion in global commitments, it has actually received only $1.8 billion from donor countries. The U.S. is part of the problem, he says, noting that President Bush has not lived up to the $15 billion commitment he made to fight AIDS around the world.

Both malaria and tuberculosis are on the list of diseases that confident post–World War II public health officials thought would be eradicated by 2000. Instead, they've become Third World scourges. Active tuberculosis cases can be treated with rounds of drug therapy if funding is available, says Ruxin, and prevention would make great strides if insecticide-treated bed nets were universally available.

Awash Teklehaimanot is director of the Center for Global Health's malaria program and an internationally known expert on the mosquito-transmitted disease. Far from being eradicated, he says, malaria is now a threat to 40 percent of the world's population in 90 countries. There are 500 million clinical cases each year, and 2.7 million deaths. Most victims are children under five.

Malaria is at the center of conservation medicine controversies. Temperature is important for mosquito breeding, so populations can soar because of climate-induced warming. Rainfall creates the pools that mosquitoes breed in, and paradoxically so does drought (by drying up flowing rivers and leaving stagnant water).

An often-touted public health treatment for malaria is long-lasting DDT, the scourge of environmentalists since its role in ecosystem poisoning was exposed in Rachel Carson's *Silent Spring*. Teklehaimanot argues that DDT, if contained in low-volume, localized indoor spraying, need not create environmental disasters. (Although banned in the United States, it is still produced in Ethiopia and other countries.) Other aerosol insecticides last only a few hours; DDT, however, has a life of nine months to a year, making it a cost-effective treatment for underdeveloped countries.

"Malaria is a disease of poverty," says Teklehaimanot, "and this is one of the cheapest treatments available." In 1999, the World Wildlife Fund called for a global ban on DDT, claiming that up to 82 percent of the pesticide escapes into the environment. More recently, however, Teklehaimanot says WWF removed DDT from its list of 12 persistent organic pollutants presented at the Johannesburg summit, having designated it acceptable for certain public health uses. It has also been endorsed for that purpose by the World Health Organization (WHO).

Despite these signs of resurgence, however, DDT will remain a very controversial chemical. And this will be one more topic under discussion in the growing and increasingly important field of conservation medicine.

Public Health at a Crossroads

In her 2000 book *Betrayal of Trust: The Collapse of Global Public Health,* the aforementioned author Laurie Garrett outlines an international crisis, a complement to her earlier *The Coming Plagues*. She cites a partially classified 2000 CIA report that predicted widespread deterioration of global health. Its key indicators were "persistent poverty in much of the developing world, growing microbial resistance and a dearth of new replacement drugs, inadequate disease surveillance and control capacity, and the high prevalence and continued spread of major killers such as HIV/AIDS, tuberculosis and malaria."

A WHO report estimates that infectious disease causes 25 percent of global deaths. The present infrastructure is ill-equipped to handle this growing burden. As Garrett points out, WHO itself, "once the conscience of global health, lost its way in the 1990s. Demoralized, rife with rumors and corruption, and lacking in leadership, WHO foundered." Taking up the slack, she reports, was the World Bank, which became the world's largest public health funder by 1997.

But this is the same World Bank whose renewable energy portfolio is approximately $200 million per year, compared to the $2.5 billion it loans for other energy projects, most of them based around global warming–aggravating fossil fuels. These loans are very much a health issue. WHO estimates that 160,000 people die annually because of the effects of climate change.

Mark Walters, author of *Six Modern Plagues and How We Are Causing Them,* sees a "perfect storm of emerging disease." Humans, he says, "are animals, and we all share the same disease grid. No longer can we pretend that we're on some kind of pedestal above creation."

Walters asks, "How late are we in combating this avalanche of new diseases?" He traces successive waves of epidemics that began 10,000 years ago, when humans first domesticated animals. "Mutant cow viruses gave rise to smallpox," he says. "Measles developed from distemper, a virus in dogs. These animal-to-human exchanges caused wave after wave of major epidemics. But then it equilibrated, as people began to live with diseases. But civilizations built up, commerce developed, and populations began to mix, precipitating what may now be the fourth great wave of historical epidemics. And for the first time, we know why these epidemics are occurring."

Our growing medical knowledge, coupled with the relative stability of the Earth's ecosystems, led to centuries of improving human life expectancy and well being. But we're upsetting that stability and disrupting those ecosystems. "We're giving up the home court advantage, upsetting the evolutionary playing field," says Walters, whose work as a journalist and author is complimented with a veterinary degree. "Over the past century or more, humans have so disrupted the global environment and its natural cycles that we risk evicting ourselves from our shelter of relative ecological stability."

This kind of talk makes Walters a Cassandra in the eyes of some conservative thinkers. *National Review* wrote that the book "resembles an age-old religious pronouncement—and a misguided one at that." Its sin? "Worshipping Mother Nature." The reviewer surveyed our progress in eradicating some former scourges, such as smallpox, and concluded, "Those suffering the most from disease epidemics need more trade and economic growth to escape from—not return to—the life of the primitive."

But victories with smallpox and other diseases are far from enough to get us to the goal so confidently outlined by the 1978 Health for All accord, which predicted a near-total victory over infectious disease by 2000. Researcher Jonathan Patz of the Johns Hopkins Bloomberg School of Public Health estimates that 2.5 million people are at risk from dengue fever infection, spread primarily by the *Aedes aegypti* mosquito, with between 250,000 and 500,000 cases of the most severe form occurring every year.

Malaria is an even deadlier scourge. And as Patz has written in his studies of the Northern Peruvian Amazon, a pattern of development, road construction and logging have resulted in a fifty-fold increase in malaria cases. It's pretty simple, really. The destruction and clearing of ancient rainforests for development encourages mosquito breeding. And when those mosquitoes are infected with the emerging diseases we thought would be long gone by now, it makes people sick.

See also

Health—*Animal Reservoirs of Human Disease*
Health—*Diseases between Animals and Humans (Zoonotic Disease)*
Health—*Epizootics: Diseases That Affect Animals*

Further Resources

Aguirre, A. A., Ostfeld, R. S., Tabor, G. M., et al. (Eds.). (2002). *Conservation medicine: Ecological health in practice.* New York: Oxford University Press.

Conservation Medicine Center of Chicago (CMCC). http://www.luhs.org/depts/cmcc. [Loyola University's (Chicago) Health Science Web site on conservation medicine. "The Conservation

Medicine Center of Chicago (CMCC) is a collaboration among the Chicago Zoological Society, which operates Brookfield Zoo; Loyola University Chicago Stritch School of Medicine; and the University of Illinois College of Veterinary Medicine. The Center, which uses facilities at the three institutions, brings together a unique team of physicians, veterinarians, researchers and clinicians in many disciplines."]

Consortium for Conservation Medicine, The. http://www.conservationmedicine.org/. [The Consortium is "a collaborative institution that studies the link between anthropogenic environmental change, the health of species, and the conservation of biodiversity."]

EcoHealth. http://www.ecohealth.net [Published quarterly in hard copy and online.]

Tufts Center for Conservation Medicine. http://www.tufts.edu/vet/ccm/. [The Tufts Center for Conservation Medicine "brings together veterinarians, physicians, ecologists, and conservation professionals to develop education and research activities that explore the relationships among animal, human, and environmental health."]

University of Minnesota—Conservation Medicine. http://www.tropical.umn.edu/TTM/Conservation/index.htm. [The University of Minnesota's conservation medicine Web site].

Wildlife Trust. http://www.wildlifetrust.org. [An international organization of scientists promoting wildlife diversity conservation.]

Jim Motavalli

Conservation and Environment
Conservation Psychology

Conservation psychology is the scientific study of the reciprocal relationships between humans and the rest of nature, with the goal of encouraging conservation of the natural world. This relatively new field is oriented toward conservation of ecosystems, conservation of resources, and quality of life issues for humans and other species. In addition to being a field of study, conservation psychology is also the network of researchers and practitioners who work together toward a common goal. Most of the research questions address the following outcome areas:

- How humans care about/value nature, with the goal of creating harmonious relationships.
- How humans behave toward nature, with the goal of creating sustainable relationships.

With the daunting environmental challenges facing humanity, people are both the source of the problems as well as the hope for solutions. As the science of human thought, feeling, and behavior, psychology has much to offer. Environmental psychology is similar to conservation psychology in that it studies interactions between humans and the environment and applies psychological approaches to the solution of environmental problems. However, conservation psychology has a stronger emphasis on human-nature relationships and a more explicit outcome orientation. Like conservation biology, conservation psychology is directed toward the mission of biodiversity conservation and environmental sustainability. It actively invites contributions from the many subdisciplines of psychology, as well as other social and natural sciences, in order to promote connections with the natural world and stewardship behaviors.

Caring About and Valuing Animals and Nature

One research area for conservation psychologists is how people emotionally connect with nature and develop a way of valuing the living world. Topics of study include human-animal relationships, empathy, environmental sensitivity, a sense of place, significant life experiences as precursors of environmental concern, the role of directed attention in nature, individual and collective environmental identities, and moral reasoning about animals and the natural environment.

Animals are a compelling part of the human experience. Because humans are social creatures, animals appeal to our propensity to interact socially. Care occurs naturally toward individual animals when they visibly respond to the caring, whereas abstract entities such as ecosystems do not provide such tangible feedback. Researchers are just beginning to explore how the human emotions of love, caring, and connection might be extended more widely to the biotic world. The "natural care" for specific animals may potentially provide a bridge to caring about the natural world in general. There is some evidence that perception can be shaped by processes of joint attention, when people notice the same thing together. Moreover, when people have satisfying experiences in nature, they are motivated to explore further and develop competence.

As we lose biodiversity, every generation has fewer possible experiences with nature and psychologists are only beginning to understand the implications for such loss. One body of research looks at the psychological benefits of experiences in nature. Studies have documented human preference for natural settings, how humans benefit from and are affected by the natural world, how interactions with nature positively affect multiple dimensions of human health, and the effect of nature on spiritual well-being. According to the biophilia hypothesis, the human species evolved in the company of other life forms, and we continue to rely on the quality and richness of our affiliations with natural diversity. Conservation psychology research explores how a healthy and diverse natural environment is an essential condition for human lives of satisfaction and fulfillment. It also studies how to address the despair associated with the loss of natural environments.

Understanding our relationship to the natural world well enough so that we have a language to celebrate and defend that relationship is another research area for conservation psychology. A fundamental challenge for many cultures is developing ways to talk about humans as part of nature, not separate from nature. A new language of conservation will be supported if there are abundant opportunities for meaningful interactions with the natural world in both urban and rural settings. Such a language will offer new tools for education and communication outreach efforts. Psychological research has already produced an extensive set of strategies for persuasive communication, how to talk about risk and loss, how to talk about threats that are temporally and spatially distant, the role of the messenger for different audiences, the use of metaphors and stories, and approaches for framing and developing messages that connect to existing values within a culture. Such strategies are beginning to be more widely applied by environmental educators and communicators.

Conservation Behaviors

In addition to the longer-term goal of developing an environmental ethic, the transition to global sustainability will require more immediate changes in human behavior. A key assumption underlying environmental sustainability is the need to decrease the negative impact of humans, as well as the need to encourage environmentally friendly behaviors. Collectively, any activities that support sustainability, either by reducing harmful behaviors or by adopting helpful ones, can be called conservation behaviors. Achieving more sustainable relationships with nature will require that large numbers of people change their reproductive and consumptive behaviors.

An increasing number of conceptual models related to pro-environmental behaviors have been developed by psychologists and other social scientists. Many look at the relationships between environmental knowledge, concern, attitudes, beliefs, values and behaviors, and they often link to well-established theories and constructs from social psychology, cognitive psychology, and other disciplines. Although it is difficult to directly compare these models and their underlying assumptions, they each suggest behavior changing strategies that can be used by practitioners. More studies are needed that directly compare the explanatory power of the various models empirically, as well as studies that allow us to match models to critical contextual variables.

Some of the obstacles to behavior change that have been identified include the direction and strength of attitudes, lack of competence or knowledge of what to do, social norms and cultural beliefs, incentives or disincentives, structures such as laws, regulations, technology, and the broader socioeconomic and political context. Past experiences, knowledge, and fundamental motivations related to control and belonging can all influence behavior by changing a person's interpretation of the current context.

Emotions are known to contribute to nature protection behaviors but they have not been as well studied. New measures of "connectedness to nature," environmental identities, and the emotional dimensions of human-nature experiences are allowing studies of the links between a psychological connection with nature and environmental sustainability. Nature experiences have been shown to influence environmental action, especially in everyday situations with low task complexity. In comparison, there has been more research on environmental values. In psychology, studies have tended to distinguish egoistic (focus on self interest), humanistic (focus on other humans), and biocentric (focus on nature) values. Humanistic and biocentric values are both related to environmentalism, and the conditions under which these values are expressed continue to be explored. Psychologists are also investigating the base of universal biocentric valuing across cultures.

Interventions to change behavior typically use one or more of the following approaches that have emerged from the research literature:

- Show that the benefits outweigh the costs
- Arouse emotions that are appropriate for the situation
- Change a person's appraisal of the situation based on the seriousness of the problem and their perceived role
- Fit messages to personal values and demographic characteristics
- Create opportunities for action
- Provide social support for action utilizing techniques such as incentives, prompts, and commitments

The success of such interventions can be documented through evaluation.

In sum, conservation psychology is the science and practice of connecting people to various aspects of nature and promoting conservation behaviors in order to create a more harmonious and sustainable world for all living beings. Theories, principles, and methods from psychology and other social sciences are used to address the human dimensions of conservation. The research areas are theoretical (developing conceptual models), applied (identifying effective strategies), and evaluative (measuring success), and the research process is an iterative one. Practitioners in the field, such as environmental educators and communicators, play a strong role in helping shape research questions. Collaborations with natural scientists are also necessary to ensure that the combined efforts have positive effects on high-priority ecological functions and features. The ultimate success of conservation psychology will be based on whether its research results in programs and applications that increase environmental sustainability and improve the quality of human-nature relationships.

See also

Conservation and Environment—*Environmental Sociology and Animal Studies*
Ethics and Animal Protection—*Environmental Philosophy and Animals*

Further Resources

Bögeholz, S. (2006). Nature experience and its importance for environmental knowledge, values and action: Recent German empirical contributions. *Environmental Education Research, 12*(1), 65–84.

Clayton, S., & Brook, A. (2005). Can psychology help save the world? A model for conservation psychology. *Analyses of Social Issues and Public Policy, 5*(1), 87–102.

Clayton, S., & Opotow, S. (Eds.). (2003). *Identity and the natural environment: The psychological significance of nature.* Cambridge, MA: MIT Press.

Conservation Psychology. http://www.conservationpsychology.org. [A central location for research and practice associated with conservation psychology.]

Gardner, G. T., & Stern, P. C. (2002). *Environmental problems and human behavior* (2nd ed.). Boston: Pearson Custom.

Kahn, P. H., Jr., & Kellert, S. R. (Eds.). (2002). *Children and nature: Psychological, sociocultural, and evolutionary investigations.* Cambridge, MA: MIT Press.

Saunders, C. D., & Myers, O. E., Jr. (Eds.). (2003). Special issue: Conservation psychology. *Human Ecology Review, 10*(2), 87–193.

Schmuck, P., & Schultz, W. P. (Eds.). (2002). *Psychology of sustainable development.* Boston, MA: Kluwer Academic Publishers.

Winter, D. D., & Koger, S. M. (2004). *The psychology of environmental problems* (2nd ed.). Mahwah, NJ: Lawrence Erlbaum Associates.

Zelezny, L. C., & Schultz, P. W. (Eds.). (2000). Promoting environmentalism. *Journal of Social Issues, 56*(3), 365–578.

Carol D. Saunders

Conservation and Environment
Coyotes, Humans, and Coexistence

Coyotes and humans shared the same environment long before European settlers arrived in North America. To many Native American cultures, coyotes were powerful mythological figures endowed with the power of creation and venerated for their intelligence and mischievous nature. The Aztec name for the coyote was *"coyotyl,"* which loosely translates to "trickster," whereas Navajo sheep and goat herders referred to the coyote as "God's dog."

European settlers, however, viewed coyotes as a threat to livestock and as a competitor for game species, a view that unfortunately still persists in many areas of North America. As a result, the coyote remains the most persecuted native carnivore in the United States.

Coyotes typically weigh 20 to 45 pounds and look like a tan shepherd-type dog. Originally found in the grasslands and prairies of North America, the coyote has expanded its range throughout most of the continent, largely due to human-wrought changes to the environment and to the eradication of wolves, cougars, and grizzly bears. At least nineteen sub-species of coyote now roam throughout North America, from California to Newfoundland and from Alaska to Panama, occupying a broad range of habitats including grasslands, deserts, woodlands, agricultural lands, parks, and the urban/wildland fringe.

Coyotes occupy the biological niche between foxes and wolves and may live in pairs, extended family groups, or solitarily. A strong social hierarchy generally limits reproduction to the pack's leaders: the "alpha" male and female. As the top predator in many ecosystems, coyotes play an integral role in maintaining the health of a variety of habitats, primarily by regulating the numbers of smaller predators, such as foxes, raccoons, and skunks, through competitive exclusion or killing. Research in the fragmented urban habitats of coastal southern California indicates that the absence of coyotes allowed smaller predators to proliferate, leading to a sharp reduction in the number and diversity of scrub-nesting bird species. Other studies have found that coyotes have similar indirect positive effects on songbirds and waterfowl.

As opportunistic omnivores, coyotes feed on a wide variety of mammals, insects, vegetables, and fruit, though rodents are their main food source. Indeed, the success of coyotes is a testament to their ability to survive, and even thrive, on whatever food is available. Their feeding habits can place them in conflict with humans, especially in agricultural and suburban areas.

Historically, conflicts between humans and coyotes have been addressed through lethal means. Between 1916 and 2000, the United States Department of Agriculture's Wildlife Services program (formerly Animal Damage Control) killed nearly six million coyotes, largely at taxpayers' expense for the benefit of a small number of sheep and cattle ranchers. In recent years, Wildlife Services has killed approximately 75,000 coyotes annually as part of their "livestock protection program." Primary killing techniques include aerial gunning, snaring, trapping, and poisoning, methods which have been widely criticized as inhumane, ineffective, and indiscriminate. In addition, hundreds of thousands of coyotes are killed each year for their fur, for "sport," and in "body-count" contests where prizes are awarded for killing the most coyotes. Thousands more are killed by private ranchers, though the total number is unknown since most states do not require individuals to report the number of coyotes they kill. Most states have no laws regulating by what means or how many coyotes may be killed, and some states still offer bounties to encourage coyote killing. Despite decades of systematic poisoning, trapping, and shooting campaigns aimed at eradicating coyote populations, coyotes have persevered and even expanded their range and populations in some areas.

The coyote's remarkable success appears to be directly related to lethal attempts to reduce its populations. Years of intense persecution have selected for coyotes that are more adaptable, resilient, and wary of people. They have learned to spring traps without being caught, to avoid poison baits, to hide their dens from prying human eyes, and to hunt during times of little human activity. To further avoid humans, coyotes have become more active during the night in urbanized areas.

Widespread attempts to control coyote populations have had little long-term impact because coyotes' strong compensatory responses—such as increased litter size and pup survival—allow them to replenish their numbers and reoccupy vacated habitat. Further, while lethal control may produce a short-term reduction of coyotes in a particular area, the vacuum is soon filled by coyotes emigrating from surrounding areas and by shifts in neighboring packs.

Despite clear scientific evidence demonstrating the futility and counterproductiveness of lethal coyote control, many state and federal wildlife managers continue to promote killing as the best method to address conflicts. An increasing number of scientists, however, have begun to speak out against indiscriminate lethal control. They argue that to suppress a coyote population over the long-term, 70–90 percent of the coyotes would need to be removed continually. Aside from the ethical questions such intense control efforts raise, the practicality of removing large numbers of coyotes on a sustained basis is beyond the bounds of most wildlife management agencies and may even be counterproductive by stimulating growth in the remaining coyote population.

But scientific evidence is not enough. What is needed is a new paradigm for the way humans treat native carnivores—indeed all wildlife—one that recognizes the ecological importance of these species as well as their intrinsic value as individuals. If the money and efforts used to kill coyotes, and other predators, were redirected toward cost-effective, nonlethal methods, such as public education, better landscape development, improved fencing, and guard animals, conflicts could be significantly reduced without the need to kill.

Time and again, coyotes have proven themselves remarkably resilient animals; it's little wonder that the Navajo called this cunning and resourceful species "God's dog." Coyotes have much to offer humans, not only by keeping ecosystems healthy and diverse, but also by providing inspiring examples of ingenuity and adaptability in an ever-changing world.

See also

Conservation and Environment—*Wolf and Human Conflicts: A Long, Bad History*
Ethics and Animal Protection—*Wolf Recovery*
Living with Animals—*Wolf Emotions Observed*

Further Resources

Bekoff, M. (Ed.). (1978). *Coyotes: Biology, behavior, and management.* New York: Academic Press (reprinted 2001, West Caldwell, NJ: The Blackburn Press).

Ellins, S. (2005). *Living with coyotes: Managing predators humanely using food aversion conditioning.* Austin: University of Texas Press.

Fox, C. H., & Papouchis, C. M. (2005). *Coyotes in our midst: Coexisting with an adaptable and resilient carnivore.* Sacramento, CA: Animal Protection Institute.

Robinson, M. (2005). *Bureaucracy: The extermination of wolves and the transformation of the West.* Boulder: University of Colorado Press.

Camilla H. Fox

Conservation and Environment
The Earth Charter

The Earth Charter is a declaration of fundamental principles for building a just, sustainable, and peaceful global society in the twenty-first century. It seeks to inspire in all peoples a new sense of global interdependence and shared responsibility for the well-being of the human family and the larger living world. It is an expression of hope and a call to help create a global partnership at a critical juncture in history.

As its name suggests, the Earth Charter places a particular emphasis on the world's environmental challenges, but the document's key message is that the issues of environmental protection, human rights, equitable human development, and peace are interdependent phenomena that demand integrated solutions and are indivisible. As such, the Earth Charter provides a fresh conception of sustainable development.

Drafting the Earth Charter: 1987–2000

In 1987, the United Nations World Commission on Environment and Development issued a call for the creation of a new charter that would set forth fundamental principles for sustainable development. The drafting of an Earth Charter was part of the unfinished business of the 1992 Rio Earth Summit. In 1994, Maurice Strong, the Secretary General of the Earth Summit and Chairman of the Earth Council, and Mikhail Gorbachev, President of Green Cross International, launched a new Earth Charter Initiative with support from the Dutch Government. An Earth Charter Commission was formed in 1997 to oversee the project and an Earth Charter Secretariat was established at the Earth Council in Costa Rica.

The Earth Charter is the product of a decade-long, worldwide, cross-cultural conversation about common goals and shared values. The drafting of the Earth Charter has involved the most open and participatory consultation process ever conducted in connection with an international document. Thousands of individuals and hundreds of organizations from all regions of the world, different cultures, and diverse sectors of society have participated.

Early in 1997, the Earth Charter Commission formed an international drafting committee. The drafting committee helped to conduct the international consultation process, and the evolution and development of the document reflects the progress of the worldwide dialogue on the Earth Charter. Beginning with the Benchmark Draft issued by the Commission following the Rio+5 Forum in Rio de Janeiro, drafts of the Earth Charter were circulated internationally as part of the consultation process. Meeting at the United Nations Educational, Scientific and Cultural Organization (UNESCO) headquarters in Paris in March 2000, the Commission approved a final version of the Earth Charter.

Together with the Earth Charter consultation process, the most important influences shaping the ideas and values in the Earth Charter are as follows: contemporary science, international law, the wisdom of the world's great religions and philosophical traditions, the declarations and reports of the seven UN summit conferences held during the 1990s, the global ethics movement, numerous nongovernmental declarations and people's treaties issued over the past thirty years, and best practices for building sustainable communities.

The Earth Charter has been constructed as a layered document with a preamble that describes in general terms the basic challenge, sixteen main principles with supporting principles, and a conclusion that contains a call to commitment and action. This approach makes possible an abbreviated version that includes the preamble and sixteen main principles only. (The accompanying sidebar contains the sixteen main principles.) As the text grew in length, the main principles were divided into four parts in order to make the organization and main themes of the principles easily understood. Part I contains four very broad main principles that can serve as a short summary of the Earth Charter vision. The sixty-one supporting principles that follow the sixteen main principles deal with critical issues and clarify the meaning of the main principles.

As one aim was to keep the Earth Charter as short and concise as possible, the document is limited to fundamental ethical values and principles that set forth major strategies for achieving a just, sustainable, and peaceful world. The Earth Charter does not attempt to identify the mechanisms and instruments required to implement its ethical and strategic vision. The full text of the Earth Charter, together with supporting documentation and background information about the Earth Charter Initiative, can be found at the Web site of the International Secretariat (www.earthcharter.org).

The sixteen main and sixty-one supporting principles are organized around four key themes:

- respect and care for the community of life
- ecological integrity
- social and economic justice
- democracy, nonviolence, and peace

The Main Principles of the Earth Charter

1. Respect Earth and life in all its diversity.
2. Care for the community of life with understanding, compassion, and love.
3. Build democratic societies that are just, participatory, sustainable, and peaceful.
4. Secure Earth's bounty and beauty for present and future generations.
5. Protect and restore the integrity of Earth's ecological systems, with special concern for biological diversity and the natural processes that sustain life.
6. Prevent harm as the best method of environmental protection and, when knowledge is limited, apply a precautionary approach.
7. Adopt patterns of production, consumption, and reproduction that safeguard Earth's regenerative capacities, human rights, and community well-being.
8. Advance the study of ecological sustainability and promote the open exchange and wide application of the knowledge acquired.
9. Eradicate poverty as an ethical, social, and environmental imperative.
10. Ensure that economic activities and institutions at all levels promote human development in an equitable and sustainable manner.
11. Affirm gender equality and equity as prerequisites to sustainable development and ensure universal access to education, health care, and economic opportunity.
12. Uphold the right of all, without discrimination, to a natural and social environment supportive of human dignity, bodily health, and spiritual well-being, with special attention to the rights of indigenous peoples and minorities.
13. Strengthen democratic institutions at all levels, and provide transparency and accountability in governance, inclusive participation in decision making, and access to justice.
14. Integrate into formal education and life-long learning the knowledge, values, and skills needed for a sustainable way of life.
15. Treat all living beings with respect and consideration.
16. Promote a culture of tolerance, nonviolence, and peace.

The Earth Charter presents a consensus vision of an integrated agenda for the pursuit of peace, social and economic justice, and the protection of cultural and biological diversity. It affirms that each of these important goals can only be achieved if all are achieved. Justice, peace, and ecological integrity are inextricably intertwined. We can only care for people if we care for the planet. We can only protect ecosystems if we care for people by providing freedom, eradicating poverty, and promoting good governance. The Earth Charter identifies, in a succinct and inspiring way, the necessary and sufficient conditions for promoting a just and sustainable future.

The Earth Charter in Action: 2000–2006

The Earth Charter was formally launched in ceremonies at The Peace Palace in The Hague in July 2000. Over the next five years, a formal endorsement campaign attracted over 2,400 organizational endorsements, representing millions of people, including numerous national and international associations and ultimately global institutions such as UNESCO and IUCN. Many thousands of individuals also endorsed the Earth Charter. Efforts to have the Earth Charter formally recognized at the World Summit on Sustainable Development in Johannesburg in 2002 came very close to success, resulting in numerous public statements of support from world leaders and heads of state.

The Earth Charter is now increasingly recognized as a global consensus statement on the meaning of sustainability, the challenge and vision of sustainable development, and the principles by which sustainable development is to be achieved. It is used as a basis for peace negotiations, as a reference document in the development of global standards and codes of ethics, as a resource for governance and legislative processes, as a community development tool, as an educational framework for sustainable development, and in many other contexts. The Charter was also an important influence on the Plan of Implementation for the UNESCO Decade for Education on Sustainable Development.

The Earth Charter Initiative is a broad-based, voluntary, civil society effort, but participants include leading international institutions, national government agencies, university associations, NGOs, cities, faith groups, and many well-known leaders in sustainable development.

The Initiative is served and coordinated by a very small staff, working in Centers located in Stockholm, Sweden, and in San José, Costa Rica. These Centers are both regional and topical: the Center in Stockholm manages overall strategy and communications, and the Center in San José focuses on the Initiative's extensive work in the field of Education for Sustainable Development in partnership with the UN–chartered University for Peace. Other Centers are envisioned for other regions of the world.

The Earth Charter Initiative is funded, at the international level, by public grants and generous individual donors (who donate their time as well as their financial resources). The Initiative does seek or accept commercial sponsorship.

The mission of the Earth Charter Initiative is, "To establish a sound ethical foundation for the emerging global society and to help build a sustainable world based on respect for nature, diversity, universal human rights, economic justice and a culture of peace."

The long-term objective of the Earth Charter Initiative is the universal adoption and implementation of the Earth Charter as a statement of common values and principles for a sustainable future. It continues to seek endorsement by individuals, organizations, governments, and the United Nations.

Programmatic activity is organized around the achievement of four interrelated goals:

1. Expanded awareness of the Earth Charter—so that more and more people become aware of it.
2. Engaged endorsement of the Charter—so that endorsement translates into action for a sustainable world.
3. Education for a sustainable way of life—in formal as well as nonformal education settings.
4. Ethics-based assessment and governance—the use of the Charter in decision making and evaluation.

The Earth Charter and Animals

Steven C. Rockefeller, the Chair of the Earth Charter Drafting Committee, points out, "A major objective of the Earth Charter is to promote a fundamental change in the attitudes toward nature that have been predominant in industrial-technological civilization, leading to a transformation in the way people interact with Earth's ecological systems, animals, and other nonhuman species. Humanity must, of course, use natural resources in order to survive and develop. However, the Earth Charter rejects the widespread modern view that the larger natural world is merely a collection of resources that exists to be exploited by human beings. It endeavors to inspire in all peoples commitment to a new ethic of respect and care for the community of life" (Rockefeller, 2004).

The Earth Charter is the first major international document that makes the humane treatment of individual animals a necessary condition for sustainable development. Principle 1.a. reads, "Recognize that all beings are interdependent and every form of life has value regardless of its worth to human beings." What this principle acknowledges is that sustainability is not exclusively about the human situation but must address all beings and acknowledge our mutual interdependence with all living beings and ecological systems.

In addition to its general affirmation of a nonanthropocentric worldview, the Earth Charter also includes a major principle and three sub-principles focused on animal protection, 15.a, b, and c. Sub-principle 15.a is an assertion that domesticated animals should be treated humanely, without mentioning specific contexts. Sub-principles 15.b and 15.c address wild animals, asserting that methods of hunting, trapping, and fishing should neither cause unnecessary suffering to the target animals, nor be so inexact that non-targeted animals are killed.

One significant contribution of the Earth Charter is to make respect for and the protection of individual animals a necessary condition for sustainable development. Another contribution is to challenge those of us focused on our particular interests to work together with others for a larger integrated agenda. Those who care about animal protection must recognize that this agenda cannot be achieved without alleviating poverty, empowering women, and protecting ecosystems. So, too, must those who care about poverty alleviation, women's health, climate change, and other issues recognize that acknowledging animal sentience and treating all animals humanely are essential dimensions to a sustainable future.

Further Resources

Rockefeller, S. C. (2004). Earth charter ethics and animals. *Earth Ethics, 12*(1), 5. Washington, DC: Center for Respect of Life and Environment.

Richard M. Clugston

Conservation and Environment
Earthworms: The "Intestines of the Earth"

"Earthworm" is the colloquial name given to approximately 10,000 species of segmented worms in the phylum *Annelida*. The annelids are the largest of the wormlike invertebrates ranging from 1 millimeter long to the 1-meter-long Grippsland Giant earthworm. Fossil traces of annelids are found in pre-Cambrian strata some 700 million years old. Three major groups of annelids are recognized: *Polychaeta* ("many hairs"), *Oligochaeta* ("few hairs"), and *Hirudinae*. Earthworms can also be classified based on whether they live close to the surface (epogeic) or burrow deep beneath the soil (endogeic). Worms come in a variety of colors including yellow, green, and orange and can live for up to six years.

Earthworms are bilaterally symmetrical and have a definite anteroposterior orientation. They do not have lungs or gills and must breathe through the skin. A major evolutionary advance of annelids is a digestive system consisting of a tubular gut running from a well-developed mouth to an anus. The gut allows food to be digested in stages, independent from the worms' movement. An equally important advance is an internal body cavity called a coelom. The coelom is a fluid-filled cavity that separates the gut from the body wall and is involved in locomotion. The fluid serves as a hydraulic skeleton with which the muscles act to modify the shape of the worm.

In addition to what earthworms tell researchers about digestion and movement, they also help researchers understand the process of learning and memory. Starting in the 1950s, earthworms were popular subjects for studies on Pavlovian conditioning and other types of learning. The use of earthworms to study learning has declined in recent years because new techniques that tap into the biochemistry and physiology of learning in complex animals are available. Nevertheless, the learning of worms is still of interest from a comparative perspective, wherein researchers compare the similarities and differences in the behavior of a wide range of animals.

The most important contribution of earthworms is that they help to maintain our planet. In recognition of the importance of earthworms, Aristotle called them the "intestines of the earth," and Charles Darwin noted in 1881 that "It may be doubted whether there are many other creatures which have played so important a part in the history of the world."

Earthworms on a compost heap; the earthworm was called one of the most important creatures in the history of the earth by Charles Darwin. ©Scott W. Smith/Animals Animals.

Earthworms aid in the decomposition of plant, animal, and leaf litter and are one of nature's finest recyclers. Farmers and gardeners appreciate what earthworms can do for their gardens and crops. Earthworms digest litter rich in nutrients, including calcium, potassium, and organic minerals, which are deposited on the top layer of soil in the form of excrement known as castings. These castings provide excellent fertilizer necessary for healthy plant growth and provide food for a variety of animals, including microorganisms. It is estimated that there are about 50,000 earthworms per acre of moist soil producing 30,000 pounds of castings.

In addition to castings, earthworms are also beneficial because their burrows aerate the soil. Aeration brings oxygen into the soil and serves as a conduit for organic and inorganic nutrients to reach deep into the soil to stimulate plant and root growth. Burrows also improve drainage and provide channels for roots to grow. Without earthworm burrows, soil becomes impenetrable to air and water.

Although there are many benefits to earthworms, there are some limitations. Chief among them is that their activities, although good for gardens and farms, can be disastrous for forests. Earthworms eat so much that the forest floor may become exposed to weathering, and they can disrupt the delicate balance between soil structure and chemistry. Earthworms are also a favorite food for moles and can attract them to an area.

Further Resources

Darwin, C. R. (1881). *The formation of vegetable mould through the action of earthworms with observations on their habits*. London: Murray.

Dyal, J. A. (1973). Behavior modification in annelids. In W. C. Corning, J. A. Dyal, & A. O. D. Willows (Eds.), *Invertebrate learning Vol. 1.: Protozoans through annelids*. (pp. 225–90). New York: Plenum.

Edwards, C. A., & Lofti, J. R. (1977). *Biology of earthworms*. London: Chapman and Hall.

Lee, K. E. (1985). *Earthworms: Their ecology and relationships with soil and land use*. New York: Academic Press.

Charles I. Abramson

Conservation and Environment
Ecosystems and a Keystone Species, the Plateau Pika

A central observation made by ecologists is that everything in nature is connected to everything else. For this generality, like most things in life, the devil is in the details. How connected are species, and are all species equally important? We now know that the removal or extinction of some species from natural ecosystems is accompanied by insignificant change. However, the loss of other species sets in motion a cascade of events that may drastically alter the form or function of an ecosystem. These are called "Keystone Species"—and it is important to identify and conserve them if we want to preserve nature's bounty and the services that it provides for humans.

Many keystone species are top carnivores. When humans exterminated wolves (*Canis lupus*) from the greater Yellowstone ecosystem about 70 years ago, the region underwent a significant alteration. With their major predator gone, the elk (*Cervus elaphus*) population soared and reduced Yellowstone's aspens, cottonwoods, and willows down to stubs in many areas. Wolves were reintroduced to Yellowstone in 1995, and within five years the elk population decreased. They developed behaviors to avoid wolves, and in the process the browsing pressure on willows was relaxed. Yellowstone willows now grow three meters tall, which has been a boon for beavers (*Castor canadensis*, who were down to a single colony before the wolf reintroduction). As beaver increased, their ponds provided additional habitat for fish. The newly luxuriant groves of trees beside beaver ponds have become home to many more songbirds. Another endangered species, the grizzly bear (*Ursus arctos*), enjoys the leftovers from wolf kills. During the wolf's absence populations of coyotes (*Canis latrans*) increased dramatically, but with the return of wolves the numbers of coyotes have fallen. This decline of coyotes has increased survivorship of the region's pronghorn (*Antilocapra americana*) fawns. Yes, the health of one species is tied to that of other members of the ecosystem. The complexion of the Yellowstone ecosystem with and without wolves demonstrates the keystone status of this top carnivore (Smith et al., 2003).

In the North Pacific the sea otter (*Enhydra lutris*) has also been classified as a keystone species. Sea otters dine on sea urchins and in doing so keep populations of urchins in check. When sea otter populations crash, sea urchins proliferate and devastate the productive coastal kelp forests, thereby greatly altering interactions in this ecosystem. Apparently this system has cycled through several phase shifts. Sea otter populations were initially overhunted but then rebounded when they became legally protected, and recently they have again declined precipitously. The kelp forests also recovered during the otter's recovery phase, only to decline again once the otter population began to fall. This recent decline of sea otters may be due to predation by killer whales (*Orcinus orca*) that were "fishing down" the food chain after their earlier prey, the great whales, were decimated by commercial whaling (Estes et al., 2004).

But not all keystone species are charismatic large carnivores. The plateau pika (*Ochotona curzoniae*), a small (~175 grams) rabbit relative that inhabits the high alpine grasslands of the Tibetan plateau, is in every sense a keystone species. The pika creates habitat for endemic birds and lizards, serves as the primary food source for nearly all the region's carnivores, and provides other essential ecosystem services (Smith and Foggin, 1999).

The Tibetan plateau, an area of approximately 2.5 million square kilometers, encompasses roughly a quarter of China's land area. About 70 percent of the Tibetan plateau is open rangeland occupied by pikas and Tibetan pastoralists living side by side. Environmental conditions on these high grasslands (elevations ranging from 3,000–5,000 m) are generally too severe to support growing crops, making pastoralism the only viable form of agriculture. The pastoralists live in their characteristic black tents and graze domestic yak and sheep as they have done for an estimated 4,000 years. The fate of the both the pikas and the pastoralists hinge on the health of these grasslands.

Plateau pikas occupy family territories on the alpine meadow, each territory encompassing about 500 square meters. The families are comprised of two to five breeding adults and their young from three or more large litters born during the summer. The accumulation of young over the course of the summer boosts the population density to very high levels (100–300/ha). Family members are extremely social and remain together in an interlinked labyrinth of underground tunnels; the surface of the meadow is pocketed with numerous burrow entrances. The high summer population of pikas disappears with the onset of the harsh Tibetan winter. Winter is a time of hardship for the pikas and most perish; the few animals that survive form the foundation for the next year's families (Smith and Wang, 1990; Dobson et al., 1998, 2000).

There are no trees across the high wind-swept plateau meadows, and nearly all the endemic Tibetan birds preferentially nest in pika burrows. These species include the Tibetan snowfinch (*Montifringilla adamsi*), white-winged snowfinch (*M. nivalis*), plain-backed snowfinch (*Pyrgilauda blanfordi*), small snowfinch (*P. davidiana*), rufous-necked snowfinch (*P. ruficollis*), white-rumped snowfinch (*P. tacazanowskii*), and Hume's groundpecker (*Pseudopodoces humilis*). Often one can observe a pika dart into its burrow only to have a fuzzy snowfinch chick emerge instead. There are only two native lizards on the high plateau (*Phrynocephalus vlangalii, Eremas multiocellata*), and they also live in the burrows of pikas. In this respect, the plateau pika is an ecosystem engineer—its construction project, the burrows, creating habitat for these other species (Smith and Foggin, 1999; Lai and Smith, 2003).

The burrowing activity of pikas also creates a mosaic of habitat favored by some plant species over the continuous mat of thick alpine meadow sod. Correspondingly, different species of plants can be found on pika colonies than in areas without pikas. One recent study determined that plant species richness was actually 50 percent higher in areas with pikas (Bagchi et al., 2006).

Although the Tibetan plateau is home to many small- to medium-sized herbivorous mammals, the pika is by far the most continuously distributed and numerous. As a result, most of the region's carnivores (mountain weasel, *Mustela altaica*; steppe polecat, *M. eversmannii*; fox, *Vulpes ferrilata, V. vulpes*; Pallas's cat, *Felis manul*; wolf, *Canis lupus*; brown bear, *Ursus arctos*) dine on pikas. The steppe polecat is a pika specialist, much like its close relative the North American black-footed ferret (*M. nigripes*), who specializes in prairie dogs (*Cynomys*)—the pika's ecological equivalent. Detailed studies have shown that pikas comprise 50–60 percent of the diet of wolves and brown bears, respectively. Early Russian explorers on the plateau recognized how important pikas are in the diet of brown bears; Nikolai Prejevalski gave the bear a Latin scientific name that means "bear pika eater." Pyotr Kozlov opened one bear and found 25 pikas in its stomach. The region's avian predators (golden eagle, *Aquila chrysaetos*; upland buzzard, *Buteo hemilasius*; saker falcon, *Falco cherrug*; northern goshawk, *Accipiter gentiles*; black-eared kite, *Milvus lineatus*; little owl, *Athene noctua*) appear even more dependent on pikas for survival; 90–100 percent of the nests of buzzards, saker falcons, and kites contain pika remains (Smith et al., 1990; Schaller, 1998; Smith and Foggin, 1999).

Plateau pikas may also contribute significantly to other aspects of ecosystem dynamics, such as nutrient cycling. No investigations have been completed on the ecosystem dynamics of plateau pikas (work is under way), but an ecologically equivalent species (the Daurian pika, *O. dauurica*, an inhabitant of the northern edge of the Tibetan plateau) has been studied. It was determined that Daurian pika burrows contained more humus, nitrogen, calcium, and phosphorus than nearby areas without burrows. Correspondingly, plants growing on or near pika burrows had greater root biomass and grew taller than those in areas without burrows. Additionally, certain plant species grew only in areas with Daurian pika burrows (Smith et al., 1990).

The Chinese call the Tibetan plateau their "Water Tower" because most of Asia's great rivers (Huang He, Yangtze, Mekong, Salween, Brahmaputra, Indus) originate on these highlands. The Chinese have a love-hate relationship with these rivers, particularly the Huang He and Yangtze, as they are both the lifeblood of the country and the source of rampant flooding that causes substantial loss of life and property (the Huang He is nicknamed "China's Sorrow"). These rivers become engorged with the rain from torrential monsoon storms that characterize summers on the plateau. But what does this have to do with pikas? Because the alpine meadow sod is thick with roots to a considerable depth, the high density of pika burrows may help to channel water from monsoon storms deep into the soil, thus minimizing runoff, erosion, and flooding downstream.

Plateau pikas thus contribute to both biodiversity and ecosystem functioning on the Tibetan plateau. Unfortunately, we now know what happens in the absence of pikas and their positive influence. Chinese authorities consider the plateau pika to be a pest species that causes rangeland degradation, consumes fodder that could otherwise be consumed by livestock, leads to biodiversity loss, and contributes to increased erosion (Fan et al., 1999). Despite the clear evidence that pikas contribute to the maintenance of biodiversity on the plateau, authorities have systematically poisoned them over huge expanses since the early 1960s (much as their ecological counterpart, the prairie dog, in North America has been eliminated from most of its former range). From 1960 to 1990 pikas were poisoned over an area of 208,000 square kilometers in Qinghai province alone, and the poisoning continues (Fan et al., 1999). The management budget for the newly formed Sanjiangyuan National Nature Reserve (encompassing 152,300 km^2 in southern Qinghai) is US$934 million, and government-issued reports indicate that much of this is targeted for developing more effective poisons and killing pikas. Where pikas have been poisoned, native birds and lizards have disappeared as the burrows on which they rely have collapsed and their nesting habitat is gone. No carnivores—birds or mammals—can be found in poisoned regions; they have either migrated or starved to death. And there are no data indicating that this poisoning regime has improved the quality of the rangelands in any way.

Plateau pikas sitting in contact or nose-rubbing (to indicate social relationships among family members). Courtesy of Andrew Smith.

An upland buzzard, the most common raptor on the Tibetan plateau; this bird disappears from areas after plateau pikas have been poisoned. Courtesy of Andrew Smith.

The relationships between pikas, grassland health, biodiversity, and flooding and erosion are intertwined and complicated. But this much is clear: pikas are not responsible alone for the perceived degradation of the alpine meadow ecosystem on the Tibetan plateau. In fact, all available evidence points to their positive contribution. The long-term prospectus for maintenance of biological diversity on the plateau, and the future sustainability of the grasslands and the pastoralist economy they support, is likely to be connected to the health and presence of the plateau pika. Everything is connected, and they are the key.

See also

Conservation and Environment—*The Endangered Species Act*

Further Resources

Bagchi, S., Namgail, T., & Ritchie, M. E. (2006). Small mammalian herbivores as mediators of plant community dynamics in the high-altitude arid rangelands of Trans-Himalaya. *Biological Conservation, 127,* 438–42.

Dobson, F. S., Smith, A. T., & Wang, X. G. (1998). Social and ecological influences on dispersal and philopatry in the plateau pika. *Behavioral Ecology, 9,* 622–35.

———. (2000). The mating system and gene dynamics of plateau pikas. *Behavioural Processes, 51,* 101–10.

Estes, J. A., Danner, E. M., Doak, D. F., Konar, B., Springer, A. M., Steinberg, P. D., Tinker, M. T., & Williams, T. M. (2004). Complex tropic interactions in kelp forest ecosystems. *Bulletin of Marine Science, 74,* 621–38.

Fan, N., Zhou, W., Wei, W., Wang, Q., & Jiang, Y. (1999). Rodent pest management in the Qinghai-Tibet alpine meadow ecosystem. In G. Singleton, L. Hinds, H. Leirs, & Z. Zhang (Eds.), *Ecologically-based rodent management* (pp. 285–304). Canberra, ACT, Australia: Australian Centre for International Agricultural Research.

Lai, C. H., & Smith, A. T. (2003). Keystone status of plateau pikas (*Ochotona curzoniae*): Effect of control on biodiversity of native birds. *Biodiversity and Conservation, 12,* 1901–12.

Schaller, G. B. (1998). *Wildlife of the Tibetan steppe.* Chicago: University of Chicago Press.

Smith, A. T., & Foggin, J. M. (1999). The plateau pika (*Ochotona curzoniae*) is a keystone species for biodiversity on the Tibetan plateau. *Animal Conservation, 2,* 235–40.

Smith, A. T., Formozov, N. A., Hoffmann, R. S., Zheng, C. L., & Erbajeva, M. A. (1990). The pikas. In J. A. Chapman & J. E. C. Flux (Eds.), *Rabbits, hares and pikas: Status survey and conservation action plan* (pp. 14–60). International Union for the Conservation of Nature: Gland, Switzerland.

Smith, A. T., & Wang, X. G. (1991). Social relationships of adult black-lipped pikas (*Ochotona curzoniae*). *Journal of Mammalogy, 72,* 231–47.

Smith, D. W., Peterson, R. O., & Houston, D. B. (2003). Yellowstone after wolves. *Bioscience, 53,* 330–40.

Andrew T. Smith

Conservation and Environment
Elephant Conservation and Welfare at the Amboseli Elephant Research Project

The Amboseli Elephant Research Project (AERP) strives to create, maintain, and make available an unparalleled body of knowledge on a single, long-lived mammal species—the African savanna elephant, *Loxodonta africana.* The data bank is based on the long-term study of the relatively undisturbed, free-ranging population of elephants inhabiting the Amboseli ecosystem in Kenya. The major aim of the project is to produce published results of the Amboseli Elephant Research Project that will provide base-line information on elephant biology to relevant authorities and individuals who are responsible for the conservation of African elephants throughout their range. At the same time, AERP's goal is to secure the future of the Amboseli elephant population by finding a balance between the needs of animals and the needs of the people with whom they share their range.

The Amboseli ecosystem, which includes Amboseli National Park, in southern Kenya was originally chosen in 1972 as a study site because the Amboseli elephants had not been heavily poached, had not been restricted in their movements by fencing or agriculture, and had not had their numbers artificially reduced through culling as part of a Park management program. The Amboseli situation provided an opportunity to study a relatively undisturbed, free-ranging population of elephants.

In the absence of poaching and culling, the Amboseli elephants have been increasing slowly since the late 1970s. The population remains relatively small, presently numbering just over 1,400, but it is a very important one to Kenya and the rest of Africa. Amboseli is one of the few places in Africa where the elephant age structure has not been drastically skewed by poaching. In Amboseli there are animals spanning the whole age range from newborn calves to old matriarchs in their 60s and, even more unusual, many large adult bulls in their 40s and 50s. In most of Africa there are few males over 40 years old, because it was the males which were killed first for their larger tusks. When they were finished the poachers turned to the adult females and finally even to teenagers and calves, disrupting and sometimes destroying the very social fabric of elephant life. Amboseli with its natural age structure and intact social organization is an important source of baseline data on elephant social, reproductive, and ecological patterns. As such it is being used as a model for assessing the status of other elephant populations in Africa and as a guide to developing ethical principles for captive elephants. A number of important research studies and publications have resulted from the Amboseli Elephant Research Project.

Amboseli Elephant Research Project Studies

- Social organization, population dynamics, calf development, estrus behavior—C. Moss: 1972–ongoing.
- Distribution and ranging patterns using radio-tracking—H. Croze: 1972–74.
- Musth and male-male competition, vocal communication—J. Poole: 1976; 1977; 1978; 1980–81; 1984–90; 2000–ongoing.
- Elephant ecology and habitat use—K. Lindsay: 1977–79; 1982–84.
- Calf development and maternal investment, elephant growth, and development—P. Lee: 1982–84; 1991–ongoing.
- Female cooperation and competition—S. Andelman: 1985–87.
- Maasai-elephant relationships—K. Kangwana: 1990–91.
- Reproductive hormone analysis; elephant movements and decision-making—H. Mutinda: 1991–92; 1998–2002.
- Communication and social organization—Karen McComb & Lucy Baker: 1993–2000.
- Satellite radio tracking—Iain Douglas-Hamilton: 1996–2000.
- DNA and social behavior—S. Alberts, E. Archie, J. Hollister-Smith: 2000–05.
- Maasai attitudes towards elephants and conservation—Christine Browne-Nunez: 2004–05.
- Elephant cognition—Richard Byrne and Lucy Bates: 2005–ongoing.
- Male elephant behavior, socio-ecology, and crop-raiding—Patrick Chiyo: 2005–ongoing.

Further Resources

Amboseli Elephant Research Project (AERP). http://www.elephanttrust.org. [This site provides a list of over 90 publications.]

Archie, E. A., Morrison, T. A., Foley, C. A. H., Moss, C. J., & Alberts, S. C. (2006). Dominance rank relationships among wild female African elephants, *Loxodonta africana*. *Animal Behaviour, 71*, 117–27.

Bradshaw, I. G. A., Schore, A. N., Brown, J. L., Poole, J. H., & Moss, C. J. (2005). Elephant breakdown. Social trauma: Early disruption of attachment can affect the physiology, behaviour and culture of animals and humans over generations. *Nature, 433*, 807.

Lee, P. C. (1986). Early social development among African elephant calves. *National Geographic Research, 2*, 388–401.

———. (1987). Allomothering among African elephants. *Animal Behaviour, 35*, 278–91.

Lindsay, W. K. (1987). Integrating parks and pastoralists: Some lessons from Amboseli. In D. Anderson & R. Grove (Eds.), *Conservation in Africa* (pp. 149–67). Cambridge: Cambridge University Press.

———. (1993). Elephants and habitats: The need for clear objectives. *Pachyderm, 16*, 34–40.

McComb, K., Moss, C. J., Durant, S., Baker, L., & Sayialel, S. (2001). Matriarchs as repositories of social knowledge in African elephants. *Science, 292*, 491–94.

McComb, K., Moss, C. J., Sayialel, S., & Baker, L. (2000). Unusually extensive networks of vocal recognition in African elephants. *Animal Behaviour, 59*, 1103–09.

McComb, K., Reby, D., Baker, L., Moss, C., & Sayialel, S. (2003). Long-distance communication of cues to social identity in African elephants. *Animal Behaviour, 65*, 317–29.

Morrison, T. A., Chiyo, P., Moss, C. J., & Alberts, S.C. (in press). Measures of dung bolus size for known age African elephants: Implications for age estimation. *Journal of Zoology* (London).

Moss, C. J. (1983). Oestrous behaviour and female choice in the African elephant. *Behaviour, 86*(3/4), 167–96.

———. (2000). *Elephant memories*. Chicago: University of Chicago Press.

Moss, C. J., & Croze, H. J. (Eds.). (in press). *The Amboseli elephants: A long-term perspective on a long-lived mammal*. Chicago: University of Chicago Press.

Moss, C. J., & Poole, J. H. (1983). Relationships and social structure of African elephants. In R. A. Hinde (Ed.), *Primate social relationships: An integrated approach* (pp. 315–25). Oxford: Blackwell Scientific Publications.

Poole, J. H. (1987) Rutting behaviour in African elephants: The phenomenon of musth. *Behaviour, 102*, 283–316.

———. (1989a) Announcing intent: The aggressive state of musth in African elephants. *Animal Behaviour, 37*, 140–52.

———. (1989b) Mate guarding, reproductive success and female choice in African elephants. *Animal Behaviour, 37*, 829–49.

———. (1996). *Coming of age with elephants*. New York: Hyperion Press.

———. (1999). Signals and assessment in African elephants: Evidence from playback experiments. *Animal Behaviour, 58*, 185–93.

Poole, J. H., & Moss, C. J. (1981). Musth in the African elephant, *Loxodonta africana*. *Nature, 292*, 830–31.

Poole, J. H., Tyack, P. L., Stoeger-Horwath, A. S., & Watwood, S. (2005). Elephants are capable of vocal learning. *Nature, 434*, 455–56.

Slotow, R., van Dyk, G., Poole, J., Page, B., & Klocke, A. (2000). Older bull elephants control young males. *Nature, 408*, 425–26.

Western, D., Moss, C. J., & Georgiadis, N. (1983). Age estimation and population age structure of elephants from footprint dimensions. *Journal of Wildlife Management, 47*(4), 1192–97.

Cynthia Moss

Conservation and Environment
Elephants and Humans in Kenya

Understanding the Past and Novel Approaches for the Future

The African elephant's future will be largely decided over the next few decades, shaped through its interactions with humans and its ability to adapt to human dominated landscapes. If the largest of terrestrial mammals is to remain in the wild, unfenced and free-ranging, novel strategies to enable the peaceful coexistence between these two species must be found. In order to understand and find solutions to the current conservation challenges facing elephants, we must understand both the history and current context of elephant-human interactions.

The relatively rare development of areas of high human densities in Africa throughout history has been a subject of great scrutiny by historians and anthropologists. The most accepted theory regarding why Africa's human population densities have been historically low in comparison to other continents postulates that the high disease burden characteristic to sub-Saharan Africa has limited population growth, causing premodern medicine population densities across the continent to remain low. Recently, researchers have added that a probable cofactor limiting the development of dense human populations, and the major agricultural development required to sustain high human density, may have been competition with African elephants (Reader, 1997). Despite records of elephant hunting and ivory use from as early as 27,000 years ago, elephants were historically and are currently a major threat to people when encountered without the use of modern weapons. Giving the prevalence of agricultural-focused conflict between elephants and humans in modern day Africa, despite the relatively low density of elephants, it is not difficult to imagine the problems agriculturalists faced when elephant densities were ten times higher.

The influx of modern weapons tilted the balance of power in favor of humans beginning in the nineteenth century. Access to weapons was initially limited to governing powers, which organized to protect villages and agricultural developments. In addition to better defensive capabilities, the procurement of modern weapons greatly enhanced the ability to acquire coveted ivory. Trade in ivory became a major revenue earner for governments as well as a key resource used in the trade for arms. The famed "scramble for Africa" was driven by colonial administrations' attempts to control trade in Africa's resources, of which ivory was one of the most important. Ivory harvesting was always lucrative, but as weapons increasingly flowed into Africa, ivory hunting exploded. Many weapons used in the fight for independence or civil wars across Africa were procured through trade in ivory. The rate of elephant killing for ivory rapidly increased, peaking in the latter half of the twentieth century. The decade of 1979 to 1989 experienced the highest rates of poaching, though regional populations across Africa were not affected equally (Douglas-Hamilton, 1987). During that decade, poaching for ivory halved Africa's elephant population from 1.3 million to around 600,000. Central and eastern Africa were the most seriously affected; Kenya lost over 75 percent of its elephant population during this period. Counts of elephant populations across central and eastern Africa showed the devastating force of the trade in ivory. In some parks, the numbers of carcasses counted were over four times greater than the numbers of live elephants. Elephant populations in forested countries are notoriously difficult to estimate, thus the full extent of the impact of ivory trade on these areas remains unknown. At the same time, populations across southern Africa were relatively stable during this period. Certain populations, such as in Kruger National Park, even demonstrated strong growth. This rapid decline of elephants across most of its range driven by the demand for ivory remains one of the best examples to date of the dangers of an uncontrolled wildlife trade to a species' persistence.

Responding to the predicted imminent elephant extirpation from many areas of its range, the 1989 Conference on International Trade in Endangered Species (CITES) agreed to set an international ban on the trade in ivory. The result was a drastic reduction of the price of ivory and, consequently, the rate of elephant poaching declined rapidly. Though low levels of poaching still occur in many parts of Africa and the state of elephant populations in the forested countries remain largely unknown, it is widely thought that ivory poaching is not currently having a major impact on most elephant populations in Eastern and Southern Africa (Blanc et al., 2002). Many conservationists, however, harbor the worry that a resumption of ivory trading may be a catalyst for uncontrollable ivory trade again.

Today, many elephant populations are beginning to recover. The decades of poaching, however, have had lasting effects. Elephants have been extirpated from many of the areas they once inhabited. Following the extirpation, many areas have been converted to agriculture or other types of land use, resulting in the fragmentation and isolation of what was once interconnected range. As the recovering elephant populations begin to increase, they attempt to move out of the confines of protected areas in which they found refuge from the poaching epidemic. The interconnecting regions between parks, however, are no longer conducive to elephant movements and conflict between humans and elephants is rising sharply across Africa. Encroachment and habitat restriction is now one of the primary threats facing elephants in Africa.

Such complexity in the threats facing elephants requires novel thinking and approaches to overcome these conservation challenges. The Save the Elephants (STE) research project in Samburu and Buffalo Springs National Reserves in northern Kenya is an example of a fresh conservation approach to resolve problems facing elephants through providing insight into the needs of elephants. In coordination with behavioral research focusing on spatial properties of a wild elephant population, the project investigates the relationship between local people and elephants to understand root causes of tensions and identify factors that precipitate peaceful coexistence. With conflict between humans and elephants being primarily the result of competition for space and limited resources, STE has pioneered research on the spatial requirements of elephants through the use of Global Positioning System (GPS) radio telemetry. The Samburu elephant population is of particular interest for such research because the elephants are free-ranging and primarily rely on regions outside protected areas for their persistence. Ecosystems, such as Samburu, in which elephants and other wild animals rely on areas outside protected areas are becoming increasingly rare across Africa.

Research in the Samburu ecosystem has demonstrated that elephants are reliant on numerous, interconnected core areas (Douglas-Hamilton et al., 2005). The regions between these core areas, though infrequently used by the elephants, contain essential corridors maintaining connectivity across the elephant population's range. A loss of corridors can have serious implications for the long-term sustainability of both elephants and the ecosystem. In regions where such corridors no longer exist, confined elephant populations may negatively impact the vegetative community, leading to changes in species composition and even major ecotone transitions, such as converting woodlands into grasslands. In addition, isolated elephant populations face the long term problems associated with small population conservation, such as susceptibility to natural disasters, inbreeding, and disease. Thus, the protection of core areas and the maintenance of elephant movements is a key concern for the long-term persistence of elephants. Current conservation objectives are attempting to address this need through the creation of large-scale, "transboundary" park systems that connect major protected areas in numerous countries. Additionally, research on elephant movement, such as the Samburu project, is being used to identify key corridors and define ecosystem boundaries as conservation units.

Elephant spatial behavior appears to be strongly related to the dynamics of pastoralist land use in the Samburu ecosystem. Aerial counts conducted by the Kenya Wildlife Service indicate that elephant density is negatively correlated with cattle density. In turn, cattle density in the Samburu ecosystem is a function of both forage quality and security. Cattle raids between competing groups of people are common, making use of raid-prone regions risky. Such regions are commonly the transition zones between areas defined on tribal lines. Depending on the political relationship between tribes, use of these transition zones may be unsafe, resulting in low or no utilization by people and livestock. The influence of such risk not only has major implications for the use of land by pastoralists in the Samburu ecosystem but also impacts elephant distributions. Radio tracking data indicate elephant use of a high-risk transition zone near our Samburu study area is related to the degree of political strife between the tribes. Initial radio tracking data indicated elephants relied on this transition zone as a corridor, moving quickly through the area rather than spending any significant amount of time. This behavior has changed over the course of our research. Between 1998 and 2002 tensions among the tribes south of the study area were high, resulting in low human and cattle density in the transition zone. During this time, elephant reliance on the transition zone increased and extended periods of use by the study elephants were recorded in the strife-prone area. As relations between the tribes improved in 2003 and 2004, people began reinhabiting this zone and regularly grazing their cattle in it. As a result, elephant use of the area has declined, and once again the area primarily serves as a corridor through which elephants pass.

Though pastoralists may spatially exclude elephants to a degree, in many places elephant and pastoralist communities share the same areas and resources. To some extent, the ability of humans and elephants to coinhabit an area relates to the low level of resource competition between livestock and elephants. For example, the elephants of Samburu are predominantly browsers and thus do not compete directly with cattle, who rely primarily on grass resources. Competition between elephants and goats, although theoretically possible because both are browsers, is minimized by the differentiation in feeding height between the two species—goats commonly feed at low heights whereas elephants utilize intermediate and high heights. As such, direct competition for food resources is limited. Water resources, however, are spatially limited in the Samburu ecosystem, particularly during the dry season when competition for water can become a source of friction between humans and elephants. Interestingly, radio tracking has demonstrated that elephant and human activities around the use of water are temporally segregated, thereby minimizing the potential for conflict interactions. Outside protected areas, humans and livestock typically spend the hottest period of the day along the main river in the ecosystem, whereas elephants access the same water sources at night when livestock and humans are confined within thorn corals to protect against predators. In contrast, within the protected areas, elephants typically spend the hottest time of day in the river and spend the nights away from the river browsing in more exposed regions during the coolest period of the day. Thus, the relationship between nonagriculturalists and elephants is primarily peaceful.

The tolerance of the Samburu people to elephants and wildlife not only stems from the mechanisms discussed, which alleviates direct competition, but also is a function of their cultural beliefs. The Samburu region remains an unconstricted range for elephants because of the livelihoods and attitudes of the local people. Research was conducted in the villages surrounding the Samburu National Park with the local Samburu communities to understand their cultural beliefs and perceptions of elephants. Strong knowledge of elephants exists among the Samburu people, which influences human behavior toward these animals (Kuriyan, 2002). For example, the Samburu people indicate that their relationship with elephants is influenced by various costs and benefits elephants bring to people as well as cultural perceptions of the species. General costs are occasional conflict over water and human or cattle deaths caused by elephants. In terms of benefits, the Samburu people indicate that elephants benefit those who live among them because they create paths to water, dig dams, and break branches that people can use for firewood. In terms of cultural beliefs, the Samburu indicate that there are many similarities between humans and elephants: elephants have a trunk that acts like a human arm, breasts similar to women, and skin that resembles human skin. Consequently, certain taboos exist that prohibit the killing or eating of elephants. There is also a Samburu legend that demonstrates that elephants are considered to be ancient "relatives" of humans and thus command much respect amongst the Samburu.

In addition to the cultural beliefs and legends, there are specific Samburu traditions and practices that entail various uses of elephant dung, including symbolically burning it during wedding ceremonies in the homes of newlyweds. The smoke from elephant dung acts as a blessing for newly married couples and brings them good luck when they enter their new homes. Furthermore practices of blessing dead elephants are conducted in a similar manner to the practices the Samburu use to pay homage to their deceased. In the Samburu culture, people respect the deceased by placing such small items as tobacco, milk, beads, and green branches of trees onto their graves. Similarly, when the Samburu see elephant carcasses or remains, they place green branches onto the elephant's grave as a symbol of honor and respect. These traditions and beliefs are changing, however, given the various factors influencing the Samburu people and their younger generations.

Community wildlife conservation programs have been based traditionally on the premise that humans and wildlife have conflicting existences and that monetary incentives can address these relations or modify behavior. However, the Samburu people have demonstrated a value for wildlife, particularly elephants, for reasons not solely based on economics, so a novel approach to community wildlife conservation was developed in the Samburu District. An elephant conservation program was created based on the strategy of incorporating traditional beliefs and perceptions of elephants within the more modern Samburu structures and contexts, such as in the education system. This program was designed to incorporate Samburu stories and customs into a wildlife education program focused on schools and local villages. The program aims to reinforce positive customs and perceptions of wildlife in order to foster tolerant attitudes toward elephants through a two-way system of education, involvement of local people in conservation and research, and small incentive projects. For example, a book specifically written for the Samburu people highlighting their beliefs about elephants, legends, and myths was distributed to schools and community members. The publication explores Samburu traditions involving elephants and personal experiences of community members that reflect a positive relationship with the species. Further activities within the school system included the creation of the first-ever film in the

Samburu language that was based on Samburu legends linking elephants to people. The film used local materials and Samburu actors to spark an intellectual interest in elephants and promote a conservation message.

Throughout its implementation, the elephant conservation program in Samburu highlights that cultural tools, such as legends, myths, and tribe-specific customs about wildlife, can interest local people in wildlife. Local systems of knowledge about wildlife and conservation have existed throughout the African continent in the past. For example, research illustrates positive cultural traditions about wildlife in pastoralist contexts, such as the Maasai living around Amboseli National Park in Kenya. Thus, there is the potential for replicability for other conservation or research programs in local contexts where cultural traditions toward wildlife are largely positive. However, replicability of such a project depends largely on the primary economic activity and culture of the people involved. Developing a similar project may be difficult in agricultural contexts that involve extensive wildlife crop-raiding, the loss and damage of crops, and/or damage of forest plantation trees.

Human-elephant relationships are inextricably linked to many factors, including the local economic and livelihood options of communities with whom elephants share their land, human and elephant population sizes, and the political and security contexts in parks and outlying areas. The conservation of elephants depends on political treaties such as CITES, local wildlife management, scientific research on the behavior and movement patterns of elephant populations, and the incorporation of this information into conservation efforts. Additionally these relationships depend on economic incentive and local attitudes and beliefs toward elephants. As firearms become more readily available, poaching affects elephant populations, and elephant habitat becomes more fragmented, understanding human-elephant relations and involving local people in elephant conservation becomes of paramount importance.

See also

Conservation and Environment—*Conflicts between Humans and Wildlife*
Conservation and Environment—*Elephant Conservation and Welfare at the Amboseli Elephant Research Project*
Culture, Religion, and Belief Systems—*India and the Elephant*
Culture, Religion, and Belief Systems—*Indian and Nepali Mahouts and Their Relationships with Elephants*
Living with Animals—*Elephants and Humans*

Further Resources

Blanc, J. J., Thouless, C. R., Hart, J. A., Dublin, H. T., Douglas-Hamilton, I., Craig, C. G., & Barnes, R. F. W. (2003). *African elephant status report 2002: An update from the African elephant database.* Cambridge, UK: IUCN.

Douglas-Hamilton, I. (1987). African elephant population trends and their causes. *Oryx, 21,* 11–24.

Douglas-Hamilton, I., Krink, T., & Vollrath F. (2005). Movements and corridors of African elephants in relation to protected areas. *Naturwissenschaften, 92,* 158–163.

Kuriyan, R. (2002). Linking local perceptions of elephants and conservation: Samburu pastoralists in northern Kenya. *Society and Natural Resources, 15,* 949–57.

Reader, J. (1997). *Africa: A biography of the continent.* New York: Vintage Books.

George Wittemyer and Renee Kuriyan

Conservation and Environment
Endangered Species Act

Overview of the Law

Out of concern for native plants and animals imperiled "as a consequence of economic growth and development untempered by adequate concern and conservation," the 93rd Congress created the strongest biodiversity protection statute in the world: the Endangered Species Act (ESA). Congress was inspired to action by the nation's brushes with species loss: the disappearance of the Passenger Pigeon and the near-extinction of the Whooping Crane, Black-footed Ferret, Gray Wolf, and American Alligator. President Richard Nixon signed the ESA into law on December 28, 1973, describing the rich array of animal life as a vital part of the country's natural heritage.

The ESA's explicit purpose is to conserve imperiled species and the ecosystems upon which they depend. To achieve this end, the statute directs the federal government to classify ("list") imperiled species as endangered or threatened, to designate critical habitat for listed species, and to develop recovery plans that actively conserve and restore listed species. The law obligates federal agencies to proactively conserve endangered and threatened species, to avoid jeopardizing them or adversely modifying their critical habitat, and to protect listed species from "take" (for example, killing, harassment, degradation of habitat) by private individuals and public agencies.

The ESA is a precautionary statute. It errs on the side of protecting wild flora and fauna in the face of scientific uncertainty. Under the law, a wide variety of life forms are eligible for protection, including species, subspecies, and distinct population segments. The ESA protects not only wildlife on the brink of extinction—"endangered" species—but also those on the road to becoming endangered—"threatened" species. Moreover, the law provides for plants and animals to be listed based on the best scientific or commercial data available rather than mandating a higher threshold of scientific certainty. The choice by the law's architects to act on the best available data makes addressing suspected risks to species the priority rather than allowing species to languish during largely unachievable quests for perfect knowledge.

Congress authorized citizen enforcement of the ESA in the event that the federal government violated the law or failed to enforce it against non-governmental violators. Under the statute, citizens can petition for the listing or delisting of a species and for the revision of a species' critical habitat. In addition, citizens can sue the Secretaries of Interior or Commerce, who share responsibility for the ESA's enforcement, or any other party alleged to be in violation of any provision within the ESA. Citizens can also compel these Secretaries to enforce the ESA against other parties who are suspected to be violating the law, after providing 60-days notice.

The need for a strong ESA is clear, considering the diversity of life in the United States and threats to this diversity. Scientists have documented 200,000 species existing in the United States, and the actual number may be twice this amount. This diverse tapestry of life derives from the nation's large size and its varied terrain and ecosystems, including tundra, taiga, deserts, prairie, boreal and deciduous forests, and temperate and tropical rain forests. The United States includes more biome and ecoregional types than any other country. Richly varied plant and animal life forms find niches in these diverse habitats. However, rapid development, including massive urban sprawl and fossil fuel extraction, is taking its toll on these life forms, as is climate change, continued widespread livestock grazing, crop agriculture, logging, mining, recreation, and over-allocation of rivers.

The U.S. biodiversity crisis is a microcosm of the global human-caused "Sixth Extinction," with current extinction rates 100 to 1,000 times higher than natural rates of extinction. Possible causes of the first five mass extinctions include volcanic eruptions, climate change, and asteroids colliding with the earth. Extinction rates during some of these periods topped 75 percent. After each extinction, it took more than 10 million years for biodiversity to bounce back. Scientists are uncertain whether it will ever be possible for the Earth to recover from human impact should the Sixth Extinction unfold with no restraint. Less than two million of the world's estimated 10 to 100 million species have been identified and some are likely going extinct before they have even been discovered.

Endangered Species Act Implementation

Passed almost unanimously by Congress in 1973, the ESA continues to be very popular. University researchers have documented that 84 percent of the American public supports the current or even a stronger ESA. Additional polls have reported similarly high levels of support. Researchers have also documented that the law is effective. A 2005 study published in *Science* found that 99 percent of species listed under the Act have been protected from extinction. Numerous other studies have also documented the effectiveness of this law.

Despite the ESA's popularity and efficacy, it has been enshrouded in controversy since the late 1970s. Supreme Court Justice Antonin Scalia described the ESA as capable of imposing "unfairness to the point of financial ruin . . . upon the simplest farmer who finds his land conscripted to national zoological use." Former president George H.W. Bush similarly characterized it as a "sword aimed at the jobs, families, and communities of entire regions."

Controversies span private and public lands. Private property rights groups have continually claimed that the law erodes property rights by restricting actions that harm species on private land. Despite the rhetoric, the ESA has been only lightly applied to private lands. However, the law's reach to private land is fundamental to the goal of preventing species extinction. A General Accounting Office report estimated 75 percent of listed species find the majority of their habitat on private land, and some 90 percent find a significant portion of their habitat on private land. In addition, the ESA can curtail destructive management activities (of land and wildlife) being permitted on federal land or by federal agencies, through its provision that federal agencies not permit actions that result in the jeopardy of a listed species or the adverse modification of its critical habitat. Moreover, federal agencies cannot themselves commit actions that would result in these harms.

Other critics of the ESA include industry interests (e.g., mining, oil and gas, logging, ranching, water developers, commercial builders), state wildlife agencies, federal agencies, and pro-industry administrators and politicians at all levels. Given the law's capacity to curb ecologically destructive economic activities, there have been sustained efforts to weaken the ESA for more than a decade, and these efforts have reached a crescendo in recent years.

Industrial interests have long been hostile to the ESA's purpose of tempering economic growth with "adequate concern and conservation." Various industries have long claimed that the ESA harms economic growth and is therefore a threat to the U.S. economy. Yet, although it has endured many sets of amendments over the past three decades, the ESA has emerged fairly intact. Meanwhile, the U.S. economy has experienced strong periods of economic growth. This is particularly true in states such as Florida and California, where there exist both large numbers of listed species and high rates of development.

State agencies often criticize the ESA because listed species are primarily managed by federal authority, thus preempting state authority. This is related to the "states' rights" movement, which accelerated under the Reagan Administration. However, state wildlife agencies are generally funded by hunting, fishing, and trapping licenses, and nongame conservation efforts are often funded at very low levels. State wildlife agencies also engage in damaging practices, such as the stocking of nonnative sportfish that often compete with, prey upon, or hybridize with native fish species.

The most focused attack in Congress has been on the critical habitat provisions of the ESA. Given that 85 percent of species listed under the law are at least partially imperiled due to habitat degradation, the ESA's strong safeguards of critical habitat have resulted in significantly increased rates of species recovery. Those plants and animals with such designations are twice as likely to be recovering as those without. Unfortunately, only 36 percent of domestic listed species enjoy critical habitat designations. Critical habitat's effectiveness lies in its protection of both occupied habitat and unoccupied habitat, the latter of which is essential for the rebound of species that have vanished from the majority of their natural range by the time they are listed. In addition, federal agencies are prohibited from adversely modifying critical habitat, which is a clearer and less discretionary restriction than the provision that federal agencies not jeopardize the survival of listed species.

Because citizens can and have used the ESA to protect endangered species, federal agencies and federal administrators themselves often seek to weaken the law. Because of the George W. Bush Administration's adherence to states' rights and deregulation, there have been, on average, only eight listings per year, all the result of citizen lawsuits. This is compared to 59 listings per year under President George H.W. Bush and 65 per year under President Bill Clinton. Meanwhile, as of May 2006, nearly 300 species awaited listing as candidates and proposed species. Some species have waited on the candidate list for 25 years, and delays in protection have lead to the extinction of several dozen species. Moreover, in the 2000 book *Precious Heritage,* 6,460 U.S. species were identified as imperiled or vulnerable, the majority of which are not even listing candidates.

The bottleneck on listings under George W. Bush is second only to that under Reagan, whose administration was also opposed to regulatory protections for endangered species. Administrative hostility to the ESA was evident in the George H.W. Bush Administration as well. George H.W. Bush's Interior Secretary, Manuel Lujan Jr.—the primary official responsible for implementing the ESA at the time—quipped, in reference to the critically imperiled Mount Graham Red Squirrel in Arizona, "Nobody's told me the difference between a red squirrel, a black one, or brown one. Do we have to save every subspecies?"

While the number of species listed per year was higher in the Clinton Administration than all administrations after Jimmy Carter's, most listings still resulted from litigation and listing delays were common. In addition, Clinton Interior Secretary Bruce Babbitt greatly advanced the use of ESA exemptions through habitat conservation plans applied to private lands. Aspects of these plans have been successfully challenged by citizens due to the threat exemptions can pose to species survival and recovery.

Other ways the George W. Bush Administration has weakened the ESA administratively include funding cuts for endangered species programs, designating far less critical habitat than biologists have recommended, overruling biologists' recommendations to list imperiled species, providing exemptions from the ESA for the Department of Defense, removing some restrictions on killing and trading in international endangered species, and exempting agencies from consulting with the U.S. Fish and Wildlife Service for certain federal land projects and for the use of pesticides.

Citizen Enforcement of the ESA

There has long been an important role for citizens to play in ESA enforcement, including petitioning for species to be listed, for their critical habitat designations to be revised, and to sue any party that violates any section of the ESA. An early high watermark for controversy over the ESA was the Tellico Dam issue in the late 1970s. A citizen lawsuit stopped this dam from being completed on the Little Tennessee River due to its threat to critical habitat of the Snail Darter. In the 1978 landmark opinion by the U.S. Supreme Court in *Tennessee Valley Authority v. Hiram Hill* (437 U.S. 153 (1978)), Chief Justice Burger declared that, "the plain language of the Act, buttressed by its legislative history, shows clearly that Congress viewed the value of endangered species as 'incalculable.'" The majority of the justices found that, given this incalculable worth of endangered species, it would be difficult and inappropriate for the court to weigh the economic costs of protection against the value of protecting a species. Consequently, the Supreme Court affirmed the injunction of the Tellico dam, a $100 million dollar project that was 90 percent complete, because its completion would jeopardize the Snail Darter, a three-inch long fish, in its critical habitat on the Little Tennessee River.

As a result, many in Congress were up in arms about the ESA. Although the statute itself remained essentially intact, the Tellico Dam was ultimately completed, via an appropriations rider heard for a mere 42 seconds on the House floor. Many observers reported that legislators appeared not to realize that they were voting for the completion of the dam despite its predicted effect on an endangered species. A similar end-run around the ESA and citizen participation unfolded in the 1990s in the context of the threat to the Northern Spotted Owl from logging in the Pacific Northwest. A nondescript rider called Section 318 was attached to a Senate general appropriations bill, and it briskly passed through Congress and overrode a court injunction that had stopped the logging of old-growth forests on U.S. Forest Service land. Section 318 was one of several riders used by the Pacific Northwest delegation to avoid logging prohibitions that were intended to minimize threats to the owl.

More recent examples of citizen enforcement of the ESA include efforts to address the impact of livestock grazing on listed species inhabiting federal lands in the Southwest, such as the Mojave Desert Tortoise, Mexican Wolf, Southwestern Willow Flycatcher, and Mexican Spotted Owl. Citizen groups are also pushing for better river management to address water needs of endangered fish in U.S. rivers, for example, salmonids in the West, the Rio Grande Silvery Minnow, the Devils River Minnow in Texas, and several sturgeon species across the country. In the Rocky Mountains, citizens continue to challenge logging and ski resort expansions because of their impacts on the Canada Lynx and other forest wildlife. In the Intermountain West, citizens seek listing of Greater Sage-grouse, to protect the bird and its habitat in the Sagebrush Sea. Hotspots of biodiversity and imperiled species include California, Hawaii, and Florida, and active citizens' campaigns are working to protect species, their habitat, and native ecosystems in these states.

The importance of citizen ESA enforcement is underscored by the fact that federal agencies are not just allowing land uses and actions by private parties that jeopardize imperiled species and harm their habitat, they are also themselves committing actions that are harmful to species on the brink. For instance, a division within the U.S. Department of Agriculture, little-known to the public and misleadingly named "Wildlife Services" (WS), kills millions of animals every year, both wild and feral. In 2004, WS killed 2.7 million animals, including gray wolves, which are listed under the ESA. Other examples include the federal Bureau of Reclamation, whose dam operations are a primary threat to many endangered fishes and birds, and the Department of Defense, whose military operations on the Barry M. Goldwater Range in Arizona continue to push the Sonoran Pronghorn to the brink of extinction (there are as few as 30 individuals left).

The ESA in Perspective

The ESA was visionary when it was passed by Congress almost unanimously over thirty years ago and it remains at the vanguard today. The law's architects and supporters argued for a strong biodiversity statute based on moral, ecological, and utilitarian reasons, and from the perspective that imperiled species represent (unwilling) canaries in a coal mine. We ignore the onward march of species extinction at our own peril, agreed most of Congress in 1973.

That warning still rings true. Two-time Pulitzer Prize winner E. O. Wilson argued in *The Future of Life* that we are literally mortgaging the Earth by continuing down the path of unsustainable economies. Researchers have established that humans are currently exceeding the Earth's biological capacity by at least 20 percent. Rather than merely living off the interest that the Earth's natural capital provides, we are drawing down the capital, and our bank account will soon be empty.

On the way to eventual economic collapse (if policies aren't changed), ecosystems will crumble and native flora and fauna will disappear. Economists estimate that intact natural systems provide us with $33 trillion annually in "ecosystem services." Whether by the maintenance of the atmosphere, creation of clean air, and recycling of rainfall by forests; filtering of water by forests and healthy watersheds; nourishing of agricultural plants and trees by microorganisms; decomposition of organic matter; waste disposal; nitrogen fixation and nutrient cycling; bioremediation of chemicals; biocontrol of species that attack crops, forests, and domesticated animals; or pollination by birds, bees, butterflies, bats, and others, components and processes of nature make the Earth habitable to humans.

Yet, estimates of the monetary value of a living planet are likely to be gross underestimates. We generally cannot replace ecosystems once they are in tatters. In addition, monetary measurements do not address the intangible aesthetic, spiritual, and moral rationales that are an important component of support for endangered species protection. In *The Value of Life,* Stephen Kellert reported his finding after two decades of research, that moralistic and humanistic (affection, bonding) attitudes toward wildlife are most prevalent among U.S. citizens. The ESA honors these widespread attitudes by requiring caution when economic growth and human activities overstep nature's bounds.

As John Muir put it, "When we try to pick out anything by itself, we find it hitched to everything else in the universe." By protecting species, their habitats, and their ecosystems, the ESA serves as a vital safety net for the rich tapestry of life in the United States.

See also

Conservation and Environment—*Exotic Species*

Further Resources

Endangered Species Act: 16 U.S.C. §§ 1531–1544. http://www.fws.gov/endangered/esaall.pdf.

National Research Council. (1995). *Science and the ESA*. Washington, DC: National Academy of Sciences Press.

Rosmarino, N. J. (2002). *Endangered Species Act: Controversies, science, values, and the law*. PhD Dissertation, University of Colorado at Boulder.

Stanford Environmental Law Society. (2000). *The Endangered Species Act handbook*. Stanford, CA: Stanford University Press.

Stein, B. A., Kutner, L. S., & Adams, J. S. (Eds.). (2000). *Precious heritage: The status of biodiversity in the United States*. New York: Oxford University Press.

Taylor, M., Suckling, K., & Rachlinski, J. J. (2005). The effectiveness of the Endangered Species Act: A quantitative analysis. *BioScience, 55*(4), 360–67.

Tobin, R. (1990). *The expendable future: U.S. politics and the protection of biological diversity.* Durham, NC: Duke University Press.

Nicole Rosmarino

Preble's Meadow Mouse and the Endangered Species Act

Carron Meaney

In the populated portion of Colorado and southeastern Wyoming, the Front Range Urban Corridor, there is a little mouse with a long tail that jumps, swims, travels great distances, and hibernates. Preble's meadow jumping mouse (*Zapus hudsonius preblei*) lives in riparian (stream bottom) areas, where shrubs and other plants are lush and provide food and cover for day nests and protection from predators, and adjacent meadows, where it forages for seeds and insects. Biologists noticed that this particular subspecies had not been seen for twenty years. A nonprofit organization petitioned the U.S. Fish and Wildlife Service (USFWS) to list it under the Endangered Species Act (ESA).

The USFWS reviewed the existing information and decided to list the mouse as threatened. A threatened listing means that the taxonomic unit, in this case a subspecies, is at risk of becoming endangered and requires federal protection. As is often the case, this federal action gave rise to opposing points of view. The riparian habitat favored by the mouse is also prime real estate for gravel mining, grazing, and water development projects, and uplands overlooking the riparian zone are valued for residential communities. In this case it was the developers who created the greatest outcry. Because this was the first endangered listing in the populated Front Range corridor, developers and other private property owners were confronted with the guidelines established by the ESA for the first time.

The ensuing controversy is simply a local example of the protection of endangered species versus the freedom of industry and private property owners to meet their financial goals on their land and be unencumbered by environmental regulations. As with many ESA listings, opposing viewpoints of what contributes to the public good become exposed: biodiversity, protected habitat, and benefits to the landscape versus short-term economic benefits (that often result in long-term environmental impacts), such as production of gravel, housing, beef, a plentiful water supply, and so on. However, this protective action is no different than a large number of regulations that are promulgated for the common good: zoning regulations, speed limits, specifications for roads and sidewalks, and air and water quality standards, to name a few. We are all part of a community and to enjoy the benefits we must accept the limitations on our freedom.

This particular controversy has been further fueled by a genetic study, defended by developers and the Bush administration, that suggested Preble's meadow jumping mouse was not distinct from a subspecies occurring 200 miles to the northeast. A subsequent genetic study has demonstrated that the two subspecies are clearly distinct. An independent panel evaluated the two studies and concluded that the evidence favored the recognition of Preble's meadow jumping mouse as a distinct subspecies. Unfortunately, questions about endangered species are not only scientific; they can become debates of science versus politics.

An added complexity is the myriad methods applied to the collection, extraction, and analysis of genetic information used to assess species or subspecies differences and a lack of agreement on what represents taxonomic uniqueness. Scientists are not yet in agreement on what defines a subspecies. The ESA does not mention genetics at all. This will likely get worked out in time and is due to the fact that genetic tools are relatively new.

The ESA is one of the most significant pieces of legislation for protection of biodiversity and rare plants and animals. Controversy aside, with ESA protective measures in place, the landscape all across the Front Range Urban Corridor in Colorado and Wyoming will benefit immensely from habitat protection for the mouse. Fifty years from now, someone will point out across one of a number of beautiful, intact riparian corridors and say, "and to think, it was all due to a little mouse."

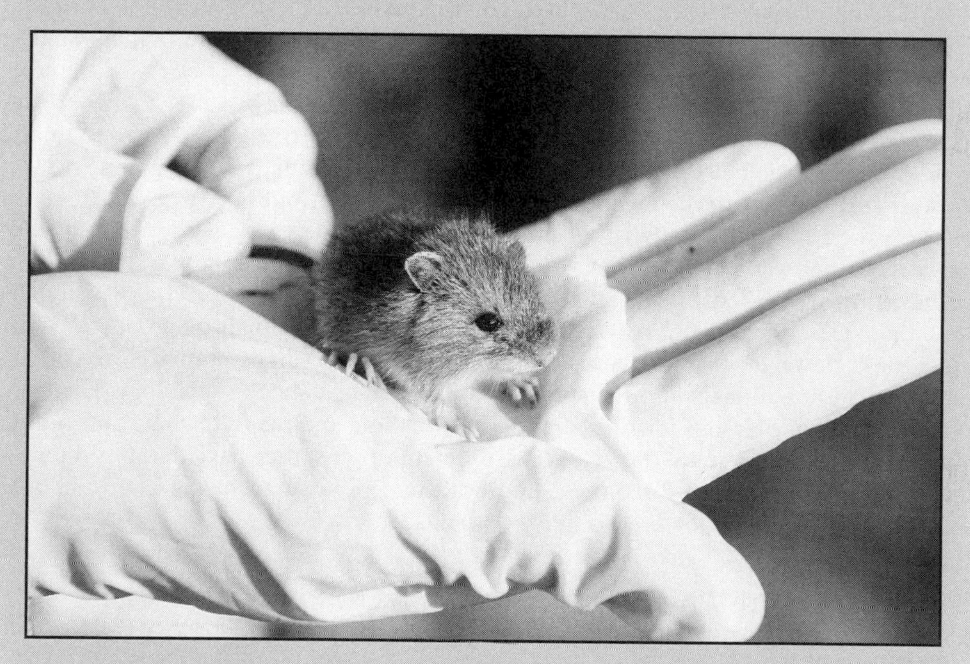

Preble's meadow mouse being held during an ecological research project in Colorado. The mouse took center stage again in 2006, as a House committee scheduled a field hearing on the debate over whether it is a separate species entitled to federal protection. Courtesy of Shutterstock.

Conservation and Environment
Environmental Sociology and Animal Studies

Environmental sociology and human-animal studies are both academic fields of study that emerged as correctives to the oversight of some critical issues by social scientific disciplines. The former field of study began to take shape in the late 1970s, while the latter developed in the 1990s. Despite many commonalities—such as critiquing the culture/nature dichotomy, pointing to the impact of the natural world on social processes and the reverse, and critiquing sociology and the social sciences more generally for overlooking these issues—the fields of environmental sociology and animal studies have remained surprisingly insulated from each other. This essay examines some of the key factors that have likely contributed to the current distance between these two fields and the potential advantages of interaction between the two. While a formidable gulf currently exists between these two fields, it can and should be ameliorated, and there are many potential advantages to doing so.

Symptomatic of the lack of engagement between these two fields, only a couple of authors have addressed it. Steven Yearly (2002) mentions the issue in his examination of what has not been included in the construction of the "environment." Likely not a coincidence, the omission of animals is the first in his discussion. The other omissions Yearly lists include the urban environment and the issue of human overpopulation. In a more focused examination of the omission of animals by environmental sociology, Hilary Tovey (2003) examines their exclusion in environmental sociology texts. She finds that in the relatively few instances where animals are addressed, it is as "wild" animals and in the form of populations, not as individual subjects. Consequently, variations within categories are overlooked, and the countless domesticated animals in contact with human societies are notably omitted. Tovey appropriately remarks, "[t]he lack of interest among environmental sociologists in domestic and domesticated animals seems paradoxical. It contrasts starkly with levels of interest among the general public" (p. 203). She also takes up the constructivist/realist tension within environmental sociology in her work and argues that while at first blush the constructivist accounts appear more likely to include animals in their examinations of the environment, they fail to live up to their potential, and, consequently, realists account more for animals than their constructivist colleagues. However, Tovey believes that neither of these camps comes close to adequately including animals into the subject matter of environmental sociology. There are numerous possible reasons for this paradoxical situation, and the following contributing factors are examined herein: the fields have different origins, they have developed different levels of analysis and discourses, and academic protectionism may be inhibiting collaboration. This list, however, is likely not exhaustive.

The Great Divide: Why Linkages between Environmental Sociology and Animal Studies Have Not (Yet) Been Formed

Divergent Origins

One factor that likely partially accounts for the gulf between environmental sociology and animal studies is that they have developed differing views of animals due to their divergent origins. Several areas have been identified in the literature as having contributed to the development of environmental sociology (for example, Buttel & Humphrey, 2002; Buttel, 1996; Dunlap & Catton, 1994); two areas in particular, rural and urban sociology, likely affected the ways in which animals are viewed by the field. The work of urban sociologists has focused on built environments and has paid negligible attention to animals in urban places (a reflection of the assumption that animals are not part of human societies). Rural sociologists, on the other hand, have paid some attention to animals, but generally as aggregates and in very instrumental and anthropocentric ways. Wildland recreation (Dunlap & Catton, 1979) and resource issues (Buttel, 1996; Dunlap & Catton, 1979) have been dominant issues of concern among rural sociologists, and both areas of concern have brought with them, to environmental sociology, rather instrumental views of nature—especially of animals.

The animal studies field, in contrast, has developed largely as a critique of the instrumental use of animals. Thus, while the origins of environmental sociology have likely contributed to a discounting and utilitarian view of animals, the animal studies field—with its strong ties to the animal rights movement—developed specifically to draw attention to the roles of animals in human societies and to challenge utilitarian views of them.

Divergent Discourses and Levels of Analysis

Environmental sociology and animal studies have developed distinct discourses and levels of analysis. It has been demonstrated that a clear relationship exists between the development and phases of the environmental movement and the emergence and development of environmental sociology (Dunlap & Catton, 1979, 1994; Dunlap 1997), and between the animal rights movement and the development of animal studies (Arluke, 2002; Beirne, 2002; Shapiro, 2002). The affinity between these academic fields and their respective social movements has serious implications for the relationship between environmental sociology and animal studies, since the environmental and animal rights movements themselves have not developed the linkages with each other that one might expect due to factors described later.

There is a notable tension between environmental sociology and the environmental movement, on one hand, and animal studies and the animal rights movement, on the other hand, regarding the appropriate level of analysis. The environmental movement, and environmental sociology relatedly, employ a holistic view, developing discourses based on ecosystems (Dowie, 1995; Gottlieb, 1993; Catton, 1980; Tovey, 2003). In contrast, the animal rights movement and animal studies tend to focus more at the individual level (Eckersley, 1992; Jasper & Nelkin, 1992; Regan, 1980). The tension between these two discourses is perhaps best illustrated with regard to the issue of hunting (Jasper & Nelkin, 1992). Examining sport hunting from a holistic or ecological standpoint, one might argue that hunting is a useful management tool to maintain ecological equilibrium (e.g., Gunn, 2001). Those who take a more individualistic animal-oriented approach generally argue that sport hunting is an unsuitable (and perhaps ineffective) means to achieve ecological equilibrium (Hawkins, 1998; Fitzgerald, 2005).

As a result of the divergent levels of analysis, discourses regarding the environment and animals are different in a fundamental way: the animal rights movement (and the area of animal studies more implicitly) has developed a discourse based on rights. For alternative conceptions to the rights-based discussion of animals, see Curtin (1996), Donovan (1990), Hawkins (1998), and Plumwood (1996). This rights discourse, which has been critiqued as being too individualistic (Curtin, 1996) among other things, has created a rift between environmental and animal-related concerns. More specifically, the language of rights generally lacks relevance in discussions of the environment, and the discourse of "animal rights" may preclude broader environmental thinking among animal rights advocates. Thus, the different levels of analysis and resultant discourses have created a formidable divide between the areas of environmental sociology and animal studies.

Academic Protectionism

Another factor that has likely contributed to the gap between environmental sociology and animal studies is academic protectionism on the part of environmental sociology, which—although still marginal within the larger discipline of sociology itself—is older and more established than animal studies. Some of the reasons for environmental sociology's continued marginalization within the discipline of sociology include taboos within sociology regarding determinism and biologism (Dunlap & Catton, 1979, 1994); the grounding of sociology in a Western worldview that discounts nature; increasing industrialization and urbanization that have made nature seem less important to studies of societies; and the theoretical sociological tradition that has overlooked such issues and has even constructed barriers to examining them (Dunlap & Catton, 1994).

The field of animal studies has encountered similar resistance from the discipline of sociology, and the social sciences more generally. In fact, all of those barriers within sociology that environmental sociology has faced have also been confronted by animal studies. It is not unreasonable to suspect that, due to its own marginalization, environmental sociology may not want to engage with the similarly marginalized field of animal studies and risk entrenching its own marginalization. Limitation of resources is a common side effect of such marginalization. Consequently, there may also be resistance to linkages between the fields, because they can be viewed as competing for the same resources. Such competition has been acknowledged as a barrier to the advancement of animal studies (Shapiro, 2002).

Another reason for academic protectionism on the part of environmental sociology could be related to the perception that environmental problems pose a greater risk to humanity and the greater general support for environmental issues. Environmental sociologists have done a good job highlighting the potential consequences of environmental degradation for humanity (e.g., Catton, 1980) and how these problems are now global in nature and a threat to everyone (Beck, 1995). They point out that addressing environmental problems is in our self-interest, to the extent that doing so mitigates environmental degradation and disasters. Conversely, animal-related issues are not perceived as endangering humans, and there may therefore be less impetus to take them seriously, by both academics and the public (although recent concern over mad cow disease and the avian flu will likely change this perception). Related to this, the environmental movement and environment-related issues receive widespread public support: the 1970 Earth Day celebration is touted as the largest single demonstration in the history of the United States (Brulle, 2000). Support for animal-related causes and issues remains more marginal. For instance, a random sample of adult residents in the United States found that only 5.2 percent reported being vegetarian, and many reported their motivation was related to environmental concerns (Kalof et al., 1999). Thus, addressing animal-related issues might be perceived as diluting environmental sociology's message and perhaps risking some of the field's moral weight and public support.

Environmental sociology may also be wary of forming a relationship with animal studies due to the backlash against ecofeminism. Numerous charges have been laid against ecofeminism, including that it essentializes women, implies biological determinism, and fails to acknowledge that women can also be agents of environmental destruction (Agarwal, 2000; Jackson, 1993a,b; Stange, 1997). It is worth noting, however, that many critics of ecofeminism fail to distinguish between the two variants of the theory: cultural ecofeminists consider the relationship between women and nature to be a natural one, whereas social ecofeminists argue that the connection between women and nature is the result of social construction, not biology. Within the social sciences, such criticisms are particularly grave. Much of the work that has been done within the field of animal studies can be considered ecofeminist in spirit if not in design (see, for instance,

Adams, 1991, 1996; Donovan, 1990; Gaard, 1997). Therefore, animal studies has largely allied itself with a heavily criticized (but mostly misunderstood) perspective. Consequently, one scholar (Kruse, 2002) has suggested that the field of animal studies pursue other theoretical avenues to enhance its own respectability. Given that some within the field of animal studies apparently want to distance the field from ecofeminism, it is entirely possible that environmental sociology may prefer to distance itself from the animal studies field for exactly the same reason.

The strong link between the animal studies field and the animal rights movement may be another reason why the field of environmental sociology has kept its distance. The animal rights movement itself has developed a reputation for radicalism, with the radical side of the movement (which James Jasper and Dorothy Nelkin (1992) refer to as the "animal rights crusade") representing the entire movement in the minds of many people. Although the environmental movement also has its own radical fringe, the entire movement has not been identified with it to the degree that the animal rights movement has. In fact, many of the large and most popular environmental organizations are reform-oriented and have become institutionalized (Brulle, 2000), and thus legitimized in the eyes of many (and impotent in the eyes of others). Unfortunately, "the undeserved charge of violence and terrorism, with all that term currently carries, readily spills over to human animal studies" (Shapiro, 2002, p. 336). The field of environmental sociology may, therefore, be wary of becoming enmeshed with a field of study that carries this liability with it.

Kenneth Shapiro continues his above statement as follows: "More insidiously, human animal studies is vulnerable to the charge that an animal rights movement agenda biases its investigations and scholarship" (2002, p. 336). This perception is not only due to the close link between animal studies and the animal rights movement, but also due to the perception that this is an area governed by emotion. The animal rights movement is dominated largely by women, and areas dominated by women are often charged with being emotional rather than rational (Munro, 2001). Thus, environmental sociologists may also choose to distance themselves from a field that may be degraded as being emotional—and, by extension, "unscientific."

Other, more philosophical, aspects of the animal rights movement may exacerbate academic protectionism on the part of environmental sociology. First of all, it can be said that the animal rights movement, and by extension animal studies, has not fully integrated environmental thinking. For instance, in his 1975 book *Animal Liberation* (widely considered the "bible" of animal rights), Peter Singer outlines an ethic for the treatment of animals based upon the claim that they are sentient beings. The perhaps unwarranted implication of this line of reasoning is that animal rights activists need not be concerned with the treatment of non-sentient beings, such as plants and rivers, which has made strategic relationships with other social movements (especially the environmental) difficult. Additionally, due to their focus on the importance of animal lives, it is not uncommon for animal rightists to accord more importance to animal lives than cultural traditions, which contrasts with the position of many left-leaning individuals cautious to avoid charges of cultural imperialism (Jasper & Nelkin, 1992). These philosophical aspects of the animal rights movement, therefore, have likely not ingratiated it into the environmental-sociology fold.

In examining the possible reasons for academic protectionism on the part of environmental sociology, the differing composition and origins of the two fields, and the different discourses both have developed, one gains a better understanding of why strong linkages have not yet developed between the two fields. Despite the divergence between the two fields, the potential exists for these fields to converge into a valuable partnership. Such a partnership would have several advantages, the most significant of which are discussed next.

Potential Advantages of a Closer Working Relationship between Animal Studies and Environmental Sociology

The first advantage of a closer relationship between these two fields would be that the sharing of knowledge and research across fields would maximize productivity. For instance, both fields have expended time and energy examining how classical social theories have addressed their respective area of interest—and how these theories could be extended and improved upon so that they better address these issues. Due to the similarities between the treatment of animal and environmental issues within classical social theory, a meaningful engagement between the work on both sides would be advantageous.

Additionally, due to the different levels of analysis employed, each of the two fields would likely benefit from attending to the insights of the other. As discussed earlier, one reason why environmental sociology and animal studies have not developed as close of a relationship as one would assume given their areas of study is that they tend to examine "nature" at different levels: environmental sociology, obviously influenced by the environmental movement, takes a holistic approach, focusing on ecosystems and populations, whereas animal studies has been impacted by the animal rights movement's focus on more individual subjectivities. As a result, each field misses something that the other one captures. Therefore, combining insights from these fields will help to fill in these gaps in their academic discourses. Further, combining insights from each may promote the development of a better ethic for dealing with animal and environmental issues. Jasper and Nelkin (1992) recommend that animal advocates adopt the environmentalists' approach for dealing with wild animals, but apply it to focus solely on domestic ones. Doing so, however, would not be as simple as implied, because definitions of "wild" and "domestic" have become increasingly blurred; and even adequately addressing issues related only to domestic animals requires a reworking of how we conceptualize society, and hence sociology.

It is the coupling of environmental sociology and animal studies that holds the promise to achieve such a reworking of sociology in particular. Within their own fields, both environmental sociologists and animal studies scholars have worked to better understand and deconstruct the culture/nature dichotomy. It is clear that, at this point in time, their efforts have not fundamentally altered the state of sociology. Instead, sociology has endeavored to "deflect critique by defining each criticism as a new specialty" (Connell, 1997, p. 163). Both environmental sociology and animal studies have been viewed as specialties and have been granted the status of sections of study by the American Sociological Association, whereby the critique of the larger discipline and cross-specialty fertilization are effectively contained. A closer relationship between these two fields would create a broader and stronger critique of the culture/nature dichotomy and the discipline of sociology in particular—which, as Ulrich Beck (1995) points out, has been structured upon said dichotomy.

One area that appears especially promising for confronting sociology's reification of the culture/nature dichotomy through its own disciplinary boundaries is the study of human "Others" who have been denigrated as being closer to nature than to culture. There is room here for a joint environment-animal studies project to argue that it is not enough to attempt to elevate those associated with nature to the level of the cultural: we must begin to address the denigration of the "natural" and "animal" in the first place.

Only then can we attend to all forms of inequality and oppression. Kruse asserts that this area presents an important opportunity for animal studies: "[t]his area holds great potential for future work. Animals, and our uses of them, play a vital part in racialization and the construction of gender. Likewise, class is, in numerous ways, tied to our relations to the animal world. It is here where those who study animals in human society have a potential truly to influence sociology" (2002, p. 376). A joint environment-animal studies project would certainly wield more influence in this regard than animal studies alone. One might reasonably ask if it is even possible for either of these two fields to persuasively argue for the dissolution of the culture/nature dichotomy and increased respect for that associated with nature without attending to the terrain covered by the other field. Together, environmental sociology and animal studies have the potential to challenge and revolutionize sociology and the social sciences more broadly.

In addition to these seemingly esoteric academic gains, there would likely be practical implications of a linkage between environmental sociology and animal studies, as well. It is likely that, as a result of the coupling of these fields, better progress could be made on issues most clearly of mutual interest (such as industrial agriculture and wildlife issues, which both have environmental impacts and impacts on animals' lives), resulting in more favorable public policies for the environment and individual animals. A closer relationship between these two fields and the blending of their holistic and individualistic views would also likely result in challenges and (perhaps) changes to policies that currently harm animals in the name of environmental sustainability—such as culling animals—and to policies that place individual animals above potential environmental harms, such as the keeping of exotic pets. Additionally, a closer relationship between the two fields would likely also result in a greater discussion of the legal techniques applied on each side, so that perhaps mutually unfavorable policies could be challenged more successfully.

Conclusion

There are numerous potential advantages to a closer working relationship between the fields of environmental sociology and animal studies, as outlined herein. Perhaps the greatest strength of this synergy would be its potential to revolutionize sociology and its reification of the culture/nature dichotomy in particular. Strategically, it would seem that efforts should be concentrated on demonstrating that undermining oppressions will require addressing the debasing of the "natural," as well as on bringing "domestic" animals into the sociological purview—a project that is already under way, but that could use the support of environmental sociology. Then, as the culture/nature dichotomy loses its salience within sociology, it would be easier to integrate the concerns of "wild" animals and the environment, but in a way that fundamentally alters sociology, instead of simply ghettoizing these concerns.

Clearly, the ways in which these two fields have developed up to this point cannot be altered, but we can begin to challenge the resultant paradoxical insulation of environmental sociology and animal studies from each other and envision new paths for the future of these two fields—paths that are meaningfully interconnected instead of simply parallel.

See also

Ethics and Animal Protection
Human Perceptions of Animals—*Sociology and Human-Animal Relationships*

Further Resources

Adams, C. (1991). *The sexual politics of meat: A feminist-vegetarian critical theory*. New York: Continuum.

———. (1996). Ecofeminism and the eating of animals. In K. J. Warren (Ed.), *Ecological feminist philosophies* (pp. 114–36). Bloomington; IN: Indiana University Press.

Agarwal, B. (2000). Conceptualizing environmental collective action: Why gender matters. *Cambridge Journal of Economics, 24,* 283–310.

Arluke, A. (2002). A sociology of sociological animal studies. *Society & Animals, 10*(4), 369–74.

Beck, U. (1995). *Ecological politics in an age of risk*. Cambridge: Polity Press.

Beirne, P. (2002). Criminology and animal studies: A sociological view. *Society & Animals, 10*(4), 381–86.

Brulle, R. J. (2000). *Agency, democracy, and nature: The U.S. environmental movement from a critical theory perspective*. Cambridge: MIT Press.

Buttel, F. H. (1996). Environmental and resource sociology: Theoretical issues and opportunities for synthesis. *Rural Sociology, 61*(1), 56–77.

Buttel, F. H., & Humphrey, C. R. (2002). Sociological theory and the natural environment. In R. E. Dunlap & W. Michelson (Eds.), *Handbook of environmental sociology*. Westport, CT: Greenwood Press.

Catton, W. R., Jr. (1980). *Overshoot: The ecological basis of revolutionary change*. Urbana: University of Illinois Press.

Connell, R. W. (1997). Long and winding road. In B. Laslett & B. Thorne (Eds.), *Feminist sociology: Life histories of a movement* (pp. 151–64). Piscataway, NJ: Rutgers University Press.

Curtin, D. (1996). Toward an ecological ethic of care. In K. J. Warren (Ed.), *Ecological feminist philosophies* (pp. 66–81). Bloomington; Indianapolis: Indiana University Press.

Donovan, J. (1990). Animal rights and feminist theory. *Signs, 15*(2), 350–75.

Dowie, M. (1995). *Losing ground: American environmentalism at the close of the twentieth century*. Cambridge: MIT Press.

Dunlap, R. E. (1997). The evolution of environmental sociology: A brief history and assessment of the American experience. In M. Redclift & G. Woodgate (Eds.), *The international handbook of environmental sociology* (pp. 21–29). Northhampton: Edward Elgar.

Dunlap, R. E., & Catton, W. Jr. (1979). Environmental sociology. *Annual Review of Sociology, 5,* 243–73.

———. (1994a). Struggling with human exceptionalism: The rise, decline, and revitalization of environmental sociology. *The American Sociologist, 25,* 5–30.

———. (1994b). Toward an ecological sociology: The development, current status, and probable future of environmental sociology. In W. V. D'Antonio et al. (Eds.), *Ecology, society, and the quality of social life*. New Brunswick: Transaction Publishers. Originally published in *The Annals of the International Institute of Sociology, 3,* 263–84, 1992.

Eckersley, R. (1992). *Environmentalism and political theory: Toward an ecocentric approach*. Albany: State University of New York Press.

Fitzgerald, A. (2005). The emergence of the figure of 'woman the hunter': Equality or complicity in oppression? *Women's Studies Quarterly, 33*(1 & 2), 86–104.

Gaard, G. (1997). Ecofeminism and wilderness. *Environmental Ethics, 19*(1), 5–24.

Gottlieb, R. (1993). *Forcing the spring: The transformation of the American environmental movement*. Washington, DC; Covelo: Island Press.

Gunn, A. S. (2001). Environmental ethics and trophy hunting. *Ethics and the Environment, 6*(1), 68–95.

Hawkins, R. Z. (1998). Ecofeminism and nonhumans: Continuity, difference, dualism and domination. *Hypatia, 13*(1), 158–97.

Jackson, C. (1993a). Women nature or gender history—A critique of ecofeminist development. *Journal of Peasant Studies, 20*(3), 389–419.
———. (1993b). Doing what comes naturally? Women and environment in development. *World Development, 21*(12), 1947–63.
Jasper, J. M., & Nelkin, D. (1992). *The animal rights crusade: The growth of a moral protest*. New York: Free Press.
Kalof, L., Dietz, T., Stern, P., & Guagnano, G. (1999). Social psychological and structural influences on vegetarian beliefs. *Rural Sociology, 64*(3), 500–11.
Kruse, C. R. (2002). Social animals: Animal studies and sociology. *Society & Animals, 10*(4), 375–79.
Munro, L. (2001). Caring about blood, flesh and pain: Women's standing in the animal protection movement. *Society & Animals, 9*(1), 43–61.
Plumwood, V. (1996). Nature, self, gender: Feminism, environmental philosophy, and the critique of rationalism. In K. Warren (Ed.), *Ecological feminist philosophies* (pp. 88–105). Bloomington; Indianapolis: Indiana University Press.
Regan, T. (1980). Utilitarianism, vegetarianism, and animal rights. *Philosophy and Public Affairs, 9*(4), 305–24.
Shapiro, K. J. (2002). The state of human-animal studies: Solid, at the margin! *Society & Animals, 10*(4), 331–37.
Singer, P. (1990 [1975]). *Animal liberation*. (2nd ed.). New York: Random House.
Stange, M. Z. (1997). *Woman the hunter*. Boston: Beacon Press.
Tovey, H. (2003). Theorising nature and society in sociology: The invisibility of animals. *Sociologia Ruralis, 43*(3), 196–215.
Yearley, S. (2002). The social construction of environmental problems: A theoretical review and some not-very-Herculean labors. In R. E. Dunlap, F. H. Buttel, P. Dickens, & A. Gijswijt (Eds.), *Sociological theory and the environment: Classical foundations, contemporary insights* (pp. 274–85). New York: Rowman and Littlefield Publishers, Inc.

Amy J. Fitzgerald

Conservation and Environment
Exotic Species

With reference to animal and plant species, the term "exotic" is used interchangeably with the terms "nonnative," "nonindigenous," "alien," "foreign," and "immigrant." Related—but not synonymous—terms are "introduced" and "invasive." Although "exotic" and the other terms are used widely and frequently in scientific, government, and popular publications, a precise definition for them remains elusive. There is general agreement, however, that the terms designate species that meet two criteria: they have spread beyond their historical native range or the habitat where they evolved, and their spread has been assisted by human activities. The term "invasive" is used particularly to designate exotic species that significantly alter the environment into which they have spread. A monitor lizard in a Minnesota apartment or a koala in a Manitoba zoo would meet the criteria for designation as an exotic species, but they would not survive outside their native habitat without the attentive care of humans. A few exotic species of animals, however, have been able to thrive without human tending and are able to reproduce in environments to which they are alien. Humans have assisted the dispersal of other animal species in several ways, both intentionally and unintentionally.

Intentional Importation, Unintentional Dispersal

When humans have traveled to new areas as colonists, they have taken with them animals to be used for food, clothing, and labor. European colonists, for example, brought cattle, sheep, pigs, donkeys, and horses to the Americas and Australia. Some of these animals escaped from human management and were able to survive and reproduce. Animals whose ancestors have a long history of domestication, but who live apart from human management, are called "feral" animals. For example, the "wild" horses and donkeys of the American Southwest are both feral and exotic. And cats and dogs, who have a long history of cohabitation with humans, have survived successfully as feral animals.

Humans have sometimes imported non-domesticated animals for economic gain but then lost control of their movement. Gypsy moths, for example, whose larvae have defoliated large areas of forest in the American East, were introduced from Europe into the Cape Cod area, in 1868, in an attempt to promote an American silk industry.

Humans frequently import wild, exotic species as household pets, and sometimes lose or abandon them. For example, parrots and parakeets, wild in their native South America but brought to North America as pets, have escaped confinement and established colonies in the urban jungles of cities such as San Francisco, New Orleans, and New York. Imported Burmese pythons and Asian walking catfish, now thriving in the Everglades and competing with native species, were originally imported into Florida as pets.

Intentional Importation and Dispersal

Humans have sometimes imported non-domesticated animals for economic gain and allowed them to disperse. In the late 1930s, for example, Edward McIlhenney imported thirteen nutrias (a type of large rodent) from Argentina to Louisiana with the intent of establishing a fur industry. After being released into a Louisiana marsh, the nutrias successfully reproduced at such an astounding rate that, within a few decades, their numbers had grown to an estimated 20 million. They have consumed enormous amounts of vegetation needed by native animals, and they have caused extensive soil erosion.

Human immigrants attempting to recreate an environment similar to the one they left have been responsible for introducing exotic species. In the nineteenth century, for example, rabbits were imported and released into Australia to provide hunters of European origin with familiar shooting targets. Similarly, the common carp was introduced into North American waterways, over 100 years ago, to provide a familiar food fish to fisherman of European descent. In the late nineteenth century, Eugene Scheiffelin devised a plan to introduce to the United States all the bird species mentioned in the works of William Shakespeare. In 1890–91, he released about 100 English starlings in Central Park. By 1940, starlings had dispersed to as far away as California. Across the North American continent, starlings, whose population now numbers about 200 million, encroach upon the nests of other birds and compete with them for food. In the reverse direction, American gray squirrels were introduced into Britain in the nineteenth century, to provide "variety" to the British landscape, and they have become competitors of the native red squirrel.

In some instances, humans have introduced one exotic species in an attempt to eradicate or control another exotic species. Such was the case when the mongoose, a native of southeast Asia, was brought to the Hawaiian Islands in 1883 to destroy the population of exotic rats that was eating cultivated sugar cane. The mongooses were brought to Hawaii from Jamaica, where they had been imported in 1872 for rat control. More recently, Cayuga ducks were introduced to Hawaii to eat another exotic species, the golden apple snail.

Unintentional Importation and Dispersal

Many animal species, both vertebrate and invertebrate, have been transported to areas alien to them through the unintentional agency of humans. The rats that arrived and thrived in Hawaii, for example, were stowaways on boats. Similarly, a host of exotic insects have hitched rides in the cargo containers of ships, trucks, trains, and airplanes. Ballast water, too, provides a vector for exotic species' dispersal. The zebra mussel, a thumbnail-sized mollusk native to the area of the Caspian Sea, reached North America in the mid-1980s, probably carried in the ballast water of a transatlantic freighter. The ballast water was discharged into Lake St. Clair, Michigan, and the zebra mussels quickly colonized the Great Lakes and the Mississippi River basin. It is estimated that, in some parts of the Great Lakes, the zebra mussel population may be as high as 70,000 per square foot.

Human-Assisted Dispersal

All of the above species can be designated as "imported" because they were, by some means, transported by humans to an area in which they did not evolve. Some species, however, can be considered *exotic,* but not *imported.* These are species whose dispersal was assisted by disturbances to the environment caused by humans. The coyote, for example, evolved in the American Southwest, but has now spread throughout North America, inhabiting urban, as well as rural, areas—as far as the Atlantic coast—where it can be considered exotic or alien. It has profited from the humans' eradication of its competitors—such as the wolf—and from the availability of food which human habitation brings. Similarly, the cattle egret migrated on its own from Africa to South America in the 1870s and, by the mid-1940s, to North America. Its dispersal, however, can be considered "human-assisted" because it benefited from the alterations that humans made to the American landscape—particularly, the dedication of vast areas of the land to raising cattle.

The majority of exotic species do not survive if deprived of human care. Of the few that do thrive (perhaps as few as 2 percent), some are of little concern to humans. For example, the opossum, an omnivorous, scavenging marsupial native to the American southeast, was imported to the San Francisco Bay area of California, around 1890, to provide a new target species for hunters. Opossums subsequently dispersed throughout the state. Since they do not stress native species as predators or competitors, their movement into areas new to them was easily accommodated. So well has the opossum coexisted with other species in its new habitat that some scientists categorize it as "naturalized." The cattle egret has also been easily accommodated by its new environment, but, in this case, the environment is one altered significantly by human activities. It has proved useful—and therefore welcome—to humans, because it feeds on insects attracted to human-managed cattle (another exotic species).

Some exotic species, however, are considered pests by humans, because they have a negative impact on human industry, economics, health, safety, and recreation. Another cause of human concern is the stress that some exotic species place on native species of animals and plants.

Industry and Economics The rabbits introduced to Australia by European colonists multiplied rapidly and began devouring both native plants and the exotic crops planted by farmers. The economic impact has been enormous in terms of crop destruction, loss of forage for (exotic) livestock, and costs of largely unsuccessful attempts to control the rabbit population through poisons, viruses, warren demolition, and fences. Zebra mussels clog the water-intake pipes of factories and utilities, causing closure of businesses. They also choke agricultural irrigation pipes and thus increase the costs of raising human food. The economic impact of this species, in market loss and pest-control efforts, is estimated to be about $5 billion annually. Feral horses in the American Southwest graze on land that ranchers want to reserve for their livestock, and rats transported to Hawaii eat into the profits of plantation owners.

Health and Safety Rats are also a concern to humans because they are carriers of disease. The proliferation of zebra mussels in American waterways is worrisome, in part, because they clog water-intake structures and reduce pumping capacity, thus threatening human water supplies and power generation. The exotic Burmese python poses a threat to visitors to the Florida everglades. In Hawaii, a bottle-cap sized Caribbean tree frog, the coqui, which was recently transported to the Pacific Islands on nursery plants, croaks at 90 decibels—a level as loud as a lawnmower. The loud, incessant noise robs people of their sleep and chases away tourists, thus affecting the economy, as well as the health, of Hawaiians.

Recreation Zebra mussels clog the engines of recreational boats. The round goby, a fish which, like the zebra mussel, is native to the Caspian Sea area and was introduced to the Great Lakes by the discharge of the ballast water of a transatlantic ship (in the 1990s), is larger and more aggressive than most fish species native to the Great Lakes and has threatened species prized by sport fishermen. (However, because the round goby feeds on zebra mussels, it may prove a helpful tool to control mussel populations.)

Environment The few species that do successfully colonize areas new to them succeed because they are resilient, have high reproductive rates, are generalist feeders (eat a wide variety of foods), and have no predators in the new area—and because their food sources or competitors for food have not yet developed defenses against them. In a relatively short period of time, therefore, they can alter an environment extensively. The voracious and aggressive Nile perch, imported into Lake Victoria in Africa in the mid-1950s as a food fish, is thought to be responsible for the extinction of about 100 species of native fish. Flora and fauna that inhabit islands are particularly vulnerable to new species; because of their long biological isolation, species have evolved that are unique to the environment of the particular island and not able to adapt to interactions with new species. For example, mammals introduced to New Zealand, not just by Europeans but also by Polynesian colonists over 1,000 years ago, caused the extinction of many native flightless, ground-nesting bird species which had evolved on the remote islands. (Prior to the arrival of Polynesian colonists, the only mammal species existing in New Zealand were two species of bats.) The rabbits in Australia have destroyed several native plant species and caused soil erosion by denuding the land; they also endanger native animal species that cannot compete with them for food, or whose habitat has been destroyed by their activities. Mongooses were introduced to the West Indies and the Hawaiian Islands for rat control. However, they preferred to prey on native species and have consequently caused the extinction of several native species of reptiles, amphibians, and birds. The brown tree snake, a native of New Guinea and Australia, reached the island of Guam about 1945, as a stowaway on a military ship. These snakes now number an estimated 13,000 per square mile and have caused the extinction of several of Guam's native bird and bat species. The extinction of these species also causes a reduction in pollination and dispersal of seeds, thus affecting plant life as well as animal life.

Categorizing animals as "exotic," "nonnative," and "invasive" is a controversial matter. Although one criterion for designation as an exotic species is dispersal beyond one's native range or place of evolution, it is, in fact, rarely possible to determine the spatial or temporal boundaries of a range or place of evolution for any species. Dispersal and colonization of new areas have always been naturally occurring phenomena. Very few species have remained strictly within the geographic area of their evolution. It is, therefore, appropriate to ask how long a species must inhabit an area before it is considered naturalized. Some scientists reserve the terms "exotic," "alien," "nonnative," and so on for species whose dispersal took place relatively recently—that is, in the modern period of European exploration and migration, beginning about 1450. For the Americas, the dividing line is the arrival of Christopher Columbus in 1492. Species inhabiting these continents in the pre-Columbian period are considered native and indigenous; those that arrived after 1492 are nonnative, alien, and exotic. However, using European migrations as the line of demarcation between native and nonnative would mean that the many species brought to the Hawaiian Islands, for example, by Polynesians from about 400 on should be considered native, a point which many biologists would dispute. Some scholars, therefore, focus on the element of human-facilitated dispersal as a key to distinguishing native from exotic. Human-facilitated dispersal is thought to be unnatural in the sense that it has moved species much farther and much more quickly than they would otherwise have moved, and has moved them across natural boundaries—particularly, oceans and mountain ranges—which they would not otherwise have crossed. In some cases, human-facilitated dispersal has resulted from the breaking down of barriers by, for example, the building of the Suez Canal, which now allows the movement of once-separated species between the Red Sea and the Mediterranean. However, calling human-assisted dispersal "unnatural" is problematic, particularly because we generally consider the migration of humans from one location to another to be a "natural" human behavior. From the beginning of their history, humans have been "on the go," usually in pursuit of a better life or economic gain. And when they move, humans take with them their possessions, many of them biological. In one sense, then, the movement of animals, even across oceans, is a natural occurrence. It is, moreover, difficult to reconcile that humans—whether Europeans in the Americas or Polynesians in New Zealand—can be considered "naturalized," but the biological items intentionally transported by them continue to be categorized as alien or exotic. And categorizing as alien and exotic the movement of animals unintentionally transported by humans is also problematic; for example, rats are opportunistic creatures that adapt well to changing circumstances. It is natural—that is, according to the nature of their species—to move in pursuit of readily available food sources. It is therefore reasonable to consider whether we can make a logical distinction, in terms of natural behavior, between a rat eating wild plants versus a rat eating human-cultivated plants. And if the rat, attracted by human-cultivated food on a ship, is transported to a place where its species has not before been, is this method of dispersal less natural than if, for example, the rat was carried on a floating tree limb?

Some scholars argue that the terms "exotic," "alien," "nonnative," and "invasive" are used inconsistently and reveal an anthropocentric bias. As noted above, the terms are not applied to humans or their domesticated animals, even though it can be argued that humans, as they have dispersed across the planet, have altered landscapes more significantly than any other species. Although it has been stated that *alien* species are second only to habitat destruction as a cause of species extinction, it is important to keep in mind that this is a far second and that habitat destruction can be traced directly to human activities, including the raising of imported animals for food. Clearly, decisions to eradicate species are based on human interests. Nonnative species that are judged to have a negative impact on human economic, health, or recreational interests are targeted for destruction, while other nonnative species, considered useful to humans or apparently benign, are not. The focus of concern thus seems to be not about the exotic origin of a species, but rather about its perceived interference in how humans want to use an area into which they have dispersed.

The use of terms such as "alien" and "invasive" influences the way people think about these species. "Invasive," for example, conjures up images of hostile armies of invaders, and it is frequently used in contexts where humans are describing species that they believe must be exterminated. Militaristic metaphors abound in these contexts. In 1999, for example, when U.S. President Clinton signed an "Invasive Species Executive Order," Agriculture Secretary Glickman declared "a unified, all-out battle" against the spread of alien species in the United States. Humans speak of undertaking "assaults" on foreign and alien species and waging "war" against invasive species. The problem with such metaphors is that they prompt us to perceive the situation as a moral issue. The presence of exotic species is represented as a struggle of "good" (humans) against "evil" (aliens). The rhetoric leads people to conclude that dispersal is, in some sense, an act of hostility and malevolence on the part of the animals. In reality, however, the animals are not our enemies; they are simply following their natural behavior in their efforts to survive. In addition, framing the issue as a war then seems to justify—and even to encourage—the harsh methods of extermination that are employed, such as poisons, traps, and guns.

Proponents of such methods argue that it is necessary to use them in order to eradicate resilient invasive species as quickly as possible. Animal protectionists, however, protest that the methods are inhumane, and, furthermore, that they are indiscriminate, because poisons and traps kill many nontargeted animals. They advocate the use of nonlethal methods of population control—such as sterilization—but also question whether there is truly a compelling argument for eradication. Among people who protest eradication, particularly by harsh methods, there is generally more sympathy for vertebrates than invertebrates, and for mammals than fishes or reptiles. Other opponents of indiscriminate poisonings include people with economic and recreational interests. When, for example, an entire lake is saturated with poisons in order to kill one nonnative fish species, all other species die, adversely affecting the fishermen and the businesspeople who are dependent upon them.

It is a natural behavior for humans to want to protect their economic and health interests and, consequently, to destroy creatures whose habits threaten those interests, although certainly every effort should first be made to confirm that eradication is the correct solution—and to ensure that damage to nontargeted species will be minimal. More controversial is the practice of killing animals in an attempt to save other species—both animal and plant—from extinction. As noted above, *exotic* species in some areas of the world, particularly on islands, have directly or indirectly caused or are threatening *native* species with extinction. The result is an increasing homogenization of the world; European and Asian species now dominate landscapes far from their original point of evolution. Some scholars contend that we humans have a moral obligation to preserve biodiversity; it is we who are responsible for introducing exotic species by transporting them across oceans and mountains, and it is we who have the intellectual capacity to recognize and reflect upon the consequences of our actions. The moral intuition that there is value in biodiversity (and, correspondingly, in landscapes that have not been altered by human activities) is a recent phenomenon, and one that conflicts with the values cherished by earlier generations of humans. Throughout our history as agriculturalists, we humans have promoted the development of monocultures—that is, cultivated areas devoted to the production of one crop, such as expansive wheat fields, rice paddies, or cattle ranges. In our efforts to alter the environment to suit our purposes, we have eliminated other species and considered that a landscape had value only if it served our needs.

A cattle egret standing on a Hereford cow, a perfect illustration of two exotic species in the Western hemisphere—one domesticated, one not. ©Carol Buchanan/Alamy.

Advocates of biodiversity, on the other hand, argue that species and landscapes have an intrinsic value that is independent of human needs. They therefore support the conservation of endangered species and of the environments in which they thrive. Many endorse restoration activities which include the eradication of exotic species that are competing with or preying on native species. They frequently claim that the exotic species "degrade" or "harm" the environment and destroy its "balance, integrity, and stability." Again, it is important to pay attention to the rhetoric of the statements. "Degrade" and "harm," like "invasive," are pejorative terms, intended to influence the way we think about a species. In truth, exotic species do not degrade or harm an environment; they *change* it (a more neutral word). They may "transform" it into a different ecosystem, with different organisms and bio-relationships. If they cause the extinction of other species, the extinction is a permanent change, but the surviving organisms and relationships continue to evolve. The argument for "balance, integrity, and stability" has been losing favor as scientists recognize that ecosystems are always in flux and that change and disturbance are persistent features of bio-communities. Nonetheless, it is undeniable that human-facilitated migrations of animals have altered ecosystems much more quickly and extensively than any nonhuman activity. Proponents of conservation and restoration defend the harsh methods they employ to kill exotic species by maintaining that they value biodiversity and are trying to ensure the very survival of native species of animals and plants. Critics of the harsh methods respond that they value compassion and are concerned about the pain, distress, and death caused by humans to each individual animal. The development of humane methods of controlling animal populations would offer a resolution to the ethical issues raised by restoration practices.

See also

Conservation and Environment—*Endangered Species Act*

Further Resources

Baskin, Y. (2002). *A plague of rats and rubbervines: The growing threat of species invasion*. Washington, DC: Island Press.

Bright, C. (1998). *Life out of bounds. Bioinvasion in a borderless world*. New York: W. W. Norton & Company.

Burdick, A. (2005, May). The truth about invasive species. *Discover, 26*(5), 33–41.

Cox, G. (1999). *Alien species in North America and Hawaii: Impacts on natural ecosystems*. Washington, DC: Island Press.

Glotfelty, C. (2000). Cold war, silent spring: The trope of war in modern environmentalism. In C. Waddell (Ed.), *And no birds sing: Rhetorical analyses of Rachel Carson's* Silent Spring (pp. 157–73). Carbondale: Southern Illinois University Press.

Larson, B. (2005). The war of the roses: Demilitarizing invasion biology. *Frontiers in Ecology and the Environment, 3,* 495–500.

McGrath, S. (2005). Attack of the alien invaders. *National Geographic, 207*(3), 92–117.

Peretti, J. (1998). Nativism and nature: Rethinking biological invasion. *Environmental Values, 7,* 183–92.

Sagoff, M. (1999). What's wrong with invasive species? *Report from the Institute for Philosophy and Public Policy, 19,* 16–23.

Simberloff, D. (2003). Confronting invasive species: A form of xenophobia? *Biological Invasions, 5,* 179–92.

Woods, M., & Moriarty, P. (2001). Strangers in a strange land: The problem of exotic species. *Environmental Values, 10,* 163–91.

Jo-Ann Shelton

Conservation and Environment
Extinction of Animals

The creation and extinction of biological species is as old as life itself; the geological record of nearly four billion years of life on earth documents long periods during which new species emerged and old species disappeared at fluctuating rates, tending to produce a gradual increase in the number of species, genera, and higher taxa over time. This record is punctuated by five recognized episodes of catastrophic extinction, when 60 to 80 percent of the then-extant species disappeared in less than a million years. The most well known of these is the end-Cretaceous event, marking the end of the dinosaurs 65 million years ago. Most biologists accept that the precipitating factor was an asteroid collision. The end-Ordovician extinctions, 440 million years ago, coincided with a drastic lowering of the sea level; the ultimate cause is uncertain. From the point of view of present policymakers attempting to learn from the lessons of geologic history, the Permian/Triassic extinctions, 250 million years ago, are probably the most instructive, both because they were the most profound and because the postulated causes most closely mirror global anthropogenic changes which are rapidly creating what Richard Leakey and others have termed "The Sixth Extinction."

During periods of geological stability, the classical Darwinian model of evolution through natural selection and survival of the fittest approximates the real world. Poorly adapted individual animals fail to reproduce, and, eventually, the numbers in a species composed of such individuals dwindle to the point where the species cannot sustain itself. During a catastrophic extinction event, on the other hand, sheer luck, rather than superior genetics, seems to be the predominant factor in species survival. This is because the entire world's biota is suddenly confronted with extreme conditions with which it has no historical experience, adaptations to which could not possibly have been selected for during the lifetime of a species.

Until modern times, humans contributed to animal extinction in much the same way other animals do—through predation and competition for finite resources. Every innovation in human technology and concomitant population increase has produced a wave of extinctions, something that has only recently been recognized. Controversy still exists concerning the relative contributions of climatic change and Mesolithic hunting to the extinction of large Ice Age mammals (including woolly mammoths, mastodons, cave bears, woolly rhinoceroses, and saber-toothed tigers). The most plausible explanation postulates that populations were already in decline as retreat of the glaciers reduced the amount of suitable grassland and that Stone Age hunters had a disproportionate effect because they were only able to hunt immature animals. Large mammal species went extinct at higher rates in North and South America than in Eurasia and Africa, where animals had millennia to adapt to man as an increasingly efficient predator. The present large-mammal fauna of North America consists entirely of "recent" arrivals from the Old World. In Australia, the disappearance of 85 percent of animal species weighing over 100 pounds coincides roughly with the arrival of humans approximately 40,000 years ago.

The Neolithic agricultural revolution introduced competition and habitat destruction as major factors in anthropogenic extinction. Overgrazing and forest depletion accompanied the rises of towns and cities. In the Middle East, this contributed to desertification. There are no documented extinctions associated with environmental degradation in the ancient Middle East or in Central America, where the rise of the Mayan civilization also produced widespread deforestation. However, many large mammals now restricted to tropical Africa disappeared from southern Europe and the Middle East in ancient times, and there were undoubtedly inconspicuous mammals, birds, and, especially, invertebrates that went extinct at the same time.

Improved shipbuilding enabled Neolithic man to colonize previously uninhabited islands, bringing his tools, crops, dogs, pigs, and rats with him. Large, ground-nesting, flightless birds, including the elephant bird of Madagascar and several species of giant moa in New Zealand, succumbed to human predation. Species narrowly adapted to lowland forests disappeared as land was cleared for agriculture. In Hawaii, the archaeologically best-studied area, over a hundred endemic species of birds—almost half the total—were eliminated before Europeans began settlement. Polynesian farmers completely deforested Rapa Nui (Easter Island), presumably destroying many species.

Between the onset of European exploration and colonization in the late fifteenth century and the first real systematic attempts to explore and categorize the biological riches of our planet in the mid-eighteenth century, a few species disappeared—among them, the proverbial dodo. The technological advantages which made Europeans superior competitors relative to native peoples had less effect on local flora and fauna; in Central America, at least, the devastating effect of introduced human diseases gave the flora and fauna a temporary advantage.

The Western scientific community became aware of extinction as a pervasive and continuing biological phenomenon in the mid-nineteenth century, as systematic paleontological research, coupled with surveys of the remotest parts of the globe, made it clear that thousands of species no longer extant had once populated our planet—and, moreover, that they had not all perished in a single biblical cataclysm. With the publication of Charles Darwin's *Origin of Species* in 1859, a picture emerged of creation and extinction of species as a gradual process occurring over many millions of years. Although increasing knowledge enabled scientists to identify endangered species, the concept of "survival of the fittest" undermined the old idea of a species' value as God's unique creation—and conservation efforts, such as they were, had to be based on utility or sentimental attachment.

The number of species of animals known to have gone extinct between 1800 and 1950 is not large. Although certainly an underestimate, it still suggests that the human contribution to the background extinction rate was not overwhelming. Many species that had either vanished from Western Europe and eastern North America or were restricted to preserves and zoos still had healthy wild populations in western North America and Russia. Tropical rainforests in Latin America, Africa, and Southeast Asia—reservoirs of most of earth's biodiversity—remained nearly untouched.

The complex causes of two well-known nineteenth-century extinctions—the passenger pigeon and the Carolina parakeet in North America—remain elusive but were certainly anthropogenic. Once exceedingly numerous, populations of these birds declined to unsustainable levels in just decades, and last-ditch conservation efforts failed to save them.

In general, human efforts to bring species back from the brink of extinction fail. By the time a species is recognized as endangered, it has already lost much of its genetic plasticity, and any population derived from captive-breeding programs will be highly inbred. The wild environment to which captive-bred individuals are returned is usually quite different from the one from which their ancestors were taken. As the natural environment is increasingly fragmented, providing enough suitable territory for large social mammals becomes impossible. The identified endangered species is part of a complex web of predators and prey, vertebrate and invertebrate herbivores, parasites, flowering plants, fungi, and microorganisms, many of them obligately bound to each other in commensal relationships. If, as is typically the case, it was a major component of its ecosystem, its disappearance from most of its range eliminates many less-conspicuous species and alters the ranges and population sizes of others.

Many of the species currently listed as endangered are, for all practical purposes, extinct. This number includes 11 percent of mammals (including all species of great apes and a third of all primates), 4 percent of birds, 8 percent of reptiles, 10 percent of amphibians, and 13 percent of fish.

Around the middle of the twentieth century, the adverse impact of humans on the environment began an exponential rise that shows no signs of leveling off in the aggregate, although valiant and sometimes-successful efforts have been made to address specific problems. World human population passed the one billion mark shortly after 1800. It increased from two-and-a-half billion in 1950 to nearly six-and-a-half billion in 2006, and, barring a global calamity, will continue to rise even if efforts at limiting birthrate find wide acceptance globally. To feed this enormous population, we have converted vast tracts of land to agriculture and stock raising, a process only possible through profligate depletion of fossil fuels. Logging to provide building materials and fuel also reduces biological diversity, even when efforts to preserve the forest as a renewable resource are successful. Often an area is reforested with a monoculture of an introduced species that the native fauna is unable to exploit.

Fossil fuels enable humans to harvest fish at an unsustainable rate. Most species of large, oceanic fishes are now declining rapidly, as a result of this direct human predation. Freshwater fish and invertebrates are even more vulnerable; a large proportion of them are restricted to single lakes or river systems, where pollution, diversion of water for agriculture, or invasion by nonnative species can rapidly produce extinction. A classic example is Lake Victoria in Africa, where deliberate introduction of Nile perch as a food fish eliminated over 200 endemic fish species in a ten-year period.

Estimates of the scope and immediacy of the current problem vary widely, but almost all environmental scientists agree that extinction due to human activities has reached crisis proportions and that officially compiled lists (such as the Red List published by the World Conservation Union) only scratch the surface. Scientists recognize just under two million species of plants, animals, and microorganisms worldwide—more than half of them insects. Estimates of the actual number of species range from five million to nearly thirty million, the bulk of the unknowns being insects, other arthropods, and fungi narrowly adapted to specific flowering-plant hosts in tropical rainforests. Engaging in practices suspected of causing mass extinction of uninvestigated species is like setting a torch to a library housing thousands of unique manuscripts—some of the information so laboriously accumulated (in the case of species, by millions of years of natural selection) is undoubtedly important, and, once it is gone, there is no way of retrieving it. Biologists' efforts to rescue nearly extinct showcase vertebrates by cloning, surrogate motherhood, or crossbreeding with closely related species (as is being done with the dusky seaside sparrow, whose only surviving representatives are both male) may conceivably succeed, but they cannot possibly preserve a species whose existence is unknown to science.

Extrapolating from the rate of disappearance of vertebrates, the proportions of various groups in diverse tropical ecosystems, and the few cases in which detailed inventories of flora and fauna were made prior to environmental destruction, the current rate of extinction of all species appears to be on the order of 100 per day, perhaps 10,000 times the background rate. Such a rate, if sustained, would reduce the world's biota to humans and their domesticated plants and animals in a thousand years, a "sixth extinction" eclipsing even the end-Cretaceous and end-Permian catastrophes in its abruptness and totality. Were the process to continue toward anywhere near its endpoint, however, humans themselves would almost certainly be numbered among the extinct species.

Some progress has been made toward preserving intact ecosystems rather than specific species. Almost 12 percent of the earth's land surface is governed by some national or international preservation protocol. Regulations and their enforcement, however, vary tremendously in their effectiveness, and may even be counterproductive. Managing large-game preserves as tourist attractions, for example, leads to artificially large populations of animal "stars," suppression of their competitors, and disruption of natural vegetation cycles.

In response to the rapid depletion of Brazil's Amazon Rainforest, which shrank in extent by 80 percent between 1920 and 2000, Brazil adopted a policy of requiring a certain percentage of the land to remain in unmanaged forest as clearance progressed. Careful monitoring of these intact parcels revealed what island biogeographers have long known: the number of species an area will support is proportional to its size. Reducing a once-extensive rainforest to isolated small parcels separated by farmland eliminated animals with large territorial requirements and, in turn, species dependent on them. In evolutionary time, islands develop their own endemic fauna, ultimately increasing total diversity; in human time, a fragmented ecosystem simply loses diversity.

Even in the developed world, preventing human exploitation of wilderness areas requires nontrivial expenditure. Expecting that near-subsistence farmers in Brazil or tropical Africa will refrain from hunting and wood-gathering in a biological preserve is wildly unrealistic. Policymakers in the countries most affected complain, with justification, that they are being asked to underwrite a disproportionate share of the costs of remediating a problem ultimately traceable to Western economic expansion. On a local level, at least, conservation efforts—including preservation of rare species—stand a better chance of success if the people most affected are compensated for the sacrifices they need to make.

Habitat loss and local environmental degradation have been the most conspicuous—and probably the most important—factors in anthropogenic extinction in recent decades. Due to increased awareness and global cooperation, the pace of both is slowing, though not enough to prevent continued, massive species loss. All current human conservation efforts, however, are likely to become irrelevant—to become what some people describe as "rearranging the deck chairs of the Titanic"—if reputable projections concerning global warming and the poisoning of the world's oceans become a reality.

The ocean is one vast ecosystem, without boundaries, "islands," or sharp environmental gradients. Some oceanic habitats—notably coral reefs—are extremely diverse biologically, but the total diversity is lower than that on land because species disperse much more readily, and because primary (plant) productivity is low in the open ocean. Local disasters and overexploitation seldom cause marine extinctions. Until very recently, it was assumed that no human activity could produce oceanwide changes. That assumption is dangerously false.

Since the beginning of the Industrial Revolution, humans have been burning three hundred million years' worth of accumulated organic carbon to fuel expanding population and material affluence (in roughly equal proportions). Some of the carbon dioxide released into the atmosphere returns to earth as acid rain, historically a serious problem in freshwater ecosystems in industrial areas; some remains in the air. Increasing atmospheric carbon dioxide creates a greenhouse effect, raising global temperatures.

The entire earth is getting warmer, at a rate that may seem gradual to humans but is precipitous on a geologic time scale. While it may be physically possible for individual species to migrate northward, reconstituting complex communities (upon which most species depend) is another matter. The survivors of massive displacement due to climatic change are likely to be the "weeds" of the vegetable and animal kingdom—species with broad ecological amplitude and high reproductive rates, such as humans, rats, cockroaches, and dandelions.

Seemingly insignificant changes in the temperature, pH, or salinity of the ocean can have massive consequences. A recent 20 percent drop in phytoplankton concentration over large areas of the Pacific Ocean, possibly related to one of the above causes, spells famine for the entire pelagic food chain. There has been a similar drop in coral growth rates in the Pacific, hampering reef regeneration and constricting available habitat for reef inhabitants.

If continued global warming results in the sudden melting of the Greenland and Antarctic ice caps, the resulting disruption of oceanic currents, decrease in salinity, and rising water levels will be catastrophic to marine life. An analogous event toward the end of the last Ice Age wiped out 70 percent of the mollusk species along Florida's Atlantic coast—in contrast to 20 percent in California, where changes were gradual.

Massive undersea volcanic eruptions are thought to have triggered the end-Permian extinctions by acidifying the oceans and releasing huge quantities of carbon dioxide into the air. The present "sixth extinction" appears to be following a similar course—and to be rapidly approaching the point when the species that initiated the process loses the ability to halt it. It has long since passed the point when preservation of individual species because of their utility or sentimental value made any sense. The existence of every species, including our own, is dependent upon an incredibly complex network of living organisms. Removing large chunks of that network, regardless of the mechanism, causes the whole to collapse.

Further Resources

Broswimmer, F. T. (2002). *Ecocide: A short history of the mass extinction of species.* London; Sterline, VA: Pluto Press.

Hoage, R. J. (Ed.). (1995). *Animal extinctions: What everyone should know.* Washington, DC: Smithsonian Institution Press.

Leakey, R., & Lewin, R. (1995). *The sixth extinction: Patterns of life and the future of mankind.* New York; London; Toronto; Sydney; Auckland: Doubleday.

Mass Extinction Resources. http://www.well.com/~davidu/extinction.html. [An extensive site with numerous links, including the Red List and the 2001 and 2006 reports of the United Nations Commission on Biodiversity.]

Swanson, T. M. (1994). *The international regulation of extinction.* New York: New York University Press.

Ward, P. D. (1997). *The call of distant mammoths: Why the Ice Age mammals disappeared.* New York: Copernicus, Springer-Verlag.

Martha Sherwood

Conservation and Environment
Extinction of Animals in Taiwan

Taiwan is located on the Tropic of Cancer and is separated from the mainland by the narrow Taiwan Strait, which is only 130 kilometers across at its widest point. It has an area of 36,000 square kilometers and is dominated by rugged, mountainous terrain with a remarkable diversity of fauna and flora. More than a century ago, naturalists Robert Swinhoe and Alfred Wallace conducted surveys of wildlife in Taiwan, and they recorded about 36 species of mammals and 187 species of birds. According to recent surveys, Taiwan harbors nearly 4,000 species of vascular plants, 61 species of mammals, 400 species of birds, 92 species of reptiles, 30 species of amphibians, 140 species of freshwater fish, and an estimated 50,000 species of insects—including 400 species of butterflies. On June 23, 1989, Taiwan enacted its Wildlife Conservation Law to give protection to endangered species. All Appendix I fauna and most Appendix II fauna listed in the Convention on International Trade in Endangered Species of Wild Fauna and Flora (CITES) are protected by the law. In addition, several species of native wildlife—including 16 mammal, 80 bird, 39 reptile, 13 amphibian, 6 fish and 19 invertebrate—have been protected by the Wildlife Conservation Law. Nevertheless, Taiwan's large mammals in general are threatened with extinction due to increasing human population pressure, ongoing habitat destruction, unregulated hunting, agricultural extension, and other developmental activities.

During the early 1600s, the sika deer (*Cervus nippon taiouanus*), which is an impressive deer with distinctive white plum blossom–like spots on its back, roamed around the coastal plains of Taiwan, numbering in the thousands. The large-scale hunting of the elegant sika deer started during the Dutch colonial time (1624–1661) and continued until the Japanese occupation (1895–1945). As a result, the sika deer became extinct in the wild during late 1960s. A restoration project was carried out to restore this species in its natural habitat. In November 1986, twenty-two deer were selected for captive breeding, and these animals were released into a large enclosure with forest habitat. By 1990, the deer population had reached sixty, and the first group was reintroduced in Kenting National Park. Nonetheless, there are possibilities of future conflict between human inhabitants who live in and around the national park, because the deer have no natural enemies except humans. The park authorities need to monitor the deer population and develop wise conservation and management strategies to avoid future conflict between deer and humans.

The largest of all of Taiwan's deer, the Formosan sambar deer (*Cervus unicolor swinhoei*), was also hunted intensively before the Wildlife Conservation Law was implemented in 1989. Now this deer is found only in higher mountain regions where human accessibility is difficult. This species is vulnerable, so data on the population status is urgently needed to predict their future survival. Another game animal, the Formosan Reeve's muntjac, has been hunted by local aborigines for its meat and fur for decades. The intensive hunting pressure had increased dramatically, before 1989, to satisfy the market demand that ultimately resulted in the decline of the species' population.

The only megabat in Taiwan, the Formosan flying fox (*Pteropus dasymallus formosus*), was previously common on a 15-square-kilometer island known as Green Island. This species has become extinct in recent years, due to severe hunting, deforestation, and habitat alteration. Although Chiroptera represents the largest order of mammals in Taiwan, the extinct flying fox is the only species in that order included in the wildlife protection list. Very little is known about the conservation status, distribution, and ecology of nineteen other species of insectivorous bats that occur in Taiwan, so priority should be given to protect the other neglected bats from nearing extinction.

The Formosan clouded leopard (*Neofelis nebuosa brachyurus*) is a subspecies of clouded leopard that was endemic to the island of Taiwan and is now believed to be extinct. Its tail is slightly shorter than that of other subspecies of clouded leopard. It was the second largest native wildlife recorded in Taiwan, after the Formosan black bear. After extensive loggings at its natural habitat and severe hunting pressure from local poachers, these secretive leopards were forced to retreat into the wilderness of Yushan (or "Jade") Mountain and Tawu Mountain. Since the 1980s, there have been no sightings of these elusive cats. The Rukai tribe of local aborigines believes that their ancestors transformed from these leopards.

Bears have been valued in Asia for centuries as medicine and food. The Formosan black bear (*Selanarctos thibetanus formosanus*) is native to Taiwan, and this species inhabits the rugged mountains to the east. The present population status of the black bear is unknown, but they have been sighted occasionally by hikers and naturalists in the Central Mountain Range, in elevations between 200 and 3,500 meters. The demand for bear body parts in traditional Chinese medicine is still growing globally; therefore, it is essential to estimate the population status and density of wild bears throughout Taiwan, so that effective protective measures can be implemented.

The Formosan macaque (*Macaca cyclopis*) is the only nonhuman primate to inhabit Taiwan. Before the Wildlife Conservation Law was passed, monkeys were captured from the wild to meet the demand in pet trade. Since 1989, the trapping of monkeys has been prohibited by law. Unlike other countries in Asia, where macaques can be seen roaming in towns and cities, the Formosan macaque is generally confined to forest areas and seldom intrudes upon human habitations. However, they are still being hunted and trapped illegally—several monkeys with wounds inflicted by poachers' traps and snares have been seen at Mount Longevity, in southern Taiwan, where the author has been conducting long-term field research on the population dynamics, ecology, and conservation of wild Formosan macaques for over a decade.

With the increases in the development of the industrial sector, economic growth, the expansion of agricultural activities, and the exploitation of natural resources resulting from the limited conservation activities of recent decades, the government must realize the fact that animals inhabiting the small island of Taiwan are extremely vulnerable for extinction; therefore, urgent measures are needed to strengthen efforts to decelerate the loss of wildlife species and their habitats, and, ultimately, save species from human-influenced extinction.

Further Resources

Hsu, M. J. (1997). Population status and conservation of bats (Chiroptera) in Kenting National Park, Taiwan. *Oryx, 31,* 295–301.

Hsu, M. J., & Agoramoorthy, G. (1997). Wildlife conservation in Taiwan. *Conservation Biology, 11,* 834–36.

Wallace, A. R. (1880). *Island life.* London: MacMillan & Co.

Minna J. Hsu

Conservation and Environment
Finland and Fishery Conservation Issues

In general, people's attitudes toward fishing and nature conservation are positive in Finland. The attitudes towards fish culture are, however, more negative. Fish culture is thought to make the water bodies eutrophic. Nevertheless, the products of fish culture are usually highly valued. Recreational and commercial fishermen are usually in favor of conservation, although fishermen who own water areas tend to have more critical attitudes. The attitudes among local water owners resemble those of coastal residents in the United States toward the conservation of Oregon coastal salmon.

Finland is a country in the northern part of Europe, between Sweden and Russia. Lakes and rivers cover about 9 percent of the country—large lakes number 56,000, and the total shoreline is about 130,000 kilometers. Finland is partly bordered by the Baltic Sea, and thus there is a good opportunity for fishing in most parts of the country. Fishing (and the fishing industry) is usually highly regarded in the minds of the citizens.

The population of Finland is five million, and two million of its people are recreational fishermen, who fish by angling and netting on lakes, rivers, and the Baltic coast. In addition to this, about 1,000 commercial fishermen harvest in inland waters. They fish vendace (*Coregonus albula*) especially, chiefly with trawls and seine nets. Finland also has about 3,000 commercial marine fishermen; approximately 75 percent of their catch includes Baltic herring (*Clupea harengus*). There are about 220 fish farms that produce fish for human consumption, mostly rainbow trout (*Salmo gairdneri*) and salmon (*Salmo salar*). Half of these farms are in the Baltic, with the rest being in the lakes.

In Finland, fishing is often enjoyed alone, but it is also an opportunity for people to socialize as "fishing friends." No age, gender, or kinship restrictions exist on who can be "fishing friends"—though fishing is a predominantly male activity.

For centuries, fish stocks have been protected to ensure sustainable use of fish. Protective measures have consisted mainly of restrictions on fisheries (e.g., closed seasons during the spawning period); these measures have been widely accepted. In the middle to late 1990s, people tended to be confused—and sometimes even irritated—by the discussion concerning the protection of biodiversity. Currently, however, attitudes are more positive in this respect.

In general, people's attitudes toward nature conservation are largely determined by their own dependence on the natural resources. Usually, this means that people tend to oppose conservation measures if their socioeconomic position is threatened. For example, in fishermen's minds, the gray seal (*Halichoerus grypus*) population, which grew rapidly in the second half of the 1990s, has now almost destroyed the salmon and whitefish (*Coregonus lavaretus*) fishery in the Gulf of Bothnia. Many Finnish citizens are perplexed about this because during the second half of the twentieth century, the gray seal in the Baltic Sea was not far from extinction. People were told for decades about the problems DDT and other toxicants caused, for example, to the breeding of seals. Now, fishermen say that people should consider the gray seal as a problematic species for fishery in the Baltic Sea. This poses a dilemma: should we protect the seals or the fishermen's livelihood?

At the moment, Baltic Sea fishermen are concerned about the high PCB content in herring and the enrichment of poisonous substances in the food chain. People are skeptical about the usability of sea fish as food for humans. This attitude strongly affects the national fish markets in Finland. Issues such as overfishing and the role of bycatches in regard to Finnish fisheries in the Baltic Sea are seldom discussed; such topics are usually brought up only in regard to salmon fisheries in the area.

Fish species that are considered valuable, such as salmon, artic char (*Salvelinus alpinus*), zander (*Sander lucioperca*), and brown trout (*Salmo trutta*), can only be fished when they reach a legal minimum size. People have accepted this as a measure to protect these fish.

Attitudes reflect general changes in society—especially, a shift from a rural and production-oriented society toward a modern society, where recreation and nature protection are more emphasized. For example, for recreational fishermen, the catch and its value are not of crucial importance. These fishermen like the tranquility of nature, and, usually, they go along with conservation measures.

Cultural changes in society are slower than economic ones; for example, old local fishing habits often conflict with modern and effective ones. In the Finnish Lake District, the attitudes of the local residents and summer cottage owners toward commercial trawling are often negative. Residents usually complain about the sustainability of the fish stocks, whereas recreational fishermen and summer-cottage owners most usually complain about the disturbance to the tranquility of nature. Attitudes of residents indicate the peasant-nature relationship in which nature is largely seen as a resource for humans, whereas recreational fishermen stress their personal enjoyment of nature. Both groups object to trawling in the lakes.

Further Resources

Finnish Game and Fisheries Research Institute. (2001). Finnish Fishery Time Series. SVT, *Agriculture, Forestry and Fishery, 60.*

———. (2004). Aquaculture 2004. SVT, *Agriculture, Forestry and Fishery, 59.*

Salmi, P., Auvinen, H., Jurvelius, J., & Sipponen, M. (2000). Finnish lake fisheries and conservation of biodiversity: Coexistence or conflict? *Fisheries Management and Ecology, 7,* 127–38.

Smith, C. L., Gilden, J. D., Cone, J. S., & Steel, B. S. (1997). Contrasting views of coastal residents and coastal coho restoration planners. *Fisheries, 22,* 8–15.

Tonder, M., & Jurvelius, J. (2004). Attitudes towards fishery and conservation of Saimaa ringed seal in Lake Pihlajavesi, Finland. *Environmental Conservation, 31*(2), 122–29.

Tonder, M., & Salmi, P. (2004). Institutional changes in fisheries governance: The case of the Saimaa ringed seal, *Phoca hispida saimensis,* conservation. *Fisheries Management and Ecology, 11*(3/4), 283–90.

Juha Jurvelius